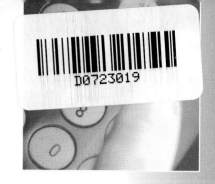

A PRACTICAL GUIDE
to Call Center Technology

by Andrew J. Waite

A Practical Guide To Call Center Technology

copyright © 2001 Andrew Waite

Published by CMP Books
An Imprint of CMP Media Inc.
12 West 21 Street
New York, NY 10010

ISBN 1-57820-094-6

For individual orders, and for information on special discounts for quantity orders, please contact:

CMP Books
6600 Silacci Way
Gilroy, CA 95020
Tel: 800-500-6875 or 408-848-3854
Fax: 408-848-5784
Email: cmp@rushorder.com

Distributed to the book trade in the U.S. and Canada by
Publishers Group West
1700 Fourth St., Berkeley, CA 94710

Manufactured in the United States of America

Table of Contents

ii

iii

iv

Preface

Total Customer Focus —
Easier Said than Done

I have just finished a meeting with a leading insurance company (USAA of San Antonio, TX) that is morphing to a full service financial services company providing life, casualty and health insurance, retail banking, investment and related services. Their mission is to serve their clients; primarily family units, up the wealth generation side of the life curve AND down the backside of the curve through their retirement years. This not only means serving the head of the household or primary breadwinner, but the surviving spouse and family in the event of the death of the head of the household.

Their business is tidily organized in "silos" reflecting the regulations, skills that apply and various products offered. When a spouse goes through the unfortunate, but inevitable life event of losing a partner, these silos within the business spring into action to provide prompt response to the survivor's financial requirements individually, and in doing so usually manage to request the death certificate...four or five separate times!

On reviewing their processes from a customer point of view, this business leader decided that "silo-ed" operation worked to its benefit and not its customers. The result has been the development of a czar-like "survivor relationship team" whose job is to take an absolute customer focus: cross-silo, cross-product, cross-channel and cross-application. Now, the customer drives their customer management, the convenience or organization of the business does not.

USAA, a recognized customer focused business leader for over a generation, very elegantly expresses the spirit of customer focus with their action: looking at customer data from a customer point of view.

Most important in this USAA illustration is their recognition of:

- the strategic objectives of customer focus and service,
- the need to examine and correct processes,
- the necessity of ensuring that the underlying parts or technologies were in place, and
- the power of training and maintaining staff commitment and human capital necessary to execute the strategy.

At various points in this book, we will revisit these values.

☐ THE CHIEF CUSTOMER OFFICER

Roy Dudley, an early thinker behind the whole customer relationship management movement, most succinctly made a vital point when he said, "A company is only as good as it's list of customers." Buried in that simple statement are the oft misunderstood disciplines of *list management, direct marketing, sales force automation and customer experience* and *relationship management* that have been made more complex by multiple channels crossing multiple disciplines and technologies.

The technology industry has taken the concept of a list of prospects and customers, along with the vision of a sales process with continuity of treatment and fulfillment, and created an industry worth billions of dollars.

As individual business disciplines, each is destined for implementation failure at your company, IF there is no one person with an overall customer-centered view and the authority to make everything and everyone work together. This is the underlying argument for the creation of a customer czar or chief customer office.

Just take a few minutes to consider the number of points where your company can touch customers. Now look at the organizational silos. It is tough enough for you to make sense of it and you work there! What about your customers?

It is time for a fundamental corporate rebuild that breaches the walls of marketing, sales, support and service "silos," that no amount of automation and technology will materially help.

As this book is written, a new Administration is arguing for a review of the military because the current organization is structured around a 200-year-old Napoleonic war fighting model of massive field armies and naval fleets. The world no longer looks like this, so budgeting and adding to an already obsolete model is putting off the inevitable and sapping productivity now.

Companies are like that. They are organized around disciplines that reflect business models perpetuated by business schools and business structures that are way out of date. Customers are everything to a business yet a marketing organization finds them, a sales organization sells them, an installation and support organization installs them and a yet another organization services them. Different people, different systems and often, different databases. No continuity or experience "threading" and as a result, little loyalty on the part of customers.

☐ THE INFLUENCE OF THE INTERNET

"The more things change, the more things stay the same." Anon. Witness the arrival of the Internet with its ubiquitous media and ecommerce opportunity offered to all.

"Ideas are free, implementation is a bitch" Anon. Brilliance and success lie with those who can execute on the vision with the fundamental steps needed to make these ideas work.

Never before has there been so much hype or so many fortunes made and lost as have been delivered by the Internet, but the realities remain the same. Good managers, who understand business, can take the Internet and the new technologies and weave them into existing and emerging strategies, techniques and technologies to win more market share and build stronger business positions than ever before. What do they know and why are they successful?

The answer is simple: Common sense, learning and applying the lessons previous generations of marketing, media and distribution strategies and technologies have taught us - only with a twist. There is an overwhelmingly strong argument for a customer czar who has a cross discipline, cross channel and cross application mandate to manage the business from a customer point of view.

This book began life in 1987 as *The Inbound Telephone Call Center, How to Buy and Install an Automatic Call Distributor*. The plan was to help non-technical buyers introduce telemarketing and direct response telephone techniques into their companies. The book then tried to provide guidance in buying the best technologies to protect their advertising, marketing and sales investments. The book has sold over 25,000 copies and became the best selling standard text in buying call-handling equipment.

Ten years later, in 1997, integrating computer databases with telephone systems was the big trend. *The Inbound Telephone Call Center* evolved into *Customers: Arriving with a history, Leaving with an Experience*. This discussed

3

the burgeoning business of tying customer communications, mostly telephone calls, to a customer's record or computer telephony integration.

The conclusion reached then was that manufacturers and the industry at large was still being too calculating about handling calls as transactions. These communications represented individual customers with needs and wants, but more importantly, people. People with feelings and perceptions that needed something from a company. This was a manifestation of using technology to serve mass markets as customized one-to-one relationships.

These transactions were not just a bunch of engineering and telephone traffic statistics, but real customers with business that had better be addressed in a businesslike way. Powerful inexpensive computing made growing personalization and mass customization possible. The age of computer integrated telephony was coming into the mainstream.

Now the World Wide Web is upon us. The Internet has potentially inverted everything: threatening to destroy traditional cost models, accelerating business and heightening the need for near-instant customer response. Now your customers are more often than not, dotcom customers, moving at a pace and an understanding that anything less than prompt and knowledgeable support is unacceptable. Despite the tech bubble of the late Nineties, this will be the lasting legacy of the dotcom revolution.

A *Practical Guide to Call Center Technology:* is here to help. With the book, the "Chief Customer Officer" or equivalent has help to enable him or her maintain the best "front door" to your business, while navigating the choppy waters of a mixed communications media and its channels to profitability.

☐ SERVING THE CUSTOMER

In this book, we discuss building an integrated market position using existing electronic and emerging marketing, promotion, sales and customer support channels, tools and tactics.

The author's background includes selling technology, building and running sales organizations, publishing magazines and building and managing subscriber lists and all that implies. This has left the author with an overwhelming respect for each of the required disciplines. This respect is attended by a constant frustration at the inability of each discipline to seamlessly apply themselves to integrated customer support for the mutual benefit of customers and business stakeholders alike.

Since this book's first edition in 1987, core principals for serving customers arriving by phone, mail, fax, email and electronically have remained relatively constant. Your business has presented itself as available to do business and customers still expect service, within their perception of reasonable standards. If a business does not meet that expectation, these prospective customers vote by doing business elsewhere with their had won earnings.

Over the intervening ten plus years we have seen massive advances in database marketing, prospect identification and customer ranking technologies, customer "propensity modeling," printing, direct mail, coupled with predictive dialers and overall follow up technologies delivered by sales force automation and customer relationship management systems. This means companies can reach prospective customers faster, more easily and less expensively than ever before.

Other news is both good and bad. The Year 2000 census count of the United States identified over 280 million individuals resident in the United Sates. The bad news is that there are only 280,000,000 consumer prospects in the United States and provided they all have phones we can deliver them all a telemarketing or email message within three months with the technology available today. With all this new marketing technology you can reach out and touch these folks as never before. Any market is increasingly finite.

☐ THE EVOLUTION

This book is written for the business technophobe so that essential communication with the "computer guys" is maintained. Failure to remain on the same page produces unintended and undesirable results. "Plumbing" all these data and systems together to produce a cohesive and responsive customer-centric "face" is a discipline that requires the skill and knowledge to look at business objectives first, then articulate them to the technologists to meet the business goal in an integrated customer contact systems environment.

One of the first things we are going to do is to redefine the call center as the customer contact center. Because the term "call center" perpetuates the notion that customer contact is limited to a live person answering a telephone call to and from a customer or prospect. Mail, electronic and personal contact are ignored as alternative, integrated or complementary processes. This will be one of the first issues dismissed.

Customers have communication choices and they will contact a business whenever, wherever and however they choose. Our mission is to anticipate and manage any communication with the same sensitivity, quality and concern for success.

☐ WINNING AND KEEPING CUSTOMERS

Keeping a customer happy and maximizing the lifetime value of your relationship with that customer is more important than it has ever been. Yet often we only emphasize new customer acquisition.

The customer contact center is there when our customer wants to:

- Get more information.
- See a demonstration.
- Place an order.
- Check order or delivery status.
- Ask questions on how the product works (or is meant to).
- Troubleshoot remotely (manual or remote diagnostics).
- When all else fails, dispatch a service person and manage a service call.

This book examines the customer contact process, not only from an inbound and outbound perspective, but also integration with the any other market contact techniques a customer or company may use to reach and serve pending and existing customers. We'll look at philosophies, applications, and the strategic and tactical tools and uses of a customer contact process. We will discuss the techniques, tools and technologies available to aid the customer contact center owner and manager in realizing their expectations of winning and keeping sophisticated dotcom customers more successfully and more profitably.

Andrew J. Waite
Phoenix, AZ
September 1, 2001

Introduction

The Customer Contact Center is the Focal Point

Strategically, the call or customer contact center is the point of entry for most customer communication. This is where a customer can make any type of inquiry or contact, oblivious to the type of connection, content and context of the communication, and expect a meaningful response. The processes, technology and people with the skills, training and motivation, all exist to serve this relationship.

Once available to do business, a company establishes and advertises an entry point where an inquiry from a prospect or customer can expect to be answered. This communication can arrive at the company via any medium. Here the connection is defined in the broadest sense:

By telephone:
- advertising or response via 800 call
- an outbound telephone solicitation
  Electronic: Internet inquiry
- assisted browsing
- request a callback, "click to call" or even a
- web chat or kiosk visit (ATM etc.)
  By text:
- mail solicitation and any response
- correspondence delivered by mail or
- FAX
- Email

In person:
- visit to a retail store
- visit by a salesperson

The options are many; and, as long as choices exist, customers will find a way to use them all. However, are you able to serve any customer request arriving via any media with the same level of service quality?

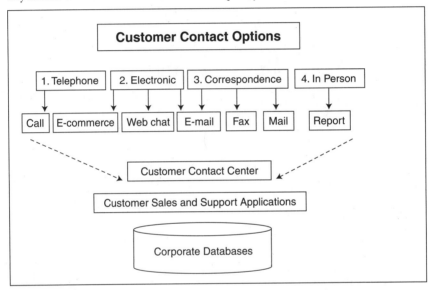

Figure 0.1

□ WHY USE A CUSTOMER CONTACT CENTER

At some stage, nearly every company on the planet decides their sales process needs some formality and organization. The simple reason for this is simplicity: A single business focus with one phone number or web address to remember and to contact.

The next stage is fulfilling the expectation that is created, that is to receive orders and deliver answers to customer questions. To do this a number of customer communications mechanisms are usually established. This can be as simple as opening a retail store on a busy thoroughfare. If a prospect wishes to buy or an existing customer desires help, they simply go to the store. Thus, all of the customer contact center elements are met: an advertised location where the business is prepared to serve customers.

This is the simplest manifestation of customer contact and probably the most mature execution as the staff has corporate memory for particular customers, their likes and dislikes and their status in relation to the business.

As a business grows, adding more complex products, more complex disciplines, longer sales cycles, extended product life cycles, the business itself becomes more complex.

Added complexity arises when a company begins to identify alternate places prospects and customers can be served so as to deliver greater convenience for the customer and improve contact performance rates (increased sales and faster and easier customer satisfaction) for the company. This is the point where a business establishes a customer contact center where the staff are equipped to serve and treat prospects and customers as you would expect to be served. A formal customer contact center should be positioned to provide dramatic reductions in:

• cost of sales,
• length of a sales cycle,
• time to customer satisfaction, and
• speed of payment.

9

☐ A SEISMIC SHIFT IN CALL CENTER "PLUMBING" TECHNOLOGY

The great Confucian curse is "May you live interesting times!" These are now, and it will only get more complicated. We are entering a sea change in connection technology that is bringing a shift from separate telephone, data networks and address identification for physical delivery, to telephony and data converging into a single Internet protocol-based voice and data network, broadly described as the Internet.

This book relies heavily on the proven lessons of a traditional call center because most of the equitable customer focused work distribution, workflow and service level management were developed for responding to high volumes of real time customer calls in larger call centers. Updating "the plumbing" or adding different types of customer contacts and methods adds complexity, but does not negate those lessons.

Previous technical shifts were generally "plumbing" based accompanied by heavy vendor marketing efforts that most often failed to meet buyer expectations and reality during the early years of the adoption cycle. This change offers significant cost and process advantages but requires massive investment at the end customer level and changes in culture.

Two other realities remain and will take some time to change: the fact telephones, wired or wireless, are still the most ubiquitous and easily used communications devices in the world and remain fundamental to more than 80% of all customer contacts. Given the massive worldwide capital and political investment in telephone companies and infrastructure, this is not expected to be displaced fast.

Data Systems

Wide availability of solid state computing technology for business applications is about five decades old and has gone through various changes: mainframes and time-sharing, mini-computers, PCs, client server and now, Internet-based networks and databases.

Telephone Systems

PBXs and telephone switches have experienced a similar evolution, although it has occurred more slowly because of past regulation. First, there was the telco government regulated or "tariffed" crossbar analog PBX of the early sixties and the solid state analog switch of the early deregulated interconnect days (1967 to 1980). This equipment has been widely replaced by computer controlled digital switching systems running proprietary operating systems with fixed hardware address structures, proprietary signaling and proprietary computer telephony links. These systems have a life span of approximately 20 years and many are reaching the end of that life span.

This market space is ripe for rework. IP protocols allow data and voice to travel on the same cable (or wireless) network to a common address and provides common packetization and control signaling that completely eliminates the requirement for a separate PBX switch or key system (as we now know them).

Note: The earliest technology iterations still have major application today, albeit, on newer hardware.

Software and applications are far more resistant to the obsolescence that plagues hardware, mainly because hardware manufacturers have the tendency to use proprietary strategies to improve hardware and gain supposed market differentiation.

Only when operating systems and related standards became ubiquitous and programmers plentiful did "branded" computers lose sway. That's when "open" systems became widely adopted and volume computing became cheap and accessible to nearly everyone.

However, for this to occur, there may be a need for substantial upgrade in bandwidth in the corporate networks.

Voice-over-IP (VoIP)

VoIP or voice transmission over the Internet has entered the telecommunications fray. Although there are still volume and fidelity problems that need to be resolved before it can become a viable communications option for most businesses. For instance, many times the audio quality and reliability of the call connection is currently less than is required for a business transaction. Nevertheless, VoIP does have its attractions: a unified network within the company premises where IT controls ample bandwidth (voice and data capacity potential) and the integrated applications for customer contact and contact centers.

The rate of change related to the adoption of VoIP will be faster than any previous generational migration. The move from mainframes to minis to PCs, or analog to digital PBX systems, was slow as new code (and more work) was required. But few major end user advantages accompanied these plumbing improvements. VoIP brings huge efficiencies by eliminating redundant networks, eliminating the middleware required to perform protocol translation between proprietary systems, and simplifying applications development. Less work, less cost, fewer systems and faster performance.

If a new contact center at a new site is required, VoIP may soon be a foregone conclusion. Updating an older PBX (ACD) to a "newer technology" PBX is merely an "expensive paint job" when VoIP is an option.

Setting the Applications Stage

Call centers have proven themselves as successful customer sales and support tools for over a generation. The customer contact center has its roots in the call center (beginning with airline reservations and even earlier manned telephone exchange services), so initial call center philosophies, structures, references and management standards are the baseline for building the next wave of integrated customer contact centers. The roots of this next generation of integrated customer contact centers lies with the old — just more options are available with the next generation center.

If your business objective is to sell and service customers to the point of mutual satisfaction, the technology and process should be subordinate to the business. When technology goals supersede the business strategy, things get confusing and can cause failure. All of the ensuing contact center discussion, put achiev-

11

ing the business objective ahead of technical issues. This includes all applications within a customer contact center:

From order entry to CRM. During the intervening 13 years since this book was first written, the customer call center has also gone through a significant market shift. At first the customer call center was a processor of customer transactions so order entry and fulfillment systems were the standalone. Take an order and ship it, invoice and update the customer order history.

With the realization of the key role in the building and maintaining of a relationship with a customer, the shift was to full relationship documentation and management. From transactions to relationships, hence the evolution of the support system description to customer relationship management or CRM systems.

From handling customer complaints to customer support systems. Once the sale is made, a customer has to be supported. The core reason for this is the value of satisfied and referenceable customers who return. This means some customer support or help desk system had to exist to track help requests, solutions and any resource necessary to achieve this goal.

Two realities flow from this process: First, customers that may have been sold by third parties in a wholesale/retail distribution model or consumer sales environment, ideally should be identified and stored for future marketing purposes. Second, the marketing and individual customer intelligence gathered by an unthreatening non-sales transaction such as customer support, is vital to future sales.

Help desks. A note should be made here that although there is separate genera of systems for internal company customers, known as help desks we will make little distinction and reemphasize the fact that any one asking for help is "either a customer or someone serving a customer." Ironically badly run internal help desks are exponentially worse than poor direct customer service as the host corporation is paying both parties to do a poor job! And that's before they visit their ineffectiveness on a customer!

From advertising budgets to high power promotional performance analysis. As customer propensity modeling, marketing and advertising systems begin to make their way into companies, the data resident in sales and support elements of these databases is vital to finding and winning future business from existing or future customers. If you don't know where you have been, how can you improve you future direction.

Successful reengineering a customer database system depends on lots of information about process and the state of the relationship with a customer.

Data and information are no longer enough; knowledge is needed and this is being gathered well beyond the stationary customer contact center. It is necessary to add all of the data gathered from a dispersed and mobile sales organization using sales force automation (SFA) tools and integrate this data with information flowing from the stationary customer relationship and support system in the knowledge mix.

From mass marketing to mass customization and one-to-one messages to mass markets. A primary value of all of this technology is to be able to produce results in progressively smaller market increments, and manage many more increments. As granularly as one-on-one. When Henry Ford built the Model T assembly line, the goal was volume. Now that same assembly line concept can be dynamically adjusted to build a specific automobile for a specific customer and their specific choices, while not contradicting the original goal economic scale. Technology has done the same for marketing, selling and supporting the individual customer.

The Call Center Evolution

Traditionally field representatives have been tasked with account management, customer support and service contacts through individual and personal visits. They typically travel to the customer. This is personnel, labor and cost intense as well as expensive and inefficient, except for major situations or courtesy visits. Therefore a customer call center strategy was adopted. Initially many call centers were used only as a supplement or secondary strategy for the sales organization, i.e. direct response or telemarketing, while the field force continued business as usual.

Soon it was realized that the customer call center could provide consistent and predictable service, often around the clock. Today, the frequent telephone conversation with a courteous, professional, and knowledgeable sales, service or support representative is more desirable, or even preferable, to the visit by the local field sales or service representative. Now many companies and products are launched with direct marketing, direct response and customer contact centers as the core of their sales and support process.

Successful pharmaceutical companies promoting new drugs or new uses for older products took the trend mainstream: busy medical practitioners prefer a telephone call to an office visit, if the call is from a knowledgeable telephone salesperson or account service person. The contact is faster, often more frequent, more effective and less expensive for all parties concerned.

☐ THE COST OF A CONTACT

Let's concentrate a bit on the call center business. In this way we can understand the importance, cost and potential of an inbound customer inquiry or opportunity. Let's look at the actual cost breakdown of a traditional call handling process.

The transaction elements do not differ that much from any other communication though the cost of carriage and the response effort may be incrementally less (even to point of replacing live individuals with email "robots").

These inbound contacts are made up of three basic elements:

- telephone lines and their costs, (carriage)
- the staff and management necessary to process the call or contact and,
- the equipment necessary to answer, record and fulfill the request the call represents.

The Lines

Your local telephone company typically provides local telephone lines; although this is no longer an absolute rule. Your local telephone service franchise is

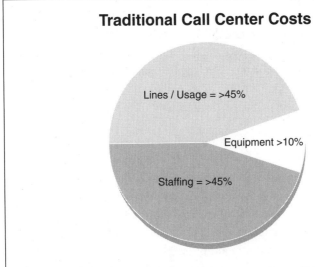

Traditional Call Center Costs

Lines / Usage = >45%

Equipment >10%

Staffing = >45%

Shown here are the three traditional cost categories in a traditional call center. Communications may be less of a percentage, but relatively speaking the equipment to serve these customers is always less than 10% of the annual investment.

Figure 0.2 But cost analysis is less than half the story. We will get to this in a moment in Figure 1.1.

absolutely necessary to your happy existence. They may even provide your connection to the Internet. We will discuss this relationship further when we discuss lines and telephone traffic in Chapter 4.

For the large telecommunications user, there are many long distance and bulk service alternatives other than the local phone company — including:

- stringing you own cable.
- hooking up with a local broadband fiber optic company.
- installing your own private microwave connection to your long distance carrier and anchoring a satellite antenna to your roof. Many companies have successfully installed such systems and many more are installing them.

These alternative broadband access technologies have become more accessible, some are of better quality and some are significantly more cost effective.

Today, the local telephone company is still NOT the long distance carrier that provides long distance service, although they can provide instate long distance. As we write this local telephone companies are on the verge of being freed to sell interstate long distance, which should further depress prices. Currently, the local phone company typically provides you with the connection to the long distance carrier who, in turn, provides long distance telephone service. The local phone company connects your central office to the long distance carrier.

What is important to understand is that there are at least TWO entities, and maybe more, involved in providing you with the necessary telephone services for a customer contact center. If your company is typical, it's using one of the major carriers such as Sprint, MCI WorldCom, AT&T — there are sound cost reasons for using more than one; not simply for prices, but also redundancy. Don't forget: you can use one or more of the alternative access methods as discussed previously. More about all of this in Chapters 4 and 12.

Telephone services come in a number of different flavors: Digital or analog, value added, such as ISDN (integrated services digital network) and DSL (digital subscriber line), bulk services via T-1 format or just "plain old telephone service" (POTS). Volume services are now delivered in digital format.

There are two different types of customer call centers — those focused at serving a local population and those that receive calls from all over the country. For example, a local electrical or municipal utility company offers service to a local customer base that does not need to call long distance to reach the center. No toll charges are incurred. Here the customer call center owner need not offer toll free long distance service. As a result customer call center costs drop substantially when compared with a national center receiving nearly all of its calls from out of the local area.

However, for many customer contact centers, the cost of making a long distance call must not become an objection to placing an order, so management offers toll free "800" service that substantially adds to operational costs. Thus the argument for 800 service is increased sales. Dropping the caller's cost of placing an order leads to increased orders, which serve to outweigh the cost of the 800 lines. Nonetheless, some very mature and successful mail order catalog houses only use 800 service for customer service not for new orders. They let the customer pay the cost of long distance call to place the order. Of course, some do it in reverse — they allow the customer to place the order toll-free but to get assistance the customer must dial in on a toll-line (one that costs the customer money). Again, more on this in Chapter 4.

The Staff

In a customer contact center serving a local-only market group, people are the largest expense at about 80%. In a national center, people to staff your center rank with telephone lines as the largest cost categories, though staff cost has become a larger proportion of the costs as the per minute cost long distance and 800 service has dropped. Staff will be at least 60%, if not more, of your annual center operating expense.

Staffing issues begin with selecting a location where continuity of staff availability is assured. Location is often determined by a continuing supply of adequate staff. More about this in Chapter 3.

Major direct response centers are setting up shop in such unlikely places as Yakima, Washington and Hobbs, New Mexico — a former oil town that has an abundant supply of labor. Hobb's large available labor force helps to keep costs low and, in addition, much of the available population is Spanish-speaking, meaning they can be used in bilingual marketing campaigns.

But also there's a growing trend in the US and even the UK to "haul the call" offshore to a predominantly English speaking country with low labor rates. The Caribbean was an early offshore call center site, now US originated calls are served as far away as India.

Determining the right amount of staff and balancing this to the incoming call demand is a significant management challenge. An important tool to cost effectively match staff to call demand are forecasting and scheduling packages. We will talk about this in Chapter 5.

Once staff availability is satisfied from a planning point of view, the next issue is management structure and staffing to the level and quality you require. There are several issues regarding labor/management relations. We will deal with management reporting in Chapter 14 and trends that affect customer contact centers in Chapter 19.

Training and quality control cannot be ignored. There is a finite cost to establishing effective programs. Skimp on this aspect of your customer call center and you invite long term problems that return to the issue of meeting the first time caller's (prospect or customer) expectation. Training also affects the retention of a customer as a satisfied customer over the reasonable lifetime of the relationship.

There are already developments in these areas, which promise to dramatically improve the call center manager's lot as well as allow insight into why calls occur by volume, value and opportunity. Identifying trends in the content of large call volumes is the next major development in customer contact centers. One value in doing this is to intercept negative trends, shed unnecessary calls by addressing upstream problems as well as identifying successful people and processes and clone them across the customer contact center. More on this in Chapter 19.

The facilities that house the contact center and staff also have a finite cost. We will discuss these in Chapter 6.

It costs $25.00 to acquire a new subscriber or customer. Of that $2.30, less than 10% is the expense of the customer call center. The cost to process calls could be reduced or completely eliminated by reverting to mail order only and doing away with the customer call center. But the statistics heavily favor satisfying the impulsive purchaser with a few minor questions on the phone. You could avoid providing toll free service, but there are heavy marketing arguments, reflected in sales increases, that make sense to maintain one number marketing presence and toll free service.

☐ THE EQUIPMENT

Finally, and much of the reason for this book, is the customer contact center equipment. This used to be simply buying a phone system and depending on your MIS department to provide you with a computer system to enter orders and track customer service inquiries. Now we have moved beyond buying telephone systems to buying a better way of doing business across every conceivable medium.

Since the original edition of this book, "computer telephony" or "computer telephony integration" (CTI) went mainstream. But, in many customer contact centers, it's imperative to weave in email and the web — adding further complexity to an already complicated workload.

The writer is heartened by the growing recognition that a customer contact may be via mail, calls, emails, web chats or call backs. But remember that these contacts are people with expectations and the ability to buy something from any company that can successfully satisfy their expectations.

A Background on Equipment Integration

The first integrators were dominated by a mainframe mentality, evidenced by the large computer, PBX (private branch exchange) and ACD systems. These major vendors recognized the need to gather important call/caller information delivered on a call-by-call basis. This data became really significant when collected and used in conjunction with the customer database information stored on the company computer. The opportunity to match the caller with their account record prior to being connected to a live agent was determined to save significant time and offer greater potential for personalized customer focus.

This linkage allowed significant gains in the speed of response, but initially remained heavily dependent on the rewriting of host code to become CTI compliant. Rewriting host code and making these links are expensive in terms of time and money. As a result far fewer CTI links were installed than were originally projected (at least until open or *de facto* standards were produced by the likes of Microsoft and Intel).

The second phase of computer telephony, however, arrived earlier than expected through the purveyors of switch to host links. The advent of powerful personal computers and desktop operating system standards, such as Microsoft Windows, gave rise to desktop call control versus the proposed peer-to-peer call control of the first generation of CTI links. Using local area networking architecture, the desktop device has become a "client" attached to a "server" gateway that talks to the switch and possibly the host.

By moving this link into the world of Microsoft Windows, a significant breakthrough occurred. Because Windows offers automatic or scripted data exchange between application programs, it provided a standard for the millions of programmers striving to find "killer applications." Many really useful call center programs have been developed that take advantage of computer telephony links without the cost and expense of peer-to-peer call control envisioned by the large computer and switch manufacturers.

When the PC or open standards arrives in any industry, the PC and its attendant culture remake the structural and economic reality of that industry in much the same way a swarm of locusts decimate a lush landscape. Many old-line companies that resisted the move to PC-based standards fell behind and often died.

For many businesses, the Internet is having the same effect, although its effect is growing more rapidly than the PC's effect on the traditional computing order. Fortunately, the aware customer contact center owner can achieve dramatic productivity gains and increases in customer satisfaction by taking full advantage of this new environment. We examine the impact of computer telephony and VoIP in the customer contact center arena throughout the book.

☐ RESPONSE CENTER

The inbound customer call center responds to demands that are beyond the control of the center. The destiny (i.e. success) of the customer call center is to satisfy caller demand when they decide to call. Caller demand may be understood to a degree, but can only be forecast within certain general parameters. And that's where a good forecasting and scheduling package can be most effective. But also there are wild card elements that must be considered which can totally change call traffic demand. There's the example of a bus line where the sudden call volume increase turned out to be an advertising campaign rolled out by a marketing department operating in a vacuum. Marketing management "forgot" to tell its customer call center management of their advertising campaign. Callers were not satisfied. Customer expectations were blunted now and maybe into the future? So much for their expensive marketing investment!

Often marketing management is unaware of the customer call center resources available to serve the expectation created by the marketing campaign. For example, how many catalogs do you still see without the 800 number on every page? How many print advertisements do you see with the telephone number buried or missing completely? Or an already worrisome problem can be compounded by not communicating what is about to happen (i.e. the marketing campaign) to the people charged with serving inbound calls. But there are many other wild card elements that can affect a customer call center's ability to serve callers that go beyond the common problems, such as inadequate facilities, lines, building space, equipment, people. These include things like adverse weather (this may create increased call volume and/or make it difficult to get your staff to and from work,) service outages (telephone lines, telephone systems, computer equipment and electricity). We will discuss these in Chapter 12, when we discuss "bullet proofing" your customer contact center.

☐ A BIT OF HISTORY

Telephones were originally seen as a tool exclusive to the business world. However, because of their intrinsic social value it is no surprise the telephone is vital to modern society and business development. Today they are a ubiquitous device and their presence considered a measure of a society's development.

Business telecommunications developed slowly from its invention until the early sixties. This is not the place to plot the development of telecommunications, other than to say high volume customer call centers were a well established fact with the development of telephone exchanges, information lines and manual central offices.

The image of legions of telephone operators manually connecting callers and answering their requests can be frequently found in contemporary literature. All the early work standards developed for customer call centers began in the telephone company's local and toll offices. Business systems were large electromechanical devices that offered little in the way of flexibility and operated without any real data about the performance of the switching system, the lines attached to the system or the personnel staffing it.

Initially, when the telephone companies, particularly those of the Bell System, began providing call distribution capabilities on PBX systems to the heavy inbound customer call centers, the obvious users (the airline companies) latched onto the technology and like Oliver Twist, wanted more — now.

As is typical of a market where the demand drives development, the incumbent vendors (the telephone companies) were slow to respond and this opened up an opportunity for the electronics industry.

By the early 1970s, it was clear the Bell System was not going to respond to demand for sophisticated call distribution systems. Enterprising airline communications managers, like Mike Huntley (then at Continental Airlines), had established relationships with Collins Radio, a vendor of high technology radio equipment. Together they explored the idea of building a high traffic, non-blocking, software controlled, user programmable call distribution system that would provide management reports. The result was the birth of the computer controlled automatic call distributor in 1974. The Collins Galaxy system (now part of Rockwell International) was a standard for twenty years and still serves many large customer call centers today.

A new era began that gave birth to no fewer than twenty vendors and twice as many systems to choose from. But, don't make the wrong choice. You can compromise your ability to integrate customer requests, manage quality of service and increase your staff requirement by up to 30% over ACD systems designed to give you complete insight and control into customer demand, no matter what channel and allow your company to intelligently balance and integrate all requests.

Now call centers are adding email and its implied workflow distribution systems to their existing customer contact system. With this comes new challenges, such as how to thread all of the cross channel, cross application opportunities back to a single customer relationship.

The same rules apply to every other medium and device that finds its way into the customer contact center to help manage its channels — Internet and all! We will begin by using the call center experience as a model.

21

☐ BROAD PROCESS REENGINEERING

Customer contact systems integration is not a new concept. What must change is the all-inclusiveness of this vision. In the early 1990's some businesses benefited from the breakthrough book, *Reengineering the Corporation*, by Dr Michael Hammer and James Champy. The whole notion of reengineering is fundamentally changing the process of business from the bottom up. Take the process in its smallest increments, use "zero based justification," cut unnecessary process and drive the whole process to reaching the desired goal, faster and with greater effect. A customer-focused philosophy dictates that process analysis through granulation, looks at customer impact first.

In the customer call, or more accurately, the customer contact center, reengineering goes way beyond throwing newer technology at a customer's request for goods or support services. Reengineering the customer contact center goes to the root of the business process. Apologies are offered to Hammer and Champy, as we are not going to go beyond the realm of the customer call center to reengineer your corporation, but rather will make note of the surrounding issues.

For instance, by any other name customer contact integration is process reengineering of the customer communications process from their point of view. Consequently there are four schools of promotional thought based on vendor/manufacturer bias:

The Telecentrics: the PBX and ACD systems vendors who generally promote switching systems such as telephone systems, private branch exchanges, automatic call distributors and CTI as a tool to save money based on circuit savings and labor reduction.

The Netcentrics: These are the web tools promoters who believe everything including business applications; voice, video and data will all be delivered over the web at a speed and in a time frame that will meet you business goals.

The Datacentrics: These are the computer hardware and software vendors or integrators who tend to want to make their databases the center of your system. These people believe that the justification for customer contact systems integration lies in cost savings and the efficiency driven by intermachine "plumbing." All are right for very superficial reasons.

The Systems and Services Providers: These are the consultants and systems integrators that grasp the vision and long term implications of customer contact systems integration as workflow reengineering. They typically understand that the real success of customer contact systems integration is in fact, a platform for workflow changes.

Just always keep in mind that there is no one right direction or answer unless the customer is the primary focus. Strategy first, then process, parts and people can follow! They are all valuable and great contributors to your future. Success is orchestrating a harmonious collection of information, tools and disciplines to serve your prospects and customers first to meet your business goals.

The Traditional
Call Center

The customer call center has been around for longer than officially acknowledged. Call centers didn't begin to take commercial definition until the introduction in 1972 (by Collins Radio, now Rockwell Communications) of the first computer-controlled automatic call distributor (ACD) at Continental Airlines.

A functional ACD is the core building block of any successful customer call center and computer telephone integration or CTI. But as their manufacturers most specifically state — the majority of ACDs are just that: telephone call distributors.

Prior to 1972 the airline and other businesses had been buying electro-mechanical call distribution equipment from the Bell System. These uniform call distributors had limited functionality and represented, in microcosm, the entire thinking that led to the breakup of the Bell monolith — they "knew what customers wanted and gave them that." Unfortunately for the Bell System, their customers believed otherwise and began to look elsewhere. This is what gave birth to a healthy industry of customer call center technology specialists.

As mentioned previously, Rockwell Communications (nee Collins Radio) built an electronic switching system for Continental Airlines that allowed changeable programmable inbound call routing and more extensive management reporting on the service levels offered by the airline — the ACD. Probably the most significant issue in the system design was philosophical — the system was built with an understanding of the mission critical nature of the airline reservation center.

Note: American Airlines states that a potential outage in any critical chain component in its reservation system can cost it $210,000 a minute in lost revenue. That's irrecoverable revenue. Today, most customer contact center technology is built with multiple levels of redundancy and "bullet proofing," almost guaranteeing non-stop availability to calling customers.

ACDs are considered the traditional mainstay of any customer contact center where volumes of homogeneous calls are handled — the principals of operation, functionality, and business insight based on the depth of data gathering and reporting are absolutes in any customer contact center environment.

Over the last 29 years these devices have matured and evolved based on user requirements. Today, ACDs can manage just about any demanding, revenue intense customer environment. The lessons learned in ACD development are well remembered in every one of the electronic and/or emerging media channels a modern Internet-enabled customer is likely to use when requesting help or placing an order.

☐ WINNING OR PROTECTING REVENUE

An incoming call (or for that matter any contact) handled by an established customer contact center has the potential to generate revenue. But at the same time, it has the potential to cost goodwill, and ultimately long term revenue, if the call is not handled effectively. Customer contact centers, therefore, exist to generate or protect revenue and market position. Not an easy task.

Most customer contact center managers face the need to accommodate at least some outbound calling, which has become an increasingly important consideration in customer contact centers. For instance, an inbound call or an incoming written communication may require telephone follow up. When reengineering became popular in the late 80's, identifying business processes and assigning an agent to a "case" became increasingly important, especially in solving a customer's complex problem that may require follow up.

But management in many call centers had to also mix outbound with caller-demand, which is dynamic. This usually requires dynamic allocation of labor so as to meet customer demand with the appropriate "agent force."

- Business in the basic inbound customer contact center is based on the whim of the customer — thus destiny lies with the callers.
- Outbound calling is dictated by the tasks at hand and how workload is internally allocated across workgroups of agents — the customer contact center

management controls workload priorities. This is sometimes defined as "deferrable work."

Blending outbound calling with inbound customer calls, and even interleaving reading and responding (by writing or phone) to traditional correspondence (and more recently email) potentially allows management some ability to smooth workload and avoid overstaffing the center to accommodate peak caller demand.

Allocation of resources to answer callers' requests is an issue of matching supply with demand. Despite increasingly accurate forecasting systems, you cannot hire "fractions of people" or anticipate demand precisely. So, assigning non-inbound call tasks to idle agents allows the opportunity to justify hiring additional staff to meet peak inbound load, since idle staff can be reallocated to making outbound calls or following up on correspondence or other non-telephone contacts. But, cross utilization of agents has many challenges, which we'll discuss later in this book.

To quote inbound customer call center guru, Gordon MacPherson, "calls arrive in clumps." Although business normally occurs in predictable patterns, the typical call patterns are not ideal for rationing good service. Therefore, a customer contact center manager tries to modulate the whole process with an automatic call distributor, staff forecasting and attendance software and the nearest they can come to "just-in-case" staffing.

Finally, most inbound customer contact centers do not control their own destiny. A marketing program they not only don't control and often are

Management Challenges

Bus Line Fare & Schedule Center: A major fare and schedule center in Los Angeles opened to an unprecedented volume of callers requesting information about the special fares being offered by the line. No one had received notice or information about this promotion — at least not until agents reading the paper on their morning break found a full-page ad in the Los Angeles Times!

IBM Hosting Services: An advertisement ran by IBM Hosting Services is set in the context of a "12-step" program. A group member gets up and explains his stupidity in launching a web campaign. The group chides him that no one is stupid and goes on to hear his tale of a wildly successful $2,000,000 web ad campaign. The punch line is he didn't tell the web guys — the server crashed. The more things change — the more they stay the same.

External forces such as a lapse in quality control or ambiguous marketing or documentation also frequently drive calls to product and customer support lines.

unaware of (e.g. spot buying of TV time) is only one of many generators of such unpredictability.

Successful customer contact center operation is understanding and orchestrating the correct proportions among all of the channels available to connect and serve customers. The traditional call center is a great place to begin to understand how this is done. Mainly because these centers are applications and systems heavy and have been working at this for nearly 30 years.

The Traditional Call Center

Many times the customer call center represents the first level of contact a customer or prospect has with a company. It's the front door to your business. With such an important mandate, it is essential that the process be thought through — well upstream and well downstream of the actual contact:

- Should the contact occur at all?
- Should the center strive to process the call as fast as possible?

- Or should the agent strive to keep the customer on the phone for as long as possible to learn more about them and how your products, documentation, support, fit their needs?
- How can it be done better?

This discussion begins the core of the next breakthrough in matching corporate resources to optimizing customer experience.

At the point in the relationship that a customer decides to pick up the phone and call, little things, like prompt and courteous attention, mean everything. The potential emotion bound up in this transaction, from the customer and company point of view, is tremendous. To treat the incoming caller with less than the attention they are due (and expect) is a reflection of the attention, concern and service they can expect from your company, should the relationship proceed.

How Does Your Company Rate?

Senior Management should call their company and blind test its call center personnel. There may be a shock in store for that manager.

Senior management should also sit in their customer contact center, at least for 30 minutes a month, to get first-hand knowledge of what customers think, what they want and how they are served.

These are easy and painless ways "to get a little dirty."

Now add the electronic media — web and email — and expectations and responses get more confusing. Does incorporating the web and email into the mix add or detract from the volume of calls to the contact center?

☐ CUTTING AN ORGANIZATION'S COSTS

Load Shedding as an Analytical Trend

Traditional call center thinking is to focus on delivering the best service level to a caller under the circumstances. The web and VRUs (voice response units) have given rise to the notion of self-service and load shedding. But, what portion of these calls could be serviced upstream of the call center?

More importantly, what percentage of these contacts should not occur at all, e.g. many are based on some imperfect process or message that has caused the customer to pick up a phone and call a live agent. Root cause analysis is gaining impetus in analyzing all customer contacts to find a method that most suits a given customer and keeps them off the phone.

27

Customer Acquisition Cost

The cost of acquiring a new customer or cost of sale has steadily increased and continues to do so at a rate that shows no sign of slowing. Yet ironically, the average company does not know what it costs them to acquire a customer. When they do figure out the costs, the amount is often a "stunner!" Many companies have found that it is very expensive to acquire a new customer, and that the process is often far less profitable than retaining and servicing an existing customer. So, it's no surprise that service has been rediscovered.

In the magazine business it costs between $25 ("free to qualified readers" or controlled circulation) and $50 (paid circulation) to acquire a new subscriber. When you figure most consumer magazine subscriptions sell for less than $25 a year (and it costs more than that to service the subscription), you can easily see the revenue burden on the advertising side of the house. Magazines, like newspapers, commonly use the telephone as their primary sales medium. Because magazine advertising rates are set on the basis of size, quality and depth of the subscriber base, winning and keeping subscribers is a major goal of a magazine publisher.

How is this $25 to $50 subscriber acquisition cost calculated? First, all the marketing and sales efforts and expenses are identified and added up and then divided by the number of new subscribers. (Think of your customers.)

Let's look at some of these cost elements and what goes into pursuing a new magazine subscriber. Typically the costs include the following:

1. Creative efforts, i.e. the defining of who and where the market is believed to be, obtaining the list of target prospects or selecting the media which best suits the market, then acquiring the lists.
2. Creating the advertising and marketing materials necessary to a specific campaign.
3. Purchasing the space and time necessary to reach the intended market. This may be traditional broadcast advertising, print advertising in magazines or newspapers, general consumer or business vertical trade papers. It may be a direct mail/direct response campaign where mail is sent to individual targets that have been identified on the lists, which have been rented or purchased.
4. Any list preparation that's necessary to optimize the data to ensure that the marketing effort is directed to only the prospects that represent a reasonable possibility of conversion to a subscriber.

    Now the campaign is underway and the market is responding by mail, email, fax, telephone and via any other medium where a subscription order can be placed.

5. A processing system must be established to actually accept the orders. This may be an individual accepting a phone call or receiving a request on paper (through the mails, email, etc.) and entering a subscription.
6. None of this process can occur in an orderly fashion unless there is a system in place (even if it's a manual system). If the system is automated in any fashion, suddenly the equipment costs and management overhead increases, although the labor costs usually decease.
7. With the introduction of any volume the costs and effort increase again whether manual or automated. Both systems may require an upgrade in location and an increase in staffing with emphasis on personnel with the right skills to serve the incoming requests for subscriptions.
8. Then there is the campaign development, testing, rollout, training and supervision for the products, offer, center and staff.
9. Housing, facilities and furnishing for the subscription order processing is also a substantial cost.
10. Then there are the general and administrative costs.

    This process, where all responses are exclusively by telephone, is illustrated in Figure 1.1.

In this business, the "penny wise, pound foolish" rule applies with a vengeance. Let's look at Figure 1.1 again. If the entire cost of a sale of is $25.00, then the cus-

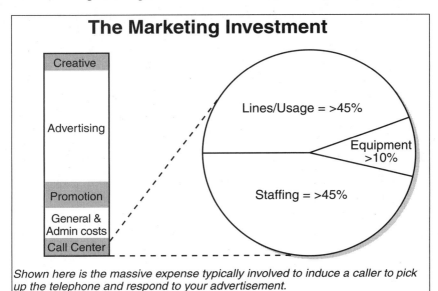

The Marketing Investment

Creative

Advertising

Promotion

General & Admin costs

Call Center

Lines/Usage = >45%

Equipment >10%

Staffing = >45%

Shown here is the massive expense typically involved to induce a caller to pick up the telephone and respond to your advertisement.

29

Figure 1.1 So often the cost of acquiring a new customer is given lip service by corporate management — little is done before the fact since it poses a risk to equip and staff up in anticipation of a busy season. Once the "season" begins and callers are not adequately serviced, however, corporate management may justify its lack of preparedness by stating, "maybe this is only a temporary condition that will soon be over." Of course, it will end — when the season closes or the business is "knocked off" by its smarter competition.

tomer contact center costs should be only $2.30. This amounts to less than 10% of the total marketing expense to acquire a new subscriber (customer). Yet, mismanaging the incoming customer contact center by not understanding how and when your business arrives and then failing to staff and equip the center to adequately handle predicted volume can be disastrous. It can defeat the marketing expenditure, no matter how effectively such marketing efforts and expenses have been applied. In fact, the more effective the campaign, the greater the potential to destroy the effectiveness of an unprepared customer contact center.

Good marketing and good products put heavy demand on your incoming lines, your equipment and your people. Maybe your callers get a busy signal or are put on "eternity hold" (with or without announcements) and suffer through boring "music," (often on an 800 WATS line). If so, they will, most likely, hang up. When established or potential customers "abandon" their calls they also may abandon both their interest and wish to business with you. Some will never call back — they're lost as prospective customers for your beautiful new product, despite your elegant, expensive marketing efforts.

Constant inbound call volume that approaches the maximum capacity of your lines, systems and staff can create more call volume, though not necessarily more business. For example, if you have a unique offer and your customer contact center is nearly always busy, much of the call volume may be "retries." That's people who could not get through to an agent or satisfy their request the first time and have been forced to call again and again and again. Even if a number of the callers do get through, many are still placed on "eternity hold" and eventually give up, but while they are on hold they're serving to displace fresh new callers.

This book encourages closer communication and cooperation between your marketing and customer contact departments. The opportunities lost daily because neither group understands their respective roles, tools, techniques, technologies and challenges of their colleagues, would simply amaze top management — if they only knew. This book is here to argue common sense and simple solutions.

No department in an organization is an island unto itself, just as no process stands alone — "upstream" and "downstream" of that department's or departmental employee's particular function are others and other steps that directly influence the success or failure of your contribution. This becomes particularly critical when the workflow implications of multiple communications media are considered. If this book achieves nothing else, your recognition of the crucial importance of this interaction, will be enough. Then the business of reconsidering the process of adding value to the customer experience becomes vital.

Reconsider the wonderfully silly example of the bus line (related previously) — the lack of interaction and communication between the marketing organization and the customer call center. The morning the center found itself swamped with inbound calls, the center's management was totally unaware of any new promotion and, therefore, could not anticipate or prepare for the increase in call volume. It was understaffed to meet the demand. It dulled a potentially good ad promotion, not to mention negating its expense.

The language of telecommunications has a wonderfully expressive term. It states in clear terms what happens to a caller who is not connected to the party that they want or intend to reach because of inadequate or insufficient lines or equipment. That caller is "blocked" or experiences "blockage." The same thing happens here...*the business opportunity is blocked from success.* Even worse, badly planned programs generate meaningless calls displacing real business.

☐ GOOD SERVICE ASSURES FUTURE SALES

So far the discussion has focused on serving the first time inbound caller responding to an advertisement for a particular product or service. What about the existing customer who calls in for service or an explanation about an existing customer relationship?

The Customer Contact — It's Valuable

Lifetime value of the relationship is a measurement that first gained recognition in the mail order business. Business-to-consumer mail order catalogers discovered their best prospect was an existing customer.

Although keeping customers happy and maximizing "the lifetime value" of that relationship is more important than it has ever been, most compensation plans used to encourage sales organizations normally emphasize new business. Thus, the stars are the sales people who bring in new accounts and customer service is viewed, more often than not, as a backwater.

However, this is slowly changing. Now many are asking, "What is a customer 'worth' over the life of the customer relationship?"

In the mail order catalog business, customers receive a set number of catalogs and promotions each year, they buy $Y, and the profit is $X. When you multiply these numbers by the expected number of years of the relationship, it is clear that effective customer service extends the life of a customer relationship. Many mature companies have built ongoing account relationships with customers after ciphering the effort/revenue/profitability equation.

Nonetheless, for many in the business community this realization is still the exception rather than the rule.

Have you ever noticed that when you are put on hold, it always appears to last longer than the actual elapsed time? Your customers have the same perception, which is compounded when they have a problem that needs resolution. Insensitive customer service is now regular grist of popular comic strips like Dilbert and Doonesbury.

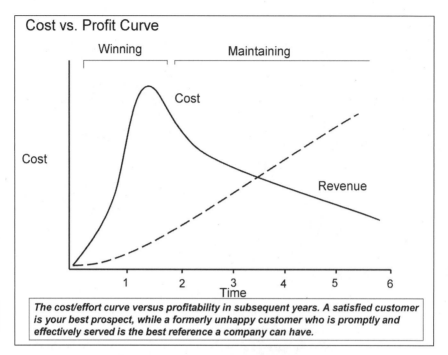

Figure 1.2 Making yourself available to your customers when they believe they need to speak to you is absolutely critical. Play hard to find, and when they do reach someone to talk to, the original problem is overshadowed by the perception you are difficult to find and difficult to do business with.

When customers have the sense that you are not particularly sensitive to their needs, there is often an increase in the number of complaints. To adequately deal with these, you need to make an additional service investment so that complaints do not escalate due to accessibility problems, thus becoming core goodwill problems. When complaints occur as telephone calls, placed to the same center that took the order, a displacement effect occurs that now "blocks" order calls as all your resources are now shared across two functions, sales and customer complaints.

The opportunity to take business is now displaced by the need to satisfy established customers with problems. This is a perfect example of superficial treatment, where partially addressing one problem creates a second and bigger problem. This "knock on" effect is significant across all aspects of a business but none so costly as customer interface disciplines.

There is an even more pervasive, yet less acknowledged problem, and that is the effect on "cash flow velocity." The timeliness of invoice payment is directly proportional to the level of satisfaction a customer has about the relationship they enjoy with your company. Sell second grade merchandise, or treat them poorly and your customers will find, even unconsciously, a reason not to pay you immediately. Maybe, not at all. Or perhaps, as it turns out, they really didn't need the product — a particularly painful experience in the business consumer mail order business, where managing the rate of product returns is critical to the profitability of the business.

What does all of this mean to your business? The whole call contact process — when a customer or prospect initiates the call — takes on a different level of importance from the customer's perception. The customer's sensitivity to careless (or callous) treatment increases. There is a major opportunity to fail if each aspect of your customer call center is not fully planned.

The first part of this book focuses on the strategies, techniques, and technology to ensure you get the most out of your telephone call portion of your customer contact center investment. The lessons here logically extrapolate into the multimedia customer contact center of the 21st century.

33

The Role of the
Customer Contact
Center

In the age of the traditional call centers, the enterprise dealt more
with the turf wars between its field organizations and its inside telesales, than try-

ing to rationalize its customer communication channels. The battle between
inside and field sales is over and — hopefully — the customers won! With the
modern customer contact center, the business community has entered a new
era, (although there are many laggards) where economics dictate the building of
corporate customer databases that make all customer relationship information
universally available. However, the key to this new customer relationship is the
orchestration of all customer contact channels and customer support resources
into a seamless centralized system.

To reach such customer relationship nirvana requires that the organization
not only set achievable goals for its customer contact center, but that it measure
the success of the center against its sales and customer satisfaction goals: both
externally to customers, and internally with other departments with whom the
center interacts. As a customer contact center manager, the key to your sanity
and personal goals are getting the executive management and client departments to set reasonable expectations and fund them accordingly. This places a
significant burden on you to unequivocally articulate the value of your center.

The principal rules for the updated call center are worth repeating: To succeed in the customer contact center business there is one absolute! Your company must establish a clear role for the customer call center. This means clear
objectives and lines of task demarcation must be laid out.

A strategic decision is usually made to establish a customer contact center when there is a demonstrable volume of contacts with particular requests initiated by your customers and prospects to your company (or internal service department), either via telephone or other media, that must be satisfied.

There are several key issues in this above statement that need explanation:

First: the volume of contacts represent PEOPLE calling with a request that can be satisfied by any one of a group of people.

Second: a successful customer call center can no longer look at each call as a stand-alone event or caller as an individual transaction, but rather as a way-point or milestone in an ongoing customer relationship.

Third: the particular request is relatively predictable and homogeneous. It also can be served by any one of several employees by the telephone, web or email. Hence, the need and logic for establishing a customer contact center in the first place.

Fourth: this service function is a defined function within your company, and your customers and all employees are made aware that the group to call for service is the group defined as the customer contact center.

Fifth: this customer contact center group recognizes they are the focal point for customer and prospect service requests and are prepared to effectively service these requests.

Sixth: the center is adequately equipped to process a particular customer's request(s) to the point of high satisfaction.

Seventh: because the center performs sales and customer support functions, it's subject to frequent change and repositioning to take advantage of the opportunities the market hands the company.

This list has undergone expansion and iteration since publication of the book's predecessors for two reasons: the addition of customer options and the ability to enrich the caller's experience through more comprehensive process integration.

The adage, "a customer arrives with a history and leaves with an experience" is a very compelling argument for giving a customer service agent all the tools necessary to respond appropriately to that one customer. Today this is less of a dream, and more of a market reality.

Is It a Formal or Informal Center?

Let's return to the telephone call center model: How many calls should a company receive before formally establishing a customer call center? What is considered volume enough to qualify as a formal customer call center and therefore warrant the attention of management?

Although you can examine total transaction volume — calls or other customer communications volume — there is no simple answer. Of course, the obvious measurement is that of the importance of the caller and the business that is represented by the incoming call.

What is the value of a typical call? How much has been invested to generate it? What is the desired call result or conclusion — a sale or a satisfied customer who will buy later on?

If it is a few important calls a day, interspersed with other non-telephone functions, then maybe there's no need for a formal customer call center. Yet add a little more call volume, revenue and/or goodwill into the formula, and suddenly all the issues and potential to succeed or fail are present. That's when it's time to take another look — is the department or application a candidate for more sophisticated telephone tools?

Also, don't be fooled by the fact that co-workers generate calls. First, this means there are now two people on the call being paid by the same company to be inefficient! Second, the adage of the request being made "by a customer or someone helping a customer" bears repeating.

Two key properties to the call volume are the homogeneity of the caller's request and the ability for the call to be served by anyone on the staff of the customer call center.

The moment a group is identified as operating in the company as a focus for customer or prospect call requests, new functions that can be satisfied by telephone, email and/or web chat are usually quickly found and added to the responsibilities of that group — the customer contact center. This is one of the factors (and there are many) that accounts for unanticipated customer contact center growth.

The Importance of Volume and Revenue Contribution

There is no such entity as a "small" customer contact center. In a small business they may represent the majority of the revenue stream. Almost at any size they represent a substantial cost to set up and maintain. Further, their business mission alone may be significant enough to warrant establishing a formal customer contact center.

The widow's mite: In smaller businesses this is compounded further by the fact the business conducted by the customer contact center is disproportionately large when compared to the number of people as a percentage of the total company head-count and their per capita sales contribution.

Add ecommerce and a formal customer contact center becomes even more essential. The irony of the Internet, the web and ecommerce and customer self-service is this: these channels were assumed to have a major potential to displace traditional telephone calls as a better solution. However, three things happened that the technical promoters of the web and ecommerce didn't anticipate:

- Websites often create more questions (often follow-on questions) than they answer. The result is more telephone calls than anticipated. Thus, many dot-coms have developed significant traditionally staffed customer service centers where all contact channels: calls, web chats, emails and other customer communications are managed.

- Often the inquiry or question has a subjective nuance that cannot be answered with an objective text answer, so a telephone call results.

- Culture. If I am going to give you money by buying something on the web, I might need an additional "feel good" to get me over the objection to placing the order. The company responds with real-time, real people interaction to provide the customer with the answer to those final questions. The web does not fulfill the need and a byproduct of shortsighted web use has expanded traditional call center call volume.

The Request for Service

Any customer contact (whatever channel is used) is technically a service request. Although it may not be a "come and fix my broken dishwasher," service demand, it will have been, nevertheless, created by the company's promotion of the existence of its customer contact center. The contact could be to buy a product or service, a request for information, a plea for help with a complex piece of software that may lead to a purchase of the latest release, an emergency call for immediate aid, or an inquiry about an invoice discrepancy. It can be any one of an infinite number of requests, all of which are requests for "service." It is typically the critical first step (but can be a later step) in a sequence of events of a larger business transaction.

Many technical managers and buyers fail to look at the "big picture" in the analysis, specification and implementation of their customer contact centers. Many opportunities for management convenience and insight and the resultant sales leverage are lost at this stage by not treating the contact as part of the larger business transaction. This is a fact that has become increasingly important in the culture of customer contact centers, and is still not recognized with the degree of importance that it deserves. It's the integration of the customer contact center systems with the ultimate fulfillment and satisfaction that this customer contact requires that's crucial. This has given rise to an industry with the omnibus description of "customer relationship management."

Seldom is the request satisfied completely with one telephone call. More calls may follow — inbound and outbound — and the dispatch of information, personnel or goods to the caller to complete the transaction. There are substantial opportunities for integration of these systems and services that result in even greater economies and efficiencies. There's more on this in Chapter 14.

A Defined Function

It's absolutely necessary that this center have a defined function and responsibility.

Often a customer call center will be established or begun timidly — so as not to upset an existing sales or service process or entities. The best and most amusing examples of this are typically inside sales departments that are set up to support (not compete with) a field sales organization. Field sales has managed these accounts for years and has done so quite well, "thank you kindly. Why should we now refer our customers to some home office type, who has never visited the account, is insensitive to the needs and the politics of the account and will be taking orders and answering inquiries that should be handled by the field organization?"

A turf battle ensues and the company loses productivity, energy, time and money. And all for lack of a clear definition, a strategic mandate and the tools and controls to make it work. The customer call center fails and the costs of doing business increase.

Corporate reengineering as espoused by Dr. Michael Hammer and James Champy in their early 90's work, "Reengineering the Corporation," challenges the types of structure, organizations and processes that object to an inside sales support group where one may not have existed. Despite the logic of such a process, the emotionally based turf battles can kill a start-up customer contact center if it doesn't have a clear top down mandate.

It's equally important that the role of the customer contact center be clearly and unambiguously communicated to both the target market and individual customers as the place where the answer and/or service can be obtained. At the same time, spend as much time communicating to the "internal audience" as you do your external market — clear recognition of the strategy along with support by peers and peer organizations are key to a successful customer contact and support process.

Take special care to ensure that the customer contact center personnel are impeccably trained. Also, coordinate their activities and take care to communi-

39

cate with any other groups that they could appear to be in conflict with. The obvious example is a field force of sales or support staff who are being refocused toward larger opportunities. Unless due care is taken to integrate these organizations, the result can be an ineffective customer contact center and a field organization subverting the center's efforts and wasting valuable time, energy and customer goodwill doing so.

A Focal Point for Customer Calls

It's essential that everyone in the organization recognize that the customer contact center is the focal point for customer and prospect contacts, calls and service requests. Everyone should understand that the center is, or is about to become, the first point of contact for many of your organization's prospects and customers.

Furthermore, it's important that you ensure (with all the correct attitudes, image and etiquette) that the email agent and/or telephone representative has access to the correct information. Well meaning but ineffectual agents are death to a customer call center. This is especially so if they are to wean the account management and support of the customer base away from a more traditional sales and support channel.

☐ ADEQUATE FACILITIES

The customer contact center should be adequately funded and equipped. This means not only budget, but staffing and facilities. When we speak of facilities, we mean "facilities" in the broadest sense:

- physical housing,
- communications services or telephone circuits that bring the emails and calls to your building and allow you to make, respond and originate outbound communications,
- the actual equipment that calls are serviced by, whether it be a lowly call sequencer, a call distributor or voice response technologies that improve caller service, and
- a complement of live agents.

Equipped to Service the Customer(s) to Their Satisfaction

To misunderstand the importance, resulting in under-funding of a customer contact center (from a staff, training and tools perspective) is just short of negligence if your company cares about customers.

Sales and customer service functions in most businesses are often considered the messy end of the business. You have heard the lament, "No, not another customer, I was just getting organized." They are not an interruption, they are your business and they are going to call, write, email and try to give you money any time, any where and any way they want. Sales processes are changing; providing poor service is another opportunity to disappoint someone who has entrusted you with their business.

Traditional Call Distribution Applications

The list of enterprises that can use call distribution equipment is extensive. It is NOT TYPES of vertical businesses that have an application for call distribution equipment, but FUNCTIONS OR PROCESSES that cross businesses horizontally. Call center applications also run diagonally across a corporation — from stockholder relations to bus fare and schedule information. These applications are increasingly candidates for self-service options, but almost universally only work well when backed up with a call center.

Customer Service and Dispatch Operations. Typically a reactive function with heavy interaction with computer aided dispatch and other outgoing communication systems. There is a need for seamless integration with voice response, web self-service, customer database profiles, cell phones, paging and radio frequency communications.

Order Entry or Inside Sales. Where customers place their orders for goods or services by regular mail, fax, telephone, email or over the Internet. This is also a reactive function that can be effectively coupled with account management strategies, such as customer databases, order entry, fulfillment systems, and credit approval to change the speed and volume of calls and orders to favor the customer contact center and center management.

Credit Authorization. Where merchants call into the authorization center as a substitute or additive use of automated point-of-sale credit authorization devices. Although voice-to-voice authorization has declined as a percentage of authorization activity, it still represents significant and valuable transaction volume.

Information Provision. Still a significant field particularly with web-based customer self-service support, FAX-back services, and voice response technologies coupled with database systems and backed by live operators.

Reservations Sales and Service. The reason for the birth of call center-related technology. Even though the web has exploded as an order placement tool, the necessity to support this with answers from live operators to "the final question,"

41

Contact Center Application Matrix	Customer Service	Order Entry	Credit Authorization	Information	Reservations	Catalog Sales	Dispatch	Claims	Shareholder Services	Technical Services	Help Desks	Appointment Desks	Registration	Circulation	Classified Ads
Banking	●		●	●					●						
Travel	●	●			●							●			
Airlines	●			●	●		●	●							
Utilites	●	●		●			●			●	●	●			
Publishing		●		●			●					●		●	●
Healthcare	●			●			●	●		●	●	●	●		
Shipping	●	●		●			●	●							
Insurance	●			●			●				●	●			
Cable TV	●	●		●			●			●	●	●			
Public Transit	●			●	●		●								
Information	●			●	●	●	●	●		●	●	●			
Ticket Sales		●	●	●	●										
Universities	●		●	●			●	●		●	●		●		
Ebusiness	●	●	●			●				●	●				
Distribution	●	●	●	●			●	●	●	●					
Manufacturing	●	●					●	●		●	●				
Retail	●	●	●	●			●	●		●					

Figure 2.1.

or after-the-fact live customer service is essential. The demise of many dotcom strategies was partially due to poor or non-existent customer service strategies.

Catalog Sales. Popping up more frequently as companies discover the business strategy of coupling mail order and telephone sales and service works as a complementary channel to existing "bricks and mortar," television and websites.

A major computer company began an experiment ten years ago to determine whether the sale of small items and consumable supplies was a viable revenue stream and business. Today, its two customer call centers (with 200 agents) deliver over 10% of this company's total revenue. A number of leading PC makers began with sales almost exclusive by catalog and call centers, then added the web as it became ubiquitous, only to expand their call centers to support increased web activity. Today, every major computer company is emulating this successful strategy. How many applications for a catalog and telephone sales strategy exist in your company?

Dispatch. Generally coupled with a customer service function unless this is a staff scheduling and dispatch center. There are a number of applications in this type of center for the integration of databases, voice response and Internet technologies. Once again, a voice channel is required as a seamless back up for "customers" who are denied technology access or during system failures.

Claims Processing. For customers who have experienced a loss, whether it is goods or services. There are increasingly persuasive arguments for using the Internet and telephone in this function. Besides the obvious reason of turn-around time, there are very critical demographic issues encouraging increased use of the telephone rather than less:

- First, aging of the population produces a concurrent reduction in mobility and the ability and desire to travel.
- Second, lower literacy standards versus higher lifestyle expectations. This latter fact ensures the existence of live call centers almost in perpetuity. Leading with customer culture is a wonderful thing!

This has an even more important implication when a company automates — it begins to deny itself to large portions of a population that may have a lower level of educational expectation but still want the goods and services a higher lifestyle expectation brings.

Also, a lower education level is generally coupled with a lack of access to technology, while there is a heighten requirement to maintain connection with the goods and services available. What all this adds up to is an increased importance of traditional channels, like phones or "retail" presence, in very diverse and very specific market segments.

Blind adherence to a consumer ebusiness strategy may effectively close off a market. Consider the huge segment of the economy that are profitably served by brick and mortar check cashing and payment remission services.

Stockholder Relations. A recently automated function in many corporations caused by the upheavals in corporate ownership and control. American Transtech is a case in point. Its customer call center service agency is an AT&T unit that began life in the early 80's, "consoling the widows and orphans" who held stock in the-about-to-be dismembered AT&T. Transtech was so successful it has continued as a profitable strategic business unit within AT&T.

Internal Technical Services or Software Support. Otherwise known as help desks, this application is a particularly complex endeavor to manage. Mainly due to the highly skilled personnel that staff such support centers and the complex nature and length of the call. Here, the call volume may not be high but the necessity to keep vital internal company applications running, account for service levels provided by the help desk and deliver productivity make it a particularly attractive application for ACD, desktop workflow technology and call management techniques.

Appointment Centers. Such as those that show up in modern health care delivery institutions. Self-service for web-enabled patients should again be backed up with a voice call option.

Registration, Permit and License Information Applications. These occur in many hospital, college, state and municipal institutions.

Circulation, Subscription Management and Customer Service in the Print Media. There are some particularly interesting phenomena that occur in the outbound aspects of circulation sales and service that allow smoothing of the inbound demand curve. These increase the flexibility for the circulation department.

Advertising. This is typically the main source of revenue for any media. The opportunities to better manage the capture and renewal of revenue from successfully managing the "running advertising bank" is absolutely dependent on the effective management of the customer call center resources from the new inbound advertiser to managing the "aging ad bank" with outbound calls offering timely advertising renewal advice.

This generic list crosses an even more extensive list of company types as illustrated in Figure 2.1. Large users have generally recognized the advantages of automatic call distribution and related technology and have gone out and acquired the systems they need. These same companies are among the early users of computer telephony integration strategies and adoption of email contact management alternatives.

Now smaller units, which were formerly unwilling or unable to pay the price of entry, want access to the benefits provided by technology. Now the Internet and standards-based technologies have caused the price of entry to be within the reach of any business.

The greatest application of call distribution equipment has just begun as the equipment and systems become less expensive to acquire. The largest markets fall under the 40-agent position and 90% of the total market are under the 100-agent position. The same can be said for computer telephony applications that have now been embraced by vendors of PC client server solutions.

☐ DEATH OF THE AUTOMATIC TELEPHONE CALL DISTRIBUTOR?

Just as sophisticated automatic call distribution (ACD) systems became available to all, email and use of the Internet took off, introducing new channels of customer contact that needs integration into the overall business workflow. This has introduced a challenge that is being answered at varying paces and degrees of success.

Nonetheless, you as the contact center business buyer, are still challenged to understand your business, your business flow and the opportunities and flexibility that must be constantly provided for. Leaving this job to the technical buyers in your company places your revenue and operations sanity at risk daily. As well meaning as these folks are, they will buy first by features, then by cost, and even for their peace of mind before considering your real day-to-day business needs. This may not result in addressing your greatest needs.

Even at this point in time, the customer contact is still dominated by live telephone calls driven by live people that want the interaction with a live telephone agent. The arrival of the web has created more calls that existed prior to widespread deployment. ACD systems, as we know them will change, but their functionality will remain vital throughout the foreseeable future.

45

The Parts and Principles of the typical Customer Contact Center

We've been jumping back and forth between the terms call and contact center. The traditional customer call center has become the customer contact center but still more than 80% of all customer contacts remain live calls. The counter intuitive example is the Internet catalog shopper who dials the phone to place the credit card call with the live agent! As illogical as this may seem technically, changing culture and breaking old habits dies hard.

The future is a blend of channels, all served by the "customer contact center." However, we do not want to diminish the importance of the traditional call center or the worthwhile lessons twenty-five years of evolution have delivered for the cross-channel, cross-media, cross-application customer contact center. Thus, we will review the components, workflow processes and lessons, while at the same time remembering the strategy behind the customer contact center — the processes, technology and people remain intimately entwined.

Initially all references made to "telephone calls" are referring to traditional telephone calls, not Internet or Voice over IP (VoIP) telephony. Later we will call out the differences between analog, digital (time division multiplexed models) and Internet (digital voice packets over IP).

☐ A BIRD'S EYE VIEW

What is a customer call center and what are the physical elements involved in such a center?

To receive telephone calls at a customer call center there must be a connection to the outside world. This is provided through a local loop consisting of phone circuits or trunks connecting the customer call center with the telephone company's Central Office, which is the telephone exchange or the nearest long distance provider's point-of-presence facility. The local loops may be digital or analog. They may be single channels or digital T-1/E-1 spans of various capacities. In North America these circuits are delivered in increments of 24 voice paths (32 in Europe). Asia and Africa generally follow Europe because of their colonial legacy.

In turn, these links are connected or terminated on the telephone switching equipment in the customer call center. Since the equipment is located on the customer premise, the termed coined is "customer premise equipment" or CPE. It may be a key system, if the application is small. If the application is larger, a PBX may be used if there is a somewhat undemanding call center environment. If it's a high volume application, there will be a specific inbound call-processing system; typically an Automatic Call Distributor (ACD).

Attached to the CPE switch are three types of devices other than the telephone trunks. They are the voice telephone instruments used by the customer call center representatives, the management information display devices, such as cathode ray tubes (CRT, VDT or supervisor screen) and the printers. These devices are the ACD management tools and are provided with the switching system.

Additionally, over the last 15 years, we have seen significant deployment of voice response and automated attendant technology primarily for the triage of callers so as to quickly provide simple information and later self-service routines. These voice devices are typically attached behind the switch so that they can be placed in the inbound call routing table like a group or a single agent. Often the application for VRUs (Voice Response Units) was the fulfillment of rote functions formally served by live agents. A caller, upon request, following a predetermined scenario or VRU process failure can be transferred to a live agent on the ACD.

This is increasing in significance as VRUs are used to request account and other caller identification in order to match customer data with the caller. Alternatively, these may provide the customer with simple self-service access. This market has evolved to a point where entire order transactions can be conducted by a VRU system gathering dialing number data, account and credit card numbers and specific caller input. The individual agents or representatives also may work at a CRT device or workstation where orders are entered or customer information inquiries made. These are typically attached to a departmental system with, more often than not, a client/server architecture supported by a legacy mainframe database behind the departmental or customer relationship management application. Upon this system, orders are placed, service dispatched or other customer facing application fulfilled.

The telephone sales or service representatives (TSR) or agents do not perform their role in a vacuum. So they, in turn, must interact with the other aspects of the company that are responsible for meeting a prospect's or customer's need. Little or no fulfillment of the customer request can be delivered over the telephone unless it is pure information, with no need of follow-up. A similar limitation applies to an order placed over the Internet.

The methods and speed in closing the fulfillment loop are critical to customer satisfaction and cash flow. A customer who has had their request satisfied quickly is much more inclined to pay promptly. What these methods are and how they interact with the customer call center represents another element in the delivery of integrated customer experience.

Case Study: *Disconnecting Order Taking from Fulfillment*

This case study gives a good example of why failure to integrate the whole customer response process from beginning to end can defeat the objective of the transaction from the outset.

A certain company introduced a clever device that assured long distance service resellers the ability to tell when a connected call was a billable call. This device "listened" to the call as it was set up. If there was a human voice present, it was a billable call. This meant that many abnormally short calls, previously thought to be circuit overhead and written off since there was doubt (which skewed in the subscribers' favor) could now be billed. The economic rationale for buying this device was self-evident as a reseller could now increase billings for previously lost calls and recover revenue on circuit overhead as productive calls.

The company advertised and began a telemarketing program. It received a substantial number of phone orders but didn't complete many sales because it had not systematized the paperwork or fulfillment process. The company wrongly assumed that once a customer said yes, and the "lead" was turned over to field sales, the deal would close and the equipment would be shipped, invoiced, etc.

The Moral: Just because an order is agreed to on the phone, the systems must be in place to close the cycle and satisfy the customer expectation created by the promotion and the call center representative. The fulfillment system interfaces must be thought through. This can vary with the type of application.

A catalog sales application may be simply satisfied with the use of an order entry package running on the company computer. This system probably would already have the fulfillment function within the software package. A more complex interface may be that of a service dispatch system for the repair of malfunctioning computer equipment. Here the system may interface to a field service, manpower management system, telephone or wireless dispatch facilities and response administration to ensure service is provided in a timely and satisfactory fashion.

To summarize, the customer call center elements are:

- the telephone circuits and services,
- the customer premise switching system,
- the telephone instruments,

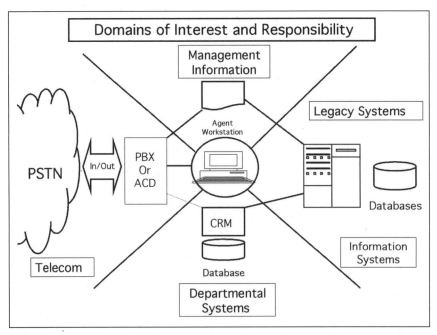

Figure 3.1. There are four domains of technical and administrative interest in a customer contact center. At the center of a customer contact center is the individual customer representative equipped with a phone and a computer terminal. If they are not happy and productive, the system is nearly meaningless. Next (to the left) is the telephone system and network. To the right is the company computer system. At the bottom of the picture is a departmental system often introduced because the mainframe can't keep up with all the needs of all the departments. At the top is the management information system that measures and reports on all the elements of the customer center.

- the customer call center management system tools,
- the staff at the workstations,
- the workstations, and
- fulfillment interfaces.

☐ SYSTEM ELEMENTS AND RESPONSIBILITY

One of the first tasks of a customer contact center manager is to determine who is responsible for critical customer contact center elements. Then the manager must decide how to ensure that minimal problems are encountered when these elements interact.

Within each of the elements in Figure 3.1 there are subsets of responsibility as different entities may provide different parts of the puzzle or different services. For example the providers of information services in your company have evolved from the MIS department controlling the mainframe, terminal users and applications development, to include local area networks, desktop and data networks. Figure 3.1 illustrates the lines of demarcation and therefore responsibility. The point of this is that internally the interests of each of these departments affect the acquisition of the ACD, email system and other contact management systems. CTI deployment, unless done with a clear vision and driven by a very senior user, can devolve into a political slugfest.

51

Technology Goes Hand-in-Hand with the Business's Objective

One of the problems confronting a customer contact center manager, and one of the reasons for this book, is that there are few sources of assistance through this minefield of opportunities. There appears to be a "mind boggling" array of technology and no one place to learn about it.

This is doubly complicated with computer telephony links, customer relationship management packages and the promised metrics promoted by vendors of customer experience management tools. Many vendors have found themselves in the customer contact management business, which they have no direct experience, they only hijack the practical with marketing hype. Despite this marketing distraction the basic systems are relatively easy to understand when the pieces and their roles are explained.

Recently a seasoned ERP professional found himself leading the North America division of a voice recording company. Only after he had dismissed inhouse call center experience as trivial, was he amazed to find telephone systems had been counting calls and plotting service levels for about 20 years. His "eureka" only occurred after he and his expensive ERP recruits had invested months and millions of dollars in trying to understand how his recording system could deliver quantitative call center call and event counts and customer satisfaction levels.

☐ THE TELEPHONE CIRCUITS AND SERVICES

With telephone service, there are two basic parts to understand. Telephone circuits usually arrive at your business as wire circuits that may carry analog or digital service. Simply put, this is the cable and central office service owned by a service provider like the phone company. On this hardware communications events occur. These represent the service that is delivered to a customer contact center enabling the center to make or receive calls, write and receive emails and receive web "hits" (assuming the web server is on premise). It's important to recognize whom has responsibility for the components associated with these telephone and data circuits and usage services.

The telephone company or service provider is responsible for supplying local access to a central office or telephone exchange or, in the case of Internet services, the web gateway. Providing local service and connection is the function of the local telephone company; long distance is the responsibility of the long distance service provider, such as AT&T, MCI WorldCom, Sprint, etc. Note that this distinction continues to blur as regional telcos, cable television and Internet service providers enter the long distance market and long distance and cable companies enter the local telephone service market.

The Traditional Customer Premise Switching System

The following discussion is predicated on traditional telephone switching system, as we have known it for the last generation. With the advent of voice calls being placed on Internet services, new opportunities arise, which we will discuss later.

Traditional voice telephone circuits are terminated on equipment at the customer call center location. Here things can get a little confusing unless we break them out a step at a time. There are five different combinations of equipment that could be used by a customer contact center, which are the responsibility of up to five different entities. These are:

Single Line Instruments: These are frequently acquired in quantity. Attached to these instruments are the associated telephone lines that connect the instrument to the telephone exchange. These instruments are unswitched, meaning that a call arriving on one instrument cannot be redirected to another.

This may or may not be a handicap. These "phone banks" are typically used by temporary inbound or outbound operations, such as temporary fund-raising or political campaigns.

Typically this type of circuit service is available from the local telephone company, who may or may not provide the instruments. Typically they will not, so they must be acquired separately.

CENTREX: Another arrangement that looks a little like the above configuration and also provides service to single line instruments at the customer call center. The features and switching (intercom, transfer, call the supervisor) are delivered by virtue of a switching service at the central office. This is called CO-based switch service or CENTREX.

Although this type of service fell out of favor during the 1980's, it regained popularity when the major central office switch manufacturers began to deploy automatic call distribution services as part of their systems. The cost of entry is attractive, however, as is typical of this platform, the features are too truncated and too late compared with the popular premise-based solutions. The local telephone company provides all CENTREX service. They take responsibility for the entire system (except for the premise equipment, such as instruments or any management information required by the user) which is a feature they sell heavily as an advantage CENTREX has over premise-based ACD systems. The advantage to the customer is the allegedly reduced need for dedicated staff to support the switch. This argument fell into disfavor in the early 80's, primarily because of the high price of lifetime ownership and the meager management data and CTI features offered by CO-based ACD products. IP telephony promises to eliminate separate voice switches and any advantage of CO-based switching.

Premise Switching Equipment, Key or Private Branch Exchange Systems: There may be a system at the customer call center location, a portion of which is dedicated to switching incoming calls. These come in two flavors, a small implementation may occur on a key telephone system (KTS) or a small to medium sized customer call center may be accommodated on a private branch exchange (PBX) with integrated call distribution features. There are now hybrid implementations of these devices integrated into a PC chassis with the capacity of approximately 100 devices. IP telephony and Internet convergence with a Window NT-based universal telephony server is the most advanced in this market segment.

Responsibility falls squarely on the constructive owner of that system. The use of the legal term "constructive" owner is to differentiate between the user and the "legal" owner, such as a leasing company. Typically, the constructive owner is the primary end user who often subcontracts maintenance to a system vendor or a qualified third party maintenance entity. In some large or specialized companies such as airlines or utilities, self-maintenance may be the chosen support strategy.

Automatic Call Distributor Systems: To date the most sophisticated implementation of inbound call management capabilities occur on standalone purpose-built automatic call distribution (ACD) systems. These can be found in all size applications. There have been many recent and aggressive developments in the ACD marketplace that are driving down the cost of acquiring sophisticated universal communications (voice, email and web) features in PC driven systems, which benefit users greatly.

The responsibility for the ACD falls on the owner and their chosen maintenance strategies.

Subsets of the ACD market are the newer entrants into the ACD marketplace — the intelligent network-based ACD-like control and routing systems. These are designed to interface with the Long Distance network or local telephone provider's high capacity digital service availability as a national (or international) control and call distribution mechanism. These systems are relatively new to the marketplace and offer to deliver much of the promise offered by CO-based systems, but on an inter-exchange basis. These systems are interoperable and compliant with many vendor switches. Most involve little in the way of hardware other than the telephone "device" and a customer premise based "telephony server," which supplies the network control logic that effects the network switching of a call to the desired agent location. They can pull service level data from complex ACD systems or the busy state of the individual SOHO agent. These systems then intelligently balance and route calls based on load and business rules across multiple customer contact centers and/or individual at home agents. ISDN or DSL provides redundancy through multiple circuits and makes the "at home" agent a viable customer call serving resource.

Voice-over-IP (VoIP): The latest added dimension to voice switching is the Internet. After early skepticism over "quality of service," bandwidth availability and mediocre experience, VoIP local switching strategies are fast becoming a respectable alternative to traditional PBX or ACD. And the major data infrastructure vendors have started to attack the traditional switch vendors with VoIP alternatives as local area bandwidth and capacity has grown.

This initial skepticism is understandable. In the early days of LANs, most data applications hogged 10-Base-T networks as was demonstrated by sluggish response times accompanied by agent comments like, "wow, my computer is really slow today!" Not good for productivity as the perception generally gets way ahead of reality. As 100-Base-T networks become ubiquitous, other applications are sought to justify the upgrade in capacity, thus both voice and video are being added to the LAN infrastructure. It's noted that gigabyte LANs are within five years of broad deployment, causing most of us will struggle to recall why we ever had separate voice and data cable plant and networks!

The interim step of computer telephony was more about making two incompatible signaling environments work in harmony. CT was about connecting phones and computers. Telephone extensions are typically fixed hardware addresses while IP addresses are logical and virtual. For a start, telephones require hardware addresses or "extension numbers." Call forwarding meant knowing where you were and where you were going so you could publish "your address" to prospective callers. Computers changed this by allowing dynamic computer sign-on with an IP address. This was automatically identified and or assigned based on the session and no longer required you to know where you were or where you are going. IP telephony follows this convenience and that is just the beginning.

VoIP allows calls to be interleaved on the same network carrier (in this case circuit not brand of LD carrier) as data signals because voice is now carried as data packets to and from each party, and in stereo. This creates some serious engineering issues that were initially seen as low fidelity and poor "quality of service" connectivity issues, although these issues are being quickly solved.

Further, do not confuse business application of VoIP with the poor audio quality of early VoIP experiments. High quality voice delivery is available on local area networks when VoIP is installed in brand new call center environments.

Prediction

We have seen major generational shifts in the past between mainframes and minicomputers, minis to PC and thence to client/server architectures. These shifts in computer architecture have reduced prices and spread computing to more users, but more importantly, these adoption cycles have accelerated as more value was gained for users at a lower price. Mainframes to minis took a decade, minis to PC less and client/server less again. One of the key drivers was the availability of application software, which has been typically behind hardware availability. Value accelerates adoption and VoIP promises greatly reduced infrastructure and the associated time it takes to bring technology to productivity.

Telephone systems have a background in regulation that has retarded rapid migration as they have not been particularly application specific. ACD systems were maybe an exception to this.

The initial customer premise equipment technology was analog space division switches. Close to steam powered! This lasted through the 1970s followed by solid-state computer controlled switches that were initially internally analog and then value engineered to be digital. The telephone network was out of synch with the CPE in that the majority of circuits remained analog until the phone companies and carriers realized it was cheaper to "haul bits" versus analog voice connections. Each of these shifts took roughly a decade and the only compelling advantages were extra features and a smaller PBX device. Application software was not a large consideration as the telephone switch architectures were proprietary and closed. No longer.

Voice over IP, although still retaining some premise equipment personality based on the manufacturer, is generally an open standard with much of the application software already developed. It is inevitable that with the adoption of integrated voice and data infrastructure, particularly in new or widely networked applications, it will be adopted far more aggressively than we have seen previously. With the arrival of VoIP, upgrading a traditional telephone switch to a newer model is just an expensive "paint job" as any new customer contact center application development will inevitably favor the integration of voice and data promised by VoIP. It will eliminate much of the inefficiency of middleware and CTI as we know it.

The Entry-level Contact Center

There are a number of combined devices that are sold as customer contact systems. Many are relatively inexpensive and offer a viable entry-level strategy for smaller customer contact centers. These systems are personal computer (NT) based and begin at simple call queuing and announcement devices all the way up to universal messaging systems.

Some include switches, while others augment existing KTS, PBX or CENTREX based switches, deriving data from those devices to assist the customer call center management.

Responsibilities in this last environment are a little more complex, since equipment is from a variety of vendors and service and support considerations.

☐ TELEPHONE INSTRUMENTS

Attached to each of the switching systems are the instruments that the telephone representatives work with. These instruments fall into one of the following categories:

- the single line instrument,
- the KTS or PBX instrument, or
- phone devices based on an IP telephony connection,
- the specialized ACD instrument, and
- the soft phone. This is a PC screen based GUI (graphical user interface) phone (software) and telephone logic (inboard or outboard hardware) that places telephony control on the PC screen or even embeds it in the application so the two operate as seamlessly as possible. For example, the screen launch (salutation screen) and it's population is driven by the arrival of a call at the agents' workstation. The agent initiates the hang-up and call release as they close the "customer farewell field."

Responsibility for maintenance of the telephone instruments lies with either the provider of the device, the local telephone company or system provider. The infrequent exception may occur when the premise instrument is owned by the customer contact center and directly attached to telco lines.

57

☐ MANAGEMENT TOOLS

There is no argument that in most modern customer call centers there is a high need to manage the center. As a result, subsystems have been added to the call switching devices, CO, network and/or premise to provide some level of reporting that, in turn, allows management a superior level of control. These systems are typically limited to gathering quantitative or productivity data. How many calls are processed, not how well they are processed. This is rapidly changing as the ACD vendors search for differentiation in an increasingly crowded marketplace.

These systems may be integrated into the call switching and distribution devices or an adjunct or supplemental system. We will deal with these as we look at the different system strategies later in this book. Responsibility for support generally lies with the system provider.

But just when it appeared that ACD data gathering and reporting systems had matured extensively, a whole new era of call center administration, reporting and management server technology is making its way into use, i.e. the arrival of volumes of customer email, web chat sessions, and assisted browsing. Where better to send it than the formal call center?

☐ STAFF AND WORKSTATIONS

Responsibility for hiring and maintaining staff rests directly upon the customer contact center management unless the campaign is being subcontracted to a third party customer call center agency. Today, there are hundreds of contract TSR (telephone service representative) staffing companies who will outsource your campaign and provide staff and management for the duration of the marketing and sales program.

Every TSR sits in a workspace or at a position. At this position there are the usual accommodations such as a work surface, a chair, a telephone instrument and almost without exception a computer workstation or CRT attached to a host computer system. There are almost no call centers encountered today without computers (though temporary political and fund raising phone banks may remain the exception). Maintaining this equipment is typically the responsibility of the entity that provides the computer system or service, which could be an in-house data processing group or a remote data center or service bureau.

With the advent of screen-based telephony and integrated desktop workstations, the system lines of demarcation and responsibility is further blurred. In a few leading cases, the telephone functionality is incorporated in the PC workstation. This becomes even more of an issue when multiple communications channels like email and assisted web chats are presented to the TSR for response.

☐ THE INTERNET

Internet connectivity in the customer call center has been through the workstation via LAN connectivity to a web server. These systems were initially put in place to deliver individual agent email as part of an internal communication network. It expanded as external Internet access became ubiquitous and specific agents tasked with answering the ensuing customer emails. Integration potential now adds:

- responding to customer email as a task,
- assisted website browsing (with or without concurrent voice),
- web chat sessions, and
- customer call-back requests via "click-to-talk" technology embedded in a website and serviced by customer contact center systems.

☐ THE FULFILLMENT INTERFACE

Fulfillment is complex as it may be a subset of the main company computer system or a remote entity that has the responsibility for promptly fulfilling the customers' expectation begun by the marketing effort and confirmed by the customer call center. It's critical to profitably closing the transaction. As a popular overnight shipping company's advertising campaign aptly states: In the world of ecommerce, there is no such thing as a "virtual package." Nothing significant has ever been delivered over the phone, short of information, ergo the importance of fulfillment.

There are often issues, which customers call about following the initial transaction that need to be handled in the context of credit status, receipt and correctness of the order fulfillment. This is a good argument for providing shipping and customer receivable status updates to the customer relationship system database.

☐ CUSTOMER CONTACT CENTER SIZES

When we talk about the size of a customer contact center and when a business can justify launching a formal contact center, there needs to be some frame of reference.

- Small ......Less than 30 TSR positions
- Medium ..30 to 80 TSR positions.
- Large ......80 TSR positions and above.

Current wisdom estimates 90% of the call centers in America are under 100 agents in size so the majority of call centers fall in the Small to Medium Enterprise (SME) market. There will be some that read this and take exception that I have termed 80 and above as large, when they are managing a customer call center with 300 positions or more. These very large centers are being recognized as less and less desirable because of the sheer effort needed to staff and maintain them. It's noted that whether the center has 30 TSRs or 100+ TSRs, the tasks and disciplines needed to establish a less than mega-sized centers are comparable in variety and complexity to launching and operating one of the gigantic centers, although they are not quite as economically intense. Add all of the newer contact channels and multimedia opportunities and the complexity is increased again so the rules are relatively universal.

People Sell to People

A significant change occurred since previous editions of this work. Voice response technology is now widely deployed and the "port wars" discussed in the first edition ended as expected. Voice response unit (VRU) vendors or interactive voice response (IVR) vendors believed they would supplant most all-live ACD applications with machine voice technologies. This has not occurred to the extent initially predicted due to one thing, people like to be assured they can speak to a live agent, and that requires switching to a live operator. The Internet and web call back features reemphasize this once again. People sell to people. Every dotcom strategy under-estimated the requirement for people. In 2001, it is estimated that nearly 3% of the US working population is working in customer service centers.

Voice response vendors remained "behind" the ACD as a switched resource for three reasons:

- an ACD switch port is cheaper than a VRU port,
- once "behind" a switch it can be "switched" from caller to caller based on demand. A VRU port dedicated to a inbound trunk cannot, and
- VRUs have generally provided poor disintegrated reporting when compared with the more sophisticated ACD systems.

I would like to pose a question that I cannot find the answer to anywhere and the voice response and Internet industry provide perfect illustrations. "Why is it when a new industry is born, even though it is a minor register shift from a previous technology (PBXs and ACDs to VRU and email), they forget all the lessons of the previous business?" In this case, a call center cannot get enough information about performance, yet the IVR and email systems initially provide lousy reporting even though the content of the customer transactions handled is of no less importance.

Importance versus Size

What is counter-intuitive to the casual observer is the fact that size may not be a true indicator of the customer contact center importance. Some businesses are so enamored with the customer contact center as a primary customer channel that the majority of their business is placed and served through that channel. The writer is aware of one case where one 90-agent center processes nearly $200 million dollars worth of a business for a company doing a total of $250 million. These ninety agents in this mid-sized call center are responsible for an overwhelmingly significant portion of the company's revenue stream. Some would consider this only a mid-sized center, others a mega center due to its revenue production capabilities.

Location

Customer call center sizing often influences the location of a customer call center and becomes an important issue in choosing a location. This book is not written, however, to help in site selection. Brendan Read's book, *Designing the*

Best Call Center for Your Business : A Complete Guide for Location, Services, Staffing, and Outsourcing and CMP's *Call Center* magazine provide significant advice and help on this ever-changing market - check them out.

☐ PLANNING A CUSTOMER CONTACT CENTER

Customer call or contact centers are often begun as pilot projects to figure out whether they will work in the company's marketing, distribution and customer support plan. If planned and supported properly, they nearly always exceed expectations and are given expanded roles with larger projects and newly discovered ones.

Be aware this can happen to you. Congratulations. But, this success can kill your customer contact center!

Planning around this is difficult from a factual and a political point of view. The accusations will ring all around. First, you're too optimistic. Then, you're unrealistic about the corporate budgeting system. There are no simple answers other than to operate and try to plan around the potential for sudden success. Educating corporate management to the potential of the customer call center and its unusual dynamics can help to avert unwelcome surprises. In short, prepare for growth.

Overcapacity is a Fact of Life

Customer call centers are unlike most telecommunications applications your in-house telecommunications departments have encountered. This is an important consideration when recruiting a phone "head" to assist in equipping the customer contact center.

In an early ACD book, *A Management Guide to ACDs*, the author, Steven Grant, best describes the philosophic differences of a customer call center to any other telecommunications application....

> " An ACD (read inbound customer call center) is more complicated to manage than other communications devices (read applications) because its operation must be tuned in response to even short variations in staffing and call volumes. These fluctuations profoundly affect so many different aspects of an ACD-based business, it requires day-by-day, hour-by-hour analysis and management judgment to operate at an optimal level. It is never adequate to staff an ACD only for a peak period like a PABX, or to configure the number of trunks simply based on the average hours of

61

highest traffic load like a tandem switch. The ACD's proper role is not to offer an arbitrary service level at random, but to perform consistently at a service level that will generate profit or save money for the profit center"....

These are the statements of a person whose professional endeavors began as a telecommunications manager. Here is recognition from an enlightened telecommunications manager that the core philosophies driving traditional (read administrative) telecommunications planning do not apply to customer call center engineering, planning and management.

When reconciling the higher costs of establishing and operating a customer call center, remember the value of the transaction represented by the call generally far outstrips the cost of the call. Typically this can be a hundred-fold. We discussed the marketing and sales effort costs of customer acquisition and keeping the phones ringing, yet the value of the transaction conducted on the phone is also of real value, and a separate opportunity.

When visiting large mail order catalog companies, it's common to see rows and rows of empty telephone representative positions, ready to be used at the slightest up tick in business. Expensive telephone lines and switching facilities lay almost idle waiting for the next catalog mailing drop or the onset of the Christmas buying season. This idleness and under-utilization is an absolute aberration to the average telecommunications manager or accountant.

Yet, it is a way of life for the customer call center manager. No call center, which reflects the market dynamics of advertising, buying and fashion seasons, can ever afford to be equipped at average capacity.

In the words of Steven Grant, the...customer call center...role is not to offer an arbitrary level selected at random, but to perform consistently at a...level...which will generate a profit or save money...!

Overcapacity is therefore a fact of life in customer call center planning. You need to recognize this early in the development of your operating plan so as to build in significant growth allowances. If your business is successful it will grow. The good news is you grow, the bad news is you move.

☐ ORDINARY & EXTRAORDINARY CUSTOMER CALL CENTER GROWTH ISSUES

Customer call centers that are begun intelligently and well integrated into the corporate structure, grow beyond expectations, at unanticipated rates.

Customer call centers live in the world of "real-time marketing." While talking to your customers and prospects you get real-time market feedback. The best illustration of this is the restaurant business. If a customer is happy, you are promptly paid, accolades accompany the payment and the waiter is presented with a generous gratuity. Great food is seldom badly served. This is good real-time market feedback and, as a result, the restaurant management can make timely incremental changes based on the diners' responses.

Until they introduced a customer call center, few companies had this consistent amount of real-time contact with their markets. Field feedback was spotty at best. Feedback becomes an important part of judging market conditions and adjusting to them, provided the customer call center is well managed and tightly integrated with an internal communications network.

Add the Internet and email and the amount and specificity of feedback grows again. Call centers are a good deal better than web channels at gathering, triaging and handling this feedback as there is a large requirement to sort and discard self-serving self-selected feedback. A live agent can gauge this better from a caller than from a context-free email, which may deserve a callback.

Here is a trick to capture email complaints in context and sort real lessons from "flame mail." Offer a complaint form on your website with a prefilled CGI form. "For better service" insist on a phone number from the complainant "so you may address their issue personally." You will be amazed at the load shedding that occurs from complaints that are simply "noise."

Another significant asset of a live customer call center is that this data gathering can be done near real-time at a lower cost than most other research techniques. It can be done impromptually, quickly and relatively informally. A huge process does not have to be set up to deal with a field organization or distribution network. The research also can be focused at small market segments. Simply ask questions on the phone.

Because all the reps are under one set of managers who require answers to a simple script, the data gathered is much more consistent. Back this up with total call recording and call content analysis as discussed in Chapter 15 and root cause analysis lifts the customer call center to yet another level of insight into customer behavior and satisfaction.

With this potential for success and the recognition that this real-time market intelligence medium exists in the customer call center, new tasks and projects are suddenly found. Test projects are created. Questions are created to ask prospect and customer as they call in. This further expands the workload of the center and, of course, all the traditional growth planning is out the proverbial window.

☐ COMPUTER TELEPHONY INTEGRATION

By any other name this is process reengineering of the customer call center. Consequently, as explained in the Introduction, there are three schools of promotional thought; telecentric, datacentric and workflow reengineering. The telecentrics (the PBX and ACD systems vendors) generally promote CTI as a tool to save money based on circuit savings and labor reduction, i.e. screen pops save call time. The datacentric crowd (computer vendors) believe that the justification for CTI lies in cost savings and the efficiency driven by intermachine "plumbing" that reduces the process time it takes to collect and write data across applications. Both are right for very superficial reasons.

Those who grasp the vision and long-term implications of CTI understand that the real success of CTI is the fact that the platform for fundamental workflow is changed. Reengineering of any workflow process is fundamental, disruptive and downright scary. To grasp this and limit CTI to the realm of "plumbing" is to miss the entire point. The secret is in orchestrating the convergence of channels, media and applications.

This channel convergence begins with reengineering the customer call center at two levels: first, to examine the work process and determine that the process is correct. This requires a fundamental rethinking of the business. And the second opportunity is attacking the customer contact handling process from the outside in and then the inside out! What do customers see?

- First optimize agent availability to serve customers. What do customers face when they contact your center? Do you make it easy and self-evident? How about the VRU options? That is outside in.
- The second aspect is to ensure that it is easy for the agent to work as productively as possible. Ensure the agent has the best tools possible to satisfy that customer. That is the inside out.

A Word of Caution: There are a number of vendors who offer voice mail style on-line surveys that give callers the ability to score their experience as they finish the call. Great care should be given to implementing such a strategy, as these feedback channels tend to attract callers who have the time and the inclination to provide some feedback. This feedback can be undesirable as it attracts self-selecting respondents who often provide statistically insignificant data and erroneous personal insight into your processes and products.

Examine the call types (although phone calls are not the only means of customer contact) and then determine the steps and process required to satisfy a customer (within the call or with subsequent follow up).

Ascertain if there is time to be saved in the entire process of meeting that request and DO NOT limit your thinking to the actual duration of the call. Now you are looking beyond the call at the fundamentals of workflow and the real opportunities to improve process.

Connection
to the Outside World

To receive email or telephone calls at a customer contact center, the center must be connected to the outside world.

Telephone calls arrive on trunks that are connected (the telephone term is
"terminated") on the customer call center switching equipment. More frequently than not, this customer call center system is located on the customer
premise. This equipment is termed customer premise equipment or CPE. It
may be a key system, if the application is relatively small. If the application is
larger, a private branch exchange (PBX) or switchboard may be used if there is
a relatively undemanding environment. For a high volume application, a sophisticated inbound call processing system, typically a standalone automatic call distributor (ACD), may be used.

An alternative to buying a telephone switching system is now being offered by
your local telephone company via a central office-based ACD service, which can
provide a similar range of features to the ACDs based on PBX systems. The main
difference is that there is little or no equipment on premise other than the agent
station equipment located at the customer's call center. Every telephone call is
made through this established and orderly set of control signals. This has expanded to include most of the long distance network providers who offer basic call
routing and distribution, which can include customer controlled network-based
services. This type of ACD service offers significant potential in far-flung enterprise-wide call distribution where small centers and "at home" agents are used.

To receive web inquiries and emails your company and customer call center
must have access to the Internet. This connection is physically identical to tele-

phone circuit(s) and may be provided by the same vendor who provides voice telephone calls. It may be connected (via some sort of firewall technology) to a local web and or email server that is, in turn, attached to the local area network. Alternatively, this service may be provided via telephone circuits to a local Internet Service Provider or ISP — essentially an Internet service bureau with the capacity and skills to manage websites, email and customer inquiries for many individual clients in a more cost-effective way.

The same physical telephone network also carries Internet traffic. Therefore, much of the infrastructure (cables, microwave and fiber links) between COs is relevant to switching the data signals that represent Internet "calls," email and web browsing. The physical bridges, routers and servers that these cables are connected to are often housed in the same facilities. The networks (plumbing) carry not only data, but also Internet, voice and video traffic as data so that the actual communication type is transparent. Today, however, the majority of incoming customer voice calls and Internet traffic arrive on different networks.

This chapter is not meant to be a comprehensive explanation of the workings of the voice telephone network or the Internet, but does provide an explanation of a few parts and principles that affect the customer call center manager. (You can acquire more knowledge from the many fine books available from CMPBooks (1-800-LIBRARY). Ask for a free catalog.

□ THE NETWORK HIERARCHY

Basically all telephony networks are the same, whether they are "circuit-switched" telephone calls or "packet-switched" data or Internet traffic. The Internet is layered onto the existing telephone network to take advantage of this almost universal standards-based network. Despite the time, investment and many technological false starts, the telephone companies of the world have done us a significant service in "wiring the world."

Back to Basics: Telephone circuits or trunks connect the customer call center switching equipment with the Central Office ("CO"). These trunks and the CO belong to the local serving telephone company or "telco."

For a CO-ACD or a network-based service these circuits are station lines from the CO to your premises and/or the individual agent stations. This central office is then connected to other central offices and the long distance networks that are available to carry inbound calls to and from a call center.

To understand how all this stuff fits together, a review of network structure is helpful. Imagine the overall telephone network as connected hierarchical elements — sort of like a "military-style" organization chart. At the highest level, the telephone on the desk in front of you is the most rudimentary and ubiquitous element representing this network. This telephone instrument is connected to all other telephones by many central office switches assembled together in

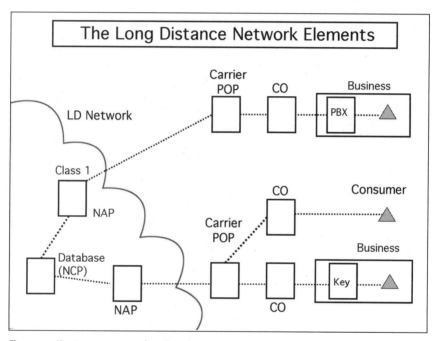

The Long Distance Network Elements

69

Figure 4.1. Key to acronyms and explanations

LD network	= dial up public long distance network, any carrier.
Carrier POP	= Long distance carrier point of presence in the local telephone companies franchise.
CO	= Local telephone company central office or phone exchange
Class 1	= Major long distance switching center
NAP	= Network Action Point is a long distance switching point
NCP	= Network Control Point is the source of control for the NAP
Database	= Network control database

All of the odd numbered boxes in this list are physical switching devices or "boxes," some the size of buildings. These devices contain a high level of intelligence that allows the system to "read and interpret" the intention of the caller by interpreting that intent from the numbers they have just dialed.

hierarchical importance. They fall into two categories:

- Local service or distribution switches serving local telephone subscribers, and,
- Switches that allow connection between these local distribution switches and the points-of-presence switches of the regional and long distance carriers. This latter category is primarily long distance transmission switches.

These traditional telephone switches have moved from space division analog devices to digital and now optical switches, switching voice energy or sound waves, digital expressions of the same voice energy, and now digital data as light pulses representing digital bits of the same voice energy. The driving forces have been the desire for increased capacity and performance out of the existing cable plant and network switching points to reduce the overall cost. Plummeting long distance costs and the entry of new players competing for voice traffic over IP-based networks are partly a result of this technology shift.

Local cable and elements of the public switched network(s) connect all these switching levels. Beginning at the highest level (individual customer) or man/machine interface (phone or voice terminal), the list and order of the elements making up a long distance call look like what's depicted in Figure 4.1.

The Network Addressing System

Whether it's for land-line phones, Internet-based phones or wireless phones, there are two different addressing systems that are need to be understood: The traditional telephone number and the Internet address.

Let's first look at a little background:

- **A Telephone Number**: Every telephone subscriber has a unique and unambiguous telephone identity or "address." This is the telephone number. The only exception to the unique numbering rule is a party line where several subscribers share the use of one line — this occurs when equipment is in short supply or it's expensive for the telco to get service to people out in the "boonies." In all but a few rural applications, party lines are very uncommon and are seldom encountered.

Numbers are assigned to subscribers under an orderly numbering plan. Telephone numbers follow a set arrangement and in doing so, reflect the switching hierarchy we have been discussing. In North America these are nine digit numbers consisting of three number groups: three for the area code (geographic or regional address), three for the exchange code (central office address) and four for the individual physical subscriber line.

- **An Internet Address**: The Internet offers a virtual addressing system that requires a similar hierarchy, although these addresses reside on a server that hosts databases containing tables or dynamic directories that allow a subscriber to access and effectively generate a "virtual" address. A user signs on to a network and generates an identification code and port address that is sent to a server with a master conversion table. This table then matches the user (passwords, access authority, email addresses etc.) to that physical port for the duration of the session. This allows a "find me — follow me" message path that was first used in ACD log-on methodology and bears a resemblance to wireless phone roaming technologies. The major difference is that all of this requires a dynamic address updating system based on users and their sign-on.

- **Wireless Phones and their Telephone Numbers**: Just because a wireless phone is mobile, it does not mean it has a "virtual number" like an Internet address. It still has a physical number that is resident on a terminal in a specific exchange in a specific area code. The difference is that the cell phone "talks back" (effectively signing on at device turn on) to the cell phone telephone network and constantly updates the cellular telephone network to which central office it is nearest to. This is how "roaming" works — the cell phone constantly "checks in" with the network and updates where it is located. The network is prepared to dynamically reroute calls and billing instructions to and from the nearest office to the cell phone as the device travels with the owner.

71

When a cell phone is email enabled, it means that somewhere in the network a directory exists that translates the IP email address to a physical cell phone number and sends the email text to the cell phone.

The Telephone Number

Back to basic telephone operation: By using the previously described numbering plans, certain subscriber capacities are attainable.

There used to be a rule that only area codes could contain a 0 or a 1 as the center digit, 212, 203, etc. More recently exchange office codes are showing up with 0 or 1 as the second digit because the phone companies are running out of numbering codes. Today, to dial long distance, the subscriber must dial 1 (one) before dialing the long distance number. This eliminates the confusion between dialing an area code and a local exchange code with the same leading digits.

Choosing a Telephone Number: Have you ever wondered why some people have simply great telephone numbers that are easy to remember. They simply ask the provider for a simple phone number. Almost always the operator will respond that they only have three to choose from, but that is never the case. This is an arbitrary policy designed to keep calls short. The operator can probably see all the next numbers available for use. AT&T Wireless has gotten so smart they will not only find an easy number but also charge you a nominal $50.00 to do this. Easy remembered numbers are worth a sales fortune.

Vanity Numbers: A number that when translated into the alpha representations on the standard touch-tone phone spells a name or action. 1-800-CALL SWA for Southwest Airlines Reservations is a good example. These may take a little more work and inquiry with your local telco or inbound 800 providers.

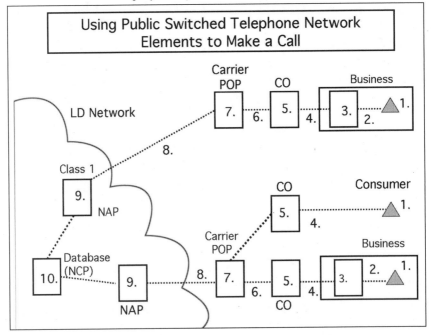

Figure 4.2.

1 Customer switching system.

2 Local telephone line to the telco exchange or central office.

3 Central office or CO.

4 Inter-exchange circuit to the long distance carrier point of presence in or near the local telephone company exchange.

5 Long distance carrier "point of presence" or POP — the long distance exchange.

6 Carrier long distance network.

7 Major long distance switching hubs or Class 1 network offices.

8 Centralized network database where number translations and routing instructions reside.

The Nitty Gritty

When a call is placed to your customer call center the caller raises the handset on the telephone (identified in Figure 4.2 as element 1) and upon receiving dial tone from the telco's CO (2), begins to dial your directory number. As this occurs (2 through 9 in Figure 4.2), necessary instructions are received in the form of the request for dial tone, then the dialed digits (pulses or tones) that indicate the intended party to be called. The call is carried into the network. This is the path the call follows "into" the network. Other elements identified as 7 through 1 in Figure 4.2 must be used "leaving" the network to reach the intended callee, or your customer call center.

To place a call, the caller must have a clear knowledge of the number of the entity they wish to call, and there are a whole series of dialing plans (local, direct dial long distance, toll free) to achieve this. We need not deal with this detail here, however, there is one exception that needs to be explained as it effects toll-free or 800 service (as it is designated in North America).

We have described a typical telephone network as it operates today. The "addressing and routing" signals (the dialed digits) travel on the same circuits the call will be conducted upon. This is called "in band" signaling because it travels in the same signal bandwidth of the voice telephone call.

73

ISDN: This same hardware and circuit structure is also used in an ISDN (Integrated Services Digital Network) network but differs in that the signaling and routing data travels on a separate parallel circuit dedicated to the administrative or control function. This is called out-of-band signaling. To run a digital high capacity network a parallel-signaling network called a Signaling System 7 (SS7) must exist between the local and long distance telephone exchanges.

The call is now separated into two basic aspects: the actual voice call, albeit digitized and the data packet that describes the call and ensures it is routed to the intended party. Basic Rate and Primary Rate service are two types of ISDN circuits that the average customer call center will encounter.

Basic rate is the service that can be provided to a residential subscriber. This is described as 2 B + D, that is the equivalent of two voice grade circuits and a data path. Primary Rate applies as volume grows or a business demands it, T1 capacity circuits or 23 B + D are typical. These, as expected, are the capacity equivalent of 23 voice grade lines and a data channel. The European standards (also as adopted in other parts of the world) are essentially the same with the circuits increasing to 32 for European driven T-1 formats.

The voice signal is digitized voice and requires ISDN compliant connections either as part of the equipment or an outboard adapter. The messages carried on the data channel allow a number of value added services that allow a customer call center to really personalize the service they are offering to their customers. Among the various services are Automatic Number Identification (ANI) and Dialed Number Identification (DNI). These two services will be described later in the context of customer call center applications.

DSL: Digital Subscriber Lines and their various manifestations have similar protocols to obtain higher bandwidth and speed out of in-place subscriber copper cables. Today, the various flavors of the DSL protocol (ADSL, HDSL and SHS-DSL) are increasingly common alternatives to ISDN.

Telephone Number Translation and Beyond

800 numbers have no "physical" or hardware address. They do not exist. An 800 number is a "software defined" or logical address that translates to an ordinary phone number.

When you dial an 800 number, the number is translated into a "physical" address identified as an area, exchange and subscriber code. Just like your residential telephone number. This is a physical address.

This is an important idea. It is important to realize the number of times software — in this case number translation software — affects customer call centers. Here is an identifiable number: 800-LIBRARY or 800-542-7279 (CMP Books). It is published and CMP book customers place calls to it. CMP is billed for usage. But, unlike a regular subscriber directory number, there are no physical trunks identified as 800-542-7279.

What is assigned to this number is a translation target directory number. The 800 number is a "logical" facility, while the subscriber number and the trunks are a "physical" facility. The 800 entity exists because of software, not hardware.

This logical versus physical location is an important consideration when you consider how wireless telephones and (ultimately) the Internet work. Even though no person-to-person communication is made, through number translation and dynamic updating of conversion tables, the cell phone updates the call routing rules. When an email user signs on to their workstation or laptop and access the Internet, the routing directories are all dynamically updated to route new emails and web chats to that device/user.

Back to the landline telephone. The "real" 800-LIBRARY phone number is 212-206-6870. A caller ultimately reaches this number in New York when they dial 800-542-7279 from California, Florida or anywhere in the U.S. Before the call proceeds through the network, the 800 database translates the called number (800-542-7279) to the final 212-number in New York City that CMP answers.

This translation ability means we cannot only translate logical 800 numbers into another physical target number, but we can use different numbers based on different conditions. These conditions may be time-of-day, day-of-week or year or when certain other things happen. These other circumstances could be based on service levels at, or calls offered to, a regional customer call center, or a center outage that invokes a backup strategy.

Once a certain parameter is reach (e.g. exceeding a call count threshold or intentional switch over to a backup strategy), the subsequent calls can be rerouted to another (customer call center) number. Incoming call services are expanding at a great rate. Some carriers can forward incoming calls to a local dealer based on knowing the number called and the number the call is originating from. New call distribution capabilities are being imbedded in networks with the intention of making that 800-service provider more competitive.

75

Customer-owned Network Control Servers

A major data networking vendor purchased a company that provided customers with the ability to interact with the long distance network on a call-by-call basis, routing callers based on time-of-day, volume, queue length, class of caller and other parameters. The customer-based server interrogates each ACD system and the inbound carrier network and dynamically matches demand to supply while gathering extensive service metrics. This type of global control approach is even

Back to Basics: Deep in the network (element 10 in Figure 4.2) is a large network control database. This is called a Network Control Point or NCP. Here resides a master table that cross-references the advertised (logical) 800 number with a regular subscriber directory (physical) number. When the 800 number is dialed, the receiving CO (element 5) communicates with this master database (element 10). A "look up" of the master cross-reference listing occurs. The destination subscriber directory number is sent back to the originating CO. This physical address follows the area/exchange/subscriber code rules so orderly call routing happens within the long distance network. This is called "number translation." This is translating the 800 number to an actual subscriber number. The CO point where the call originated and each switching step of the routing process are called Network Action Points or NAPs. The role of the NAP is determined by the route sent on a per call basis from the NCP database.

more essential when it comes to considering delivering seamless Internet and VoIP services.

Signaling

Underneath the telephone call there is a whole subset of housekeeping controls that allow the call to be made, billed for and then return the circuit path to idle for reuse by the next caller. It is important to understand this control system since there are some billing and circuit maintenance issues that are key to a customer call center's cost of operation and effective circuit use.

A telephone call occurs at two levels, the actual communication (voice or data) and the underlying housekeeping or "supervision" signaling. This signaling information falls into four categories:

1. Supervisory signals:
- Control: transmitted forward as requests for service,
- Status: transmitted backward as answers to these requests.

Supervision can be compared to a "handshake" and acknowledgment that occurs between communicating computers. "Are you there?" "Yes." "How do you wish to proceed?" Etc.

2. Address signals:
- the requested telephone number in pulse format.

3. Call Service Signals (audible tones, announcements, etc.):
- dial tone,
- busy signal,
- reorder tone,
- receiver off hook,
- recorded announcement, and
- system intercept tones.

4. Network Management Data.

It is important to understand that the entire network control system begins with a request for service by the person INITIATING the call. It seems pedantic to emphasize this, but the person making the call has the responsibility for it and the duty to pay for the usage.

This is all fine until the concept of toll free calling is introduced. Here we appear to subvert the whole notion of "user pays" and upset the physical addressing system by introducing a "fictitious" or logical number. What we do not upset, however, is the way calls are set up and the principle of the caller controlling the supervisory signaling. Only the payment responsibility is shifted to the receiving party.

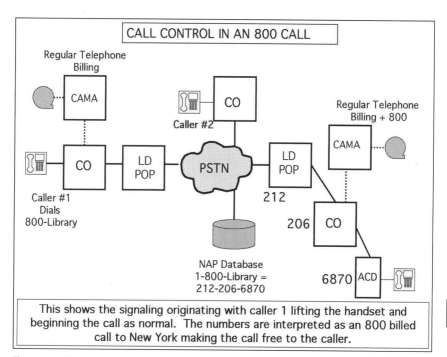

Figure 4.3. Billing Responsibility

As you would expect the notion of "user pays" is an underlying principle in establishing telephone service. With the introduction of toll-free service there was a clear recognition by sellers and buyers of 800 numbers that the market for offering free service to a caller was attractive for a much broader reason — serving prospects and customers by phone. It is no accident that today the 800 business produces over $10 billion dollars for AT&T and the other common carriers providing 800 services. AT&T, MCI WorldCom and SPRINT are offering new and innovative services with 800 service, some of which we will discuss later.

One significant change occurs to the call set up procedure on an 800 call. The billing responsibility is shifted to the receiving party. All the control and supervisory signals still originate and terminate with the caller. This is significant when we discuss call set up and breakdown control on ACD systems in Chapter 10 because the caller "owns" the link until they hang up and the disconnect signal is received by the customer call center.

Integrated Systems Digital Services (ISDN)

ISDN originated as a plan of how the world's telephone network should be since there are distinct shortcomings in the way the analog telephone network behaves due to the principles, rules and media that have evolved into the network. The limitations occur in the area of circuit capacity, supervision overhead on the network, undesirable redundancy in voice and data networks and the explosion of incompatible devices and standards. To date, digitizing this network merely reduces analog circuits and increases circuit capacity. Digitization has not reduced supervision overhead, introduced universal standards or reduced the redundancy in voice or data networks.

It was envisioned that ISDN would overcome this. It was a sweeping revision of the existent network standards and systems investments, but it's a public network standard that's struggled to gain penetration.

ISDN and its various services were targeted at call centers; but in the US, in particular, ISDN-based services have generally been bypassed in favor of higher capacity switch-to-switch based services based on the T-1 protocol.

Digital Subscriber Lines (DSL)

The initial logic for ISDN was to add capacity to a single telephone line so that the telcos could sell more services on their existing subscriber cable plant without digging up the street to lay more cable. The explosion of data requirements to the home with the advent of the web has forced telcos to look at other techniques. One of these is DSL, in particular, the asymmetric variant, ADSL.

The inventors of DSL and its derivatives, ADSL, HDSL, SHS-DSL, and others, reasoned that most of the data volumes were based on downloads to the subscriber, with commands and occasionally file uploads. Thus, they realized that the solution was to build a digital pipe within the existing copper plant that assigned more capacity to the download function than upload. This has the effect of not only increasing data transfer speed to the user, but also allows added capacity for the telcos.

Transmission Concepts

Here we must move into some technical concepts. They will make your life as a customer call center manager much simpler. Engineers will not be so quick to blind you with technical mumbo jumbo.

We have already distinguished between circuit-switched and packet-switched networks. We now need to understand how your voice gets to be a telephone call.

There are two primary types of telephone transmission signaling that a customer call center manager can encounter. These are analog and digital. Today, there are few high capacity call center switches that are analog in nature; however, a lesson in how they operate is basic to understanding the capacity of any system as every one has a finite limit.

Analog Switching and Transmission

This term is not meaningful in correct English usage; however, it is a derivative of the word "analogous" or "similar to." In telephone transmission the signal that is being transmitted is analogous to the original voice. In technical terms it is a waveform of continuously varying quantity (amplitude or loudness and frequency or tonal pitch) reflecting changes in a signal source (typically your voice) in time. In other words if you were to speak into a microphone and see your voice on an oscilloscope there would be an image or a waveform. The vertical or X coordinate expresses amplitude while the horizontal or Y coordinate expresses time. The same waveform is transmitted on a telephone line although as it is electrically transmitted over the telephone line the frequency is higher. Microphones and speakers convert this back into frequencies audible to the human ear.

Digital Telephone Transmission and Switching

Digital technology is a more recent transmission and switching technology that was brought about to economically increase the capacity and speed of switching

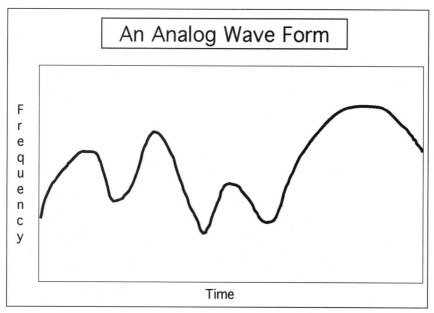

Figure 4.4.

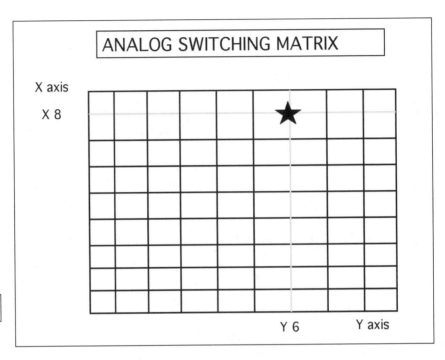

Figure 4.5. In an analog switching system there is a physical switch matrix, which is a matrix of switches. They are either physical switch relays or solid-state switches that represent physical relays. These switches are arranged in a square matrix with X and Y coordinates. Each one of the ports on the X/Y matrix has an address and a physical path to a second X/Y port. The connections are made between these ports (for example an incoming trunk and an agent instrument) by directing the switch point that allows X 8 to connect to Y 6. X 8 could be a trunk, whereas Y 6 might be an agent station. To connect between X 8 and Y 6 an available physical path (unsurprisingly called a talk path) is needed through the network matrix. If not, the call is blocked.

for transmission systems and facilities. Digital systems are less expensive and take up less space. They are infinitely more complex and software intense and require absolutely stable power environments to maintain orderly operation. They are now the norm; and with the need to integrate voice, data and emerging channels such as video, analog services and switches are no longer an option for most centers.

Digital transmission and switching technologies take the analog waveform we discussed above and plots it on an imaginary grid. Each point on this grid has a distinct numeric or digital value. The signal is then transmitted in order of occurrence (time) in the waveform (amplitude) and then reassembled at the end of the transmission. This feat is accomplished by coding and decoding (analog to digital and back to analog).

The device that performs this is called a CODEC. The most common method of coding and decoding is called "pulse code modulation or PCM. PCM coding and decoding occurs at 8000 samples a second. This means the analog wave is sampled and encoded or decoded 8000 times a second. To transmit across a digital network, tiny time slices or slots are assigned on a transmission buss to each of the samples. Attached to each of these digital samples is a destination port address so that the digital packet ends up at the right destination for decoding into the voice of the person calling this destination. The capacity test of a switch or transmission facility is the number of time slots the device will support.

To determine simultaneous conversation capacity of a switch, divide the overall time slot count by two, as a time slot only serves one side of a two-way

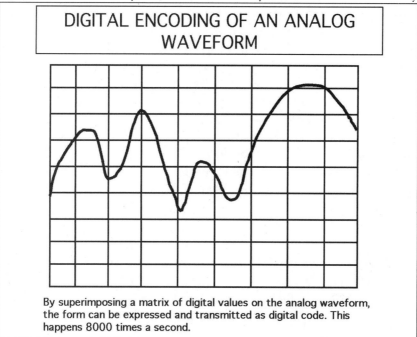

DIGITAL ENCODING OF AN ANALOG WAVEFORM

By superimposing a matrix of digital values on the analog waveform, the form can be expressed and transmitted as digital code. This happens 8000 times a second.

Figure 4.6.

conversation. One is dedicated to the caller and the other to the called party. This is analogous to the capacity of a single talk path in an analog switch that is by nature two way. The connection between two ports in an analog switch equals the capacity of two time slots in a digital switch. A manufacturer may talk about time slots, simultaneous conversations or highways, you must ask what it takes to carry one conversation between two parties, then how many simultaneous conversations can be carried.

You now have the necessary the information to understand the conversation capacity of a switch. So be sure the vendor answers your questions concerning capacity in such a way that you can understand, then document that capacity in your purchase contract.

The capacity issue also may present a problem when you are building a screen pop application where not only voice connections are important but also the capacity of the host to deliver screens over your LAN as the calls arrive. One of the greatest tests of computer systems and applications in a customer call center is the ability to keep up with the pace of the customer and agent conversation. If they do not, the system "steps all over" the business and any value becomes a negative.

Analog versus Digital

Although the world has almost completely moved to digital transmission and switching systems, analog transmission and call center switches are still in use in some older centers. Analog techniques have served us well and will continue to serve much of the world in some kind of capacity for many years to come. Mainly because analog works and technologies that have been developed to support analog transmission are extremely reliable.

Analog has built-in signal redundancy and is often powered by industrial strength telco central office power. This means that when the lights go out callers can still be served since the signal isn't just a digital signal of an analog form, rather the actual sound energy — a replication of the analog wave. But, with a digital system, if digital bits are "dropped" (lost), the message is lost. Also, analog signals are more robust; a little noise may momentarily interfere with the clarity of the signal, but the signal is still present; although more noise does equal less signal. Nevertheless, the signal still goes through unless and until the "signal to noise ratio" overwhelms the signal.

Digital switching and transmission systems are much more economical to design and build as they take up less space and power and therefore allow much larger capacity at a lower cost than their analog equivalents. Digital signals also have a much lower susceptibility to interference from noise over distance, and a digital signal needs a much "cleaner" electronic environment to travel in, otherwise the signal can be interrupted.

All signals — analog and digital — degrade over distance. An analog signal overcomes this degradation by "boosting" or "amplifying" that signal when it gets weak, but this means any circuit noise that's picked up is also amplified. Contrast this with a digital signal that is regenerated or reconstructed. A simple circuit makes a judgment: is this signal a one or a zero? It then reconstructs the signal, boosts it and retransmits it. When no one is speaking, digital signals can be so quiet some phone systems actually inject "side tone" or noise — just to make the users feel something is happening.

Voice-over-IP

This is a big change from the whole notion of switching a telephone circuit to connect two parties to a telephone call. In the traditional telco world, a connection is made and basically "nailed up" for the duration of the conversation, i.e. the two parties have an open line between them that is theirs and theirs alone. Although the telcos involved may have to do some compression, packet and switching tricks in the background, there is still an identifiable path between the parties to the call for the duration of the conversation and the circuit is effectively idle when natural breaks occur in the conversation.

83

The data processing world has long understood this and used these idle spaces in computer conversations to interleave other packets of data. This packet approach differs from the nailed or continuous link approach. The need to automatically accommodate for intermittent packets transmitted with micro time irregularities and jumbled order doesn't faze a computer. It merely takes the packet index data, performs any necessary error correction, then reassembles and presents an orderly file. The Internet has followed that data model, but now we ask it to carry a voice conversation. This represents major challenges. That's where Quality of Service enters the picture.

The term Quality of Service or QoS should not be confused with service quality (as in probability of blockage) in a call center. Rather QoS is the ability of a packet network to emulate a nailed telephone connection. The routines not only must recognize sensitive voice packets, but these packets must arrive in the right order and at the right time. This means: managing bandwidth over the entire length of the connection (the LAN, VPN or public IP network), recognizing a voice (or video) call, and assigning bandwidth based on a higher priority. This QoS discipline is a significant technology that is as important to a customer contact center manager as are issues of physical and logical blockage in a traditional telephone switch, analog or digital.

Transmission Media — Cable Plant

The most common transmission medium is a twisted pair of copper wires. These are used both on premise and in the outside cable plant.

Outside Plant

There is a vast investment in local loops and metropolitan switch-to-switch trunks (metro trunks) already in use. This medium can be a single twisted pair or occur in multiples of up to 3600 pairs per cable. The transmission characteristics of this medium are good although it can be susceptible to noise in the form of electromagnetic impulses (hum) and cross talk (noisy interference).

The intra-switch connection can be plain old voice service or a digital T-1/E-1 carrier of 24/32 voice channels per span. Although less frequently encountered, the end user might run into coaxial cable, microwave radio and optical fiber, which are frequently used by telcos and long distance carriers to bypass media to their larger contact center customers. Up until recently, the need to understand the use and characteristics of these various transmission media were not of great importance to the typical call center manager, as long as they operated flawlessly when carrying that inbound customer call.

T-1 is of absolute importance to customer call centers. Most local exchange carriers and long distance carriers such as AT&T require a local T-1 connection. There are pros and cons in buying telephone service on T-1 loops; most of these deal with reliability. Cost effective circuit sizes justifying T-1 and interconnection issues have dropped to the point that the requirement for just 10 discrete voice paths justifies the cost of a T-1. For a full explanation of T-1-based technology read *Guide to T-1 Networking: How to Buy, Install & Use T-1 From Desktop to Ds-3* by William Flanagan.

In-building Wiring

You contact center's network wiring will typically consist of Category 5 (Cat 5) cables (drops) to the workstation. This is an eight-conductor cable and is primarily used for data. If the network is a 10BaseT local area network (LAN), the cable will use 1 2 3 & 6, with 4 5 7 and 8 being idle or in use by voice transmissions. If a 100BaseT network (there is a growing use of 100 MB LANs) all 8 conductors are used for the LAN.

The ideal is to have at least three drops of Cat 5 cable per workstation — two data and one voice. In installations that are five to 10 years old, Cat 3 and even IBM Type 2 coaxial cable may be encountered. Little can be done with this limited data capacity in the new customer contact center environment, therefore, this type of cabling will need to be replaced.

VoIP relegates all of the signals to one 100BaseT cable so that LAN data traffic, voice and even video images arrive on the same cable. This cable rationalization in a new call center is one of the great arguments for early migration to VoIP, provided the LAN capacity exists.

☐ THE PLAYERS

There is a minimum of three players in the provision of telephone services to a traditional customer call center. They fall into three categories:

- The provider of the switching system (KTS, PBX, PBX/ACD or ACD).
- The provider of local telephone service and access to long distance services. The telephone company or telco may also own the provider of the premise equipment, or as an alternative to premise equipment, may provide CO-based premise switch services called CENTREX.
- The long distance carriers.

In a small to medium sized company, dealing with the local telephone company is typically the responsibility of the customer call center manager. In a larger organization this is normally the role of telecommunications department.

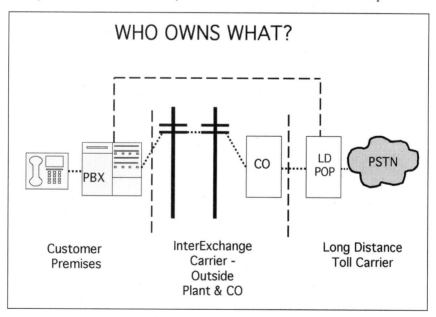

WHO OWNS WHAT?

PBX — CO — LD POP — PSTN

Customer Premises

InterExchange Carrier - Outside Plant & CO

Long Distance Toll Carrier

Figure 4.7.

The telephone company is capable of everything from absolute stupidity to sheer miracles. Your relationship with them and understanding of what they do is important to achieving your goals.

Long distance carriers are changing services and tariffs at an ever-accelerating rate. There is no substitute for up to date information from the individual carriers that serve your location. There are a few basic traffic principles we will examine in a moment that is necessary to use when you interpret this data.

☐ VOICE TELEPHONE TRAFFIC

As we have discussed, there are three economic realities associated with a telephone call to your center:

1. the cost of inducing the caller to call,
2. the potential value in the revenue opportunity offered by the call, or protection of future revenue in the case of a service call, and then,
3. the actual cost of processing the call.

A simple telephone call to your center is not just a simple event. There are two types of call:

1. A simple single dimensional transaction might be a call for rock concert tickets. Get the seat reservation, make payment arrangements and organize ticket pick-up. That's it.
2. A compound relationship where a number of transaction occur over time. An ideal example is a business-to-business account management application.

A call represents a single event in a business-to-consumer (B2C) or business-to-business (B2B) relationship. Both are made up of "milestones" marked by status changes and events that comprise the call. These are:

1. The call begins by a caller lifting the phone or going "off-hook."
2. The caller dials the number.
3. The network receives the dialed digits, interprets and "sets up" the call to the receiving central office.
4. The receiving central office receives the call and "seizes" a local trunk to the customer call center and sends a request for service.
5. The customer call center equipment responds and receives the call.
6. The call is answered with a call processing method that is transparent to the caller. That process may include an announcement, a live person and more frequently an automated attendant offering the caller treatment options.
7. The call is processed by the TSR.

8. The call ends and the customer call center "hangs up" or terminates the call.

9. The breakdown signal from the originating party makes its way (at electronic speed) through the network.

10. Returning the receiving end facilities, the CO, local loop or trunk and receiving customer call center equipment to idle, for the next call, when all this is repeated again.

All of these events represent a call that takes time to process and during that time occupies various expensive facilities at the expense of other competing calls.

It is key to understand the value of a call. What it takes to process it, and then to ensure the customer call center has enough circuit, equipment and staff capacity at the time the traffic or call volume occurs. When done correctly, the objectives of encouraging your prospects or customers to call are achieved.

Remember, as we first pointed out, we believe the customer call center business is undergoing dramatic change. Even though we are dissecting this call transaction on a very basic level, we are not negating our original message — an individual expects service and an expression of respect as an individual prospect or customer with money to spend with your business.

☐ FORECASTING CALLER BEHAVIOR

Difficult as this may sound, inbound call traffic prediction is a very stable science. The problems occur only when reality deviates from theory. An example of this is when the marketing department launches a new website, finds a great buy in spot cable or TV advertising time for a direct response ad, typically without warning the call center. Don't laugh it happens daily, in fact a major computer manufacturer is using just this scenario in promoting their web tools and hosting services.

Understanding that every telephone call represents an event in time that occupies facilities is key. Only one call can be conducted on one line (talk path or time slot) at one time. Every call has two parties to it, particularly when a conversation begins. When a call arrives it occupies an inbound trunk port. While it is in queue, it is connected to an announcement or music-on-hold source port, thereby occupying two ports and a talk path or two time slots. Even if the call is not completed because the caller hangs up and abandons the call, the call still occupied customer call center resources at the expense of other callers.

The real ugliness about an abandoned call is not just the lost caller but also the triple whammy of:

• the negative marketing impression,

- the unproductive capacity use, and
- the displacement effect for the next potentially productive caller.

All this is critical to understanding telephone call traffic and the capacity and resources required to serve caller demand. We will discuss forecasting demand in Chapter 5.

As we add computer telephony, email and web connections to the mix of customer contact center services, we also demand higher performance from computers that support these applications. There are a number of key issues in high system latency and slower response times that can have the "knock on" effect of inflating call length. Operating speed is the key difference between the way typical application computers are built and those that are found in telephone systems.

The computers found driving telephone systems are designed to operate in "ultra-real-time." That is typically not the case in the computers driving most applications. Good illustrations are the microprocessors driving telephone-switching systems.

As we commented earlier in this chapter, ACD traffic engineering does not share the luxury of engineering for average usage as occurs in PBX design and engineering. Having enough resources, network services (lines), a large enough

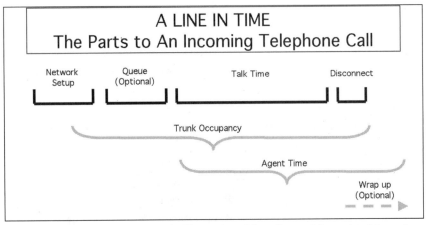

Figure 4.8. In this illustration we are looking at the call from the receiving end and the various gross states the call can be in. We are ignoring the minute signaling and status changes associated with this. The call occupies the facilities and this amounts to "occupancy time." This is not however the entire length of the actual call transaction, as we are recognizing "after call work" or "wrap up" time. Wrap up needs to be considered in telephone traffic engineering and staff scheduling. While agents are in "after call work," they are unable to answer calls, therefore calls are in queue, occupying line and switch resources.

switch and sufficient staff at the specific critical point in time when callers wish to reach you, is the decisive measurement of customer call center success.

Fortunately, in most cases, a strong case can be made for protecting the high marketing investment in inducing the caller to call and maximizing the revenue opportunity represented by the call.

□ INCOMING CUSTOMER
CALL CENTER TRAFFIC ENGINEERING

At this point, we recommend that you try to locate a copy of *The Teleconnect Guide to Automatic Call Distributors* by Steven Grant and read the twenty pages that cover rudimentary teletraffic engineering for the call center manager. Issues such as offered call load versus actual call handling are explored in depth. All the well-known traffic tables assume all calls are handled. In traditional telephone traffic engineering rules, call abandonment rates are not included in these calculations, so over staffing occurs if these tables are blindly adhered to. Alternatives and solutions to this dilemma are discussed. Customer call center traffic patterns are also explored in depth, well beyond what can be accomplished here.

Staffing Issues

This chapter touches on staffing, in general, and the impor- tance of adequate labor pools in the location you have chosen to operate your customer contact center. We will examine some of the newer staff management and motivation issues, including management strategies when we reach Chapter 14, where we will discuss not only strategy, but also the use of management reports that are available from customer contact center systems.

When there are over 150 agent positions in a customer contact center the center management become unwieldy. It's harder to maintain a full complement of staff and the center is more vulnerable to adverse weather conditions, utility problems and natural disasters. Further, with this many "pink collar"

A Great Recruitment Idea

One of the best staffing and motivation consultants in the customer call center business is George Walther who has a great interviewing idea: don't run a help wanted ad in the local paper, look at resumes and set personal appointments — you'll only hire people with a great "interview" face.

If what you are after is a great phone presence, use your answering machine as the interview tool. Set up a separate telephone number and equip it with an answering machine. Tell the job candidate to call for more information. Tell them they will encounter a machine and be prepared to leave an "audio" resume. By using this approach you hear people with great phone presence that you may not consider should you meet them in person.

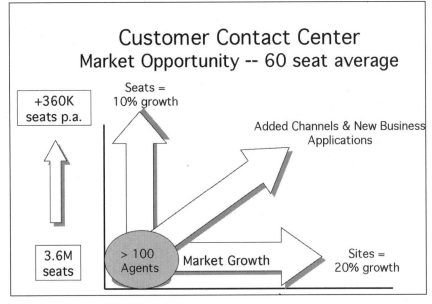

Customer Contact Center
Market Opportunity -- 60 seat average

Seats =
10% growth

+360K
seats p.a.

Added Channels & New Business
Applications

3.6M
seats

> 100
Agents

Market Growth

Sites =
20% growth

Figure 5.1.

employees in a single location, you become an ideal candidate for union organization attention.

Today, more than 3% of the US workforce is employed in stationary customer
contact centers — formal and informal. Although many of these new jobs are
being created at the expense of more labor intense support jobs, such as field
service, particularly in software and hardware (even household appliances) that
have self-diagnostic and help request routines. It's noted that help desks, remote
diagnostics, and the use of depot maintenance have reduced many field service
forces dramatically.

The issues generating the explosive growth in the customer contact center
industry occur in three dimensions:

- Wide adoption of stationary and centralized customer contact centers for
  most customer communications.
- Natural expansion and growth when customers identify a single point of contact as a service point. When a company offers easy access to a service, the
  use of this service will expand.
- Discovery and adoption of additional channels and applications that lend
  themselves to centralized contact center delivery.

A decade ago staffing was an extremely critical issue in certain northeastern states and other areas where high employment and high costs are the norm. This situation has worsened as the demographics of the 90s ran into the labor demands of a predominantly service economy.

A number of large customer call centers in the Northeast initially accepted high abandonment rates due to a lack of staff, but many eventually took the next step: establishment of secondary centers in other parts of the country. The destination in the 90's was the Sunbelt and other right-to-work states where a well-educated labor force was available to staff their call centers.

Telemarketing service bureaus were quick to capitalize on this trend and today leading call center cities like Omaha, Nebraska, Phoenix and Tucson, Arizona have now become highly competitive and high cost markets for customer contact center operators. Even remote towns like Yakima, Washington and Medford, Oregon have continuous telemarketing and contact center job openings.

Companies like British Airways have solved the problem of staffing their world-wide reservation centers by moving labor intense reservation centers to English speaking offshore locations like India and the Caribbean. Any added cost for "hauling the calls" is offset by inexpensive facilities, attractive labor rates and a dependable labor supply.

But also many university towns, resort areas and rural locales were targeted, since a significant number of highly skilled under-employed prospective agents were believed to live in those areas and also many managers court the part time employee; an extremely important strategy for many customer contact centers. But the introduction of part timers as a staffing resource means additional location issues come into play. It is nearly impossible to ask a part time employee to commute an hour each way for a four-hour shift. It then becomes almost a full

---

True Story: *The Untested Assumption*

A large hospitality corporation opened a showcase reservation center near a concentration of middle class neighborhoods in a large southern city. The assumption was that full and part time customer call center employment would be desirable to the residents of the surrounding suburbs. No one asked the target workforce candidates if this was a valid assumption. No market research was conducted. Two years after this customer call center was opened, finding adequate staff was still difficult. They have never received a job application or even an inquiry from the "model" employee they assumed would find this work desirable.

time job. Location in a university or resort town that has an abundance of "part timers" means seasonal effects have a significant impact on staff availability: spring break, finals, "powder days" or "good surf" can instantly destroy the best forecasting and scheduling plan.

Although not easy, great care still needs to be taken to ensure that the chosen customer contact center location does indeed have an abundant supply of labor that is willing to work in the customer contact center environment.

Network-based load balancing, virtual private networks and measurement systems make management of far-flung centers significantly easier. Mainly due to the ability to centrally and dynamically direct customer demand while centrally measuring and managing the supply of staff, skills and service delivery.

□ MAINTAINING ECONOMY OF SCALE WITH A DECENTRALIZED STAFF

Decentralized centers make increased sense, particularly with the advent of centralized ACD or network-based management tools. Available technology allows this to occur in three ways:

- A remote shelf off a centralized ACD system.
- Intermachine communication between similar ACD systems that allow load balancing because of features like "look-ahead" routing.
- Most recently, network-based or compatible telephony servers that allow the carrier network to distribute calls based on caller demand and agent supply.]]

Case Study: *County Funded Private Enterprise*

One example can be seen at a major California health plan. After the LA Earthquake of 1994, the company decided to expand to a satellite service site in a remote town in LA County. Lancaster is nearly an 80-mile one-way trip to the headquarters member call center. They placed a remote shelf tied to the central ACD at the Lancaster site and employed 33 customer center employees serving local (to Lancaster) and overflow member calls. By doing this for 33 employees they:

- ► reduced travel by 24,000 miles a week,
- ► reduced gasoline consumption by 1200 gallons a week,
- ► saved each employee $200 in weekly gas costs, and
- ► returned each employee 12 more hours of personal time a week!

Productivity and morale rose, average tenure increased while the actual costs of operation rose only marginally.

A less popular subset or combination of the first two alternatives is the CO-based ACD systems.

The current generation of standalone ACD and PBX-based ACD systems allow distributed operation by duplication of switching facilities or remote shelves at the second and subsequent customer call center sites. There are costs to these intermachine trunks or networking (virtual private networks for voice and data). Justification for these strategies is a matter of a payback calculation based on the business represented by successfully processing these calls versus lost business opportunity.

In some cases there is a political or public relations value. This is especially true for various public utility companies or a company that needs a market presence in a given community.

CO-based ACD systems from the two major CO vendors have been aggressively sold to the telephone companies and, in turn, to their major customers in both the public and private sector. Both vendors have embraced basic ACD features by enriching the central office to support user features, management reports and a break-through feature called at-home agents.

Now the long distance providers have gotten into the act by embedding apparent ACD features in their networks or allowing access call routing at the network control points by customer-based telephony servers. Their strong suite is nationwide distribution and load balancing with the help of companies like Cisco ICM and Teloquent. They provide customer premise telephony server-based call distribution that can "talk" to most major ACD systems, COs and high capacity services, allowing economical call distribution down to a single at-home agent's ISDN or DSL line. This is truly a breakthrough in at-home agents. The one DSL or ISDN line allows cost effective delivery of both voice and terminal data transactions on one link.

☐ HYPER-DRIVE TO AUTOMATE

The initial push in call centers was to introduce voice response systems for initial call screening and direction. These worked much in the vein of an automated attendant. The next phase was gathering account identification either through network identification from the dialing number (ANI) or customer-entered identification (either an account, membership, subscriber, telephone or social security number), thus routing a specific customer based on the relationship status and potentially the context of the call. Network derived data was not as effective since the calling number to account number matches were low, infrequent and slow to enrich the customer database. Caller derived data proved to be faster, singularly more specific and implied caller consent and support for this automated process.

Lack of staff and available skills has also heavily pushed the notion of self-service: initially via voice response technologies, more recently via the Internet and customer specific websites. This increasingly pushes the former customer call center into the requirements of a fully integrated E3 (existing*electronic*emerging media) customer contact center.

☐ STAFFING TO CALLER DEMAND: DEMAND FORECASTING

There are inexpensive telephone customer call center staff and management optimization software packages that allow the customer call center manager to better match its staffing to meet caller demand. And to do this in a manner that is both service- and cost-effective. These packages are designed to assist in solving the major challenges faced by every customer call center. How do you offer your customer base access to your center every time they call to place an order or ask for help, fulfill the callers expectation to a point of customer satisfaction and do this in a profitable manner? Much of this software is modular in structure and is designed to be flexible and easy to use by non-technical customer call center management.

These systems quickly justify their expense by allowing routine daily, weekly and monthly planning, scheduling and staffing. But more importantly, they set the customer contact center in the critical position of being able to predict the effects, response levels and costs to new marketing campaigns or service initiatives.

The benefit of these systems is simple: reduction in operating costs while increasing service to the caller. All calls presented to your customer call center fall into two categories. They offer revenue in the case of a direct response call or future revenue when it is a call for help, support or service. Now these have expanded to allow repetitive pre-campaign modeling of expected service levels and effect against short-term and long-term revenue goals, thus finding the point of diminishing return.

Failure to adequately service this incoming customer call costs your company:
- immediate revenue,
- goodwill and long-term revenue potential,
- the marketing expense invested in encouraging the caller to call, and

- the operating cost and displacement effect of carrying this call only to fail to provide service.

The major challenge in establishing an inbound customer call service center is matching your caller demand with the correct supply of telephone service or sales agents. Below the first justification layer there is added subtlety. When you miscalculate the balance of staff-to-caller-demand, you can either under staff or overstaff the customer contact center.

Understaffing: With too few agents, the worst possible scenario occurs — your callers call and don't receive the service you advertised they would receive. Therefore, the results are:

- The marketing investment made to induce the call is wasted.
- The caller is given a potentially damaging impression of your lack of care and attention to customer service.
- Any revenue potential represented by this call transaction is lost unless this prospect or customer calls back.

Overstaffing: Providing perfect service, but at the same time having more staff available than is necessary to serve the inbound callers is expensive unless the value represented by the callers and their reason for calling far outweighs the cost of overstaffing.

97

Software Solutions

Demand forecasting and staff scheduling packages are designed to help solve the staffing dilemma. The immediate benefits of such packages are:

Reduced operating costs. It is recognized by the customer call center industry that a staff forecasting and scheduling system will reduce staffing costs by up to 15% by better matching staff to caller demand.

Better staff use. If you run a center with 20 full time agents you can potentially reduce you staff head count by as many as two agents. You may wish to redeploy those staff members to better match caller demand, extend call center hours, delay asking for additional headcount or take on new projects without additional staff.

Better customer service. It is also recognized staff forecasting and scheduling allows management to reduce the number of lost calls by better matching the staff complement to caller demand thus providing better service to your callers.

Increased revenue potential. If you give better service to more callers you have a greater opportunity to capture any immediate or future revenue offered by these caller transactions.

Add in reduced management time and cost to prepare staff schedules, and you have the formula for a winning customer contact center strategy.

In many small to medium customer call centers, service level analysis, forecasting and scheduling are done manually, which requires specialized staff and/or substantial time away from active call center management by supervisory staff. It is time consuming and cumbersome to gather the data, which can lead to inaccuracy and less frequent forecasting and scheduling to match staffing to caller demand.

If these tasks are performed manually they are not performed as frequently as the dynamics of a typical customer call center demands. If you are among customer contact centers already enjoying the benefits of a PC-based forecasting system you are already using less staff time to data collection and analysis. But, such systems also position the customer call center to be more cost effective and the company more competitive since larger competitors and their customer call centers use forecasting and scheduling systems to ensure they are cost competitive in selling and serving their customers. Managing the customer call center to build sales and deliver customer service cost effectively is vital in a competitive market.

Adding Outbound Calling to the Mix. Forecasting service levels when you control the function of outbound calling is less of an issue as it smoothes the spikes by having available agents who can revert to inbound calls during spike periods. Staffing to blended calling strategies works to retard the impact of unexpected spikes.

Adding Email to the Contact Center Mix. With the addition of email requests and orders, the need to anticipate this added customer contact load into the forecasted demand adds additional complexity, especially when the differing customer response expectations can be used to smooth staffing requirements. Most of the current generation of demand forecasting and staff requirements and scheduling packages include this as a component.

Adding Assisted Web Browsing and "Click-to-Call" to the Mix. These are customer or prospect driven. The propensity for a website visitor to request assistance as a call back request or assisted browse session is based on the number of hits and exception ratio of browsers that request such service. These exception transactions then look like inbound calls to the forecasting engine and need to be treated as such.

☐ CUSTOMER DEMAND FORECASTING AND STAFF SCHEDULING SYSTEMS

These systems consist of a FORECAST module to forecast caller demand, match REQUIRED STAFF headcount to forecasted demand, SCHEDULE STAFF and allow for any necessary SHIFT BIDDING to accommodate individual staff preferences.

Historic performance files are built from customer contact call volumes. This is the basis for all forecasting of future calling activity. This data is typically gathered in monthly, daily and half-hour increments and is presented in the same way by the forecasting module.

This file may be developed in two ways. First by loading actual historic data gathered by management from the customer call center switching system or other source. These systems should provide data entry templates or forms to aid the new user in identifying and loading key data, rather than trying to generate enough meaningful historic data after installation.

If the center is a new installation and no history exists these systems should include a number of generalized models of typical customer call or contact center size and performance so the management can begin from some performance base.

Where historical data exists these systems typically provide an online guide for gathering and loading basic data. These systems provide key form fields by months, weeks, days and half-hour increments. They also allow for exception days. These account for holidays or other exceptional days brought about inclement weather, unusual labor conditions and other unprecedented developments.

Improvements have occurred in modeling of theoretical conditions such as how new marketing conditions (new product, a recall etc.) or service policy change (repricing etc.) will affect the customer contact center service levels.

Customer Contact Center Database Profile

These systems typically provide for the building of a comprehensive customer contact center database, which consists of information arranged hierarchically by contact center element. For instance: employee, supervisor, team, group, and customer contact center. The basic employee information might include:

- Proper name
- Birth date
- Social security or employee number
- Hire or seniority date

- Other seniority considerations such as skill levels and training classes attended
- Employee type such as, full time, part time, relief or floater
- ACD position or log in identity number
- Any fixed team, group and supervisor assignment
- Any shift preference
- Any lunch break schedule preference
- Agent skill set and training matrix
- Primary email abilities
- Weighted pay rate
- Other basic data

With early ACD systems, no allowance was made for personal log-on identification and therefore "free seating" or dynamic workstation assignment was used, which could create a problem. If this was the case, it was necessary to maintain a separate record of who was assigned to which ACD station to track agent work and productivity. Today, the ACD should provide a way to track agent-to-position assignment, thus producing meaningful reports by individual, even though no ACD agent sign-on is available.

The basic customer contact center database profile may include such items as:
- Company name
- Department names
- Group names
- Team names
- Assigned supervisor name
- Hours of operation for each customer contact center element
- Actual or assigned value of the typical call received by each team or group
- If revenue calls, the typical conversion rates

The Forecast

Once there is an office and employee database, it's merged with the center's historic performance data to provide the forecast. Forecasting future center performance is only possible with some historic experience. Forecasts are generated top-down: by year, month, day to half-hour increment. These are prepared and presented based on the available historical database. In the case of a new customer contact center or an existing center where poor records have been kept it is a chore to gather effective historic data.

The user may modify certain parameters to reflect expected changes based on anticipated growth brought about by an expanding business or introduction of additional sales or service campaigns. The effect of an unusual day such as a statutory holiday, calendar or special business event is automatically factored into the forecast. Special and unusual events can be forecast. All data is presented in columnar and graphical form, as a screen display or printed report.

Match Required Staff Headcount to Caller Demand

Once the forecast is built by half-hour, by day of week, and week of month, it is necessary to match the supply of available staff headcount hours to caller demand.

This module converts the forecasted half-hour call activity into the staffing required for each half-hour period. This can be done to meet a predetermined service level or may be balanced against center's expected revenue and costs. With this calculation various staffing level costs, telephone trunk and usage charges, and overhead items are balanced against expected revenue to determine maximum revenue and if and what abandoned call rate is tolerable.

Because this is an onerous task at most companies, assumptions are made which attempt to normalize data. Week one and subsequent weeks of a month look alike. In many cases this is an erroneous assumption — invoice, collection and other business cycles distort weeks and even days of a particular week. Statistical normalization and rounding ignores staffing subtleties, increases the customer contact center's staffing costs, erodes service and retards revenue opportunities.

By applying a system, these subtleties and nuances are recorded as a baseline and recognized as key data in all subsequent forecasts and explanation of the required staff headcount to meet desired service levels.

Service Levels

A service level is very much a function of what level of service a company wishes to provide callers. The received wisdom of the customer call center industry is that a live operator will answer 80% of all callers in 20 seconds. This may be adequate or, conversely, totally unrealistic.

The issue is what your company will accept. This is based on the tolerance your callers have for being queued or delayed with announcements and music before they speak to a live operator or voice response device. This is considered from an economic perspective.

It is necessary to allow the customer contact center management to experiment with a variety of service levels and determine any points of diminishing returns by pursuing completion of every possible offered transaction. If a transaction produces X revenue and costs Y marketing expense how many staff can I employ to provide potentially perfect service without it negating potential profit?

Adding Email

Forecasting systems that don't have a provision for email and other deferrable work are completely outdated, as a whole new environment exists with the addition of additional contact channels that contradict traditional thinking. The addition of electronic and emerging customer contact media channels to a customer call center mean that other customer communications can be interleaved into the general workload. This also means that overstaffing may no longer be the costly miscalculation it used to be. After saying this, however, there is not only a real need to match skill sets to call handling (inbound versus outbound), but now, there is also the need to determine an agent's ability to comprehend and respond to written email. The three skill sets are different and need to be considered before "blending" inbound and outbound calls (including media switching and actual or planned click-to-call), as well as email response.

Email servers provide some assistance as they often include "response routines" that address frequently asked questions with common responses. But, consideration also needs to be given to acknowledging the different response requirements of a caller (waiting on the line) and an email correspondent who will accept, but may not like, a less than real-time response.

Adding email response tasks to a call center is easier said than done. There are two significant issues that need to be rationalized. First, email is assumed to require a lower level of response because the communication is "batch" whereas a call is "real-time." The caller is waiting on the phone, whereas the email correspondent usually expects some delay in response.

The second major issue is breadth of customer contact center agent skill sets and what your staff tends to be good at. An acceptable telephone manner and acceptable typing ability are not necessarily concurrent skills. Adding the ability to read, comprehend (research) and articulate a meaningful email response is often not a skill found in many call center agents. The reverse is also often true. A good correspondent (email or traditional print communications) does not necessarily make a good call center agent. The problem is to rationalize across these skills to deliver good quality responses and the desired experience to the customer.

The theoretical solution is to divide your contact center staff into three skill preferences. Then hire to these skill biases. The first group (group 1) is call-centric (in and outbound preferences may be an additional sort for this group). The second group (group 2) has better correspondence and email skills, while a third group (group 3) is agnostic. Give the entire staff ACD call and email workflow features on their desktop and triage emails to group 2 first and calls to group 1 first. Group 3 is the first group to receive calls that overflow from group 1 when all agents are busy and the calls are on hold beyond desired queue time. Group 3 is also the first to receive emails that fall outside the capacity and desired response criteria (target turn-around time) of group 2. Group 2 becomes the group of last resort when calls swamp groups 1 and 3.

The good news is that as long as customers who prefer email are willing to wait up to a day for a response, the staff required to serve the email workload can be used to address caller demand spikes and smooth requirements in contact center staffing.

Take advantage of the interesting phenomenon that is currently working in favor of the customer contact center, i.e. the acceptance of less than real-time responses to emails. Nonetheless, if your email correspondents realize that email is near real-time, they might begin to demand a near real-time response. Thus, the better the service you provide, the more service customers will come to expect.

This dilemma heats up as media or channel switching becomes more viable. This is where, within a single session, what begins as a customer making a web inquiry, migrates into a decision to buy, but first there is a need to ask a question or to place the order with a live agent. The desire to keep the customer in a single session has vast implications in not interrupting the sales process and introducing an arbitrary excuse not to complete the order.

Schedule Required Head Count into the Correct Shift Slots

Once the expected customer contact load is spread across the time frame which callers typically call it's then converted to represent typical employee work schedules. It is now necessary to match the available individuals to the available shift slots. Simplistically this compares with manually sorting incoming mail (demand) with mail slots (supply) in a timely fashion. No available staff; less service. Too much staff equals over capacity.

Customer call centers fall into three categories when you consider the hours they work:

1. They may be open during typical business hours and they only deal with consumers and businesses in their local time zone. Typical hours of work are nine to five.

2. Second example is a business-to-business customer call center that works extended hours to cover the extended business day worked by a national customer base. Here a number of staggered and overlapping nine-hour shifts are necessary to cover expected customer calls.
3. Then there are those centers that serve consumers for extended periods up to 24 hours a day, 7 days a week, 365 days a year.

The scheduling component produces a set of shifts that need individual staff members to fulfill. It is necessary now to match the names with the available shift slots. This may be done by simple ASSIGNMENT of shifts to the available employees or may be done according to some plan dictated by staff seniority or union contract.

Shift Bidding and the Staff Schedule

In some cases SHIFT BIDDING is necessary to accommodate individual staff preferences. Here the SCHEDULE is published showing the shifts required to be filled. The staff is given the option to bid on their desired shifts. Beginning with the staff member with the oldest employment date or other accepted priority.

Upon completion of the bidding process each individual shift is assigned the name of the successful candidate.

Adherence

Adherence is an additional module in a number of these systems that allows an interface to the ACD data gathering function. It tracks agent sign on and sign off then provides a report that shows how closely staffing reality matched the scheduled plan.

Payroll Interface

There is a rather different approach that has been taken by some customer contact centers; particularly those that are determined to get the most from their employees. One very successful airline has taken this to the furthest extent and pays their agents off the ACD sign on/sign off reports. Agents do not get paid for being idle.

The interesting shift here is really philosophy and not simply the technical interface of the ACD time keeping to the company payroll system. The company has made it known that to work there, the agents do not sign in on traditional attendance sheets or punch a time clock. Rather they are paid from the accumulated ACD sign on time. This subtly shifts the management of personnel presence to each agent. It also reduces supervisor time and involvement in all but exceptional situations and has proved to recover an average of 30 minutes worth of agent "time shrinkage" per day! Instead of hitting the time clock and then the coffeepot, agents go to work *after* the preliminaries are attended to and *before* they go on the payroll — leading to a more powerful and effective staff.

☐ INCENTIVES AND AT RISK COMPENSATION COMPONENTS

More and more call centers are finding that additional incentive payments make employment in a call center more attractive. One hotel reservation center deems ten percent of the compensation available to be "at risk" or performance based.

This requires pulling pure quantitative data (calls offered to the agent) from the ACD, and comparing this to the number of room nights sold. The potential inequity that can occur: a single room night sales opportunity versus an individual seeking an extended itinerary.

The data is then normalized by considering the average call length and typical transaction profile against the number of room nights sold. The problem is that today no single system gathers all this data to deliver such a "gods-eye" view.

105

☐ STAFFING CHALLENGE FOR THE CUSTOMER CONTACT CENTER

As more channel alternatives are introduced into a customer contact center, it's necessary to look at the total "transaction workload" (not to contradict the core philosophy of these events being milestones in customer relationships) to gain the required understanding of staff skills, strengths and task distribution.

Further the notion of productivity management and workflow/task distribution takes on new meaning and flexibility in setting staff levels and business opportunities.

☐ CUSTOMER EXPERIENCE MAPPING AND MANAGEMENT

Much has been made of enabling a customer contact center for customer relationship management or CRM. But, the next big "term du jour" is already making its way into management consciousness and that is customer experience management or CEM. Nevertheless, most of the tools proposed today have merely opened Pandora's box and do nothing about helping manage the issues that are claimed to be identified. Still, there are significant opportunities in the customer experience improvement market space, as we will discuss later in Chapter 15.

The Budget: Building A Business Case

A typical comment following a discussion or presentation on the technologies and techniques of customer contact center management goes like this. "I really enjoyed the presentation, but we are just beginning our customer contact center. Management wants to try the concept out before they spend any money."

The problem here is not the lack of budget, but the lack of management commitment to a unified customer contact strategy. What is more important than money is the belief in the customer contact center as a key part of a company's sales or service strategy. Management is not committed to a unified vision that the customer contact center can provide immense leverage to traditional sales, support distribution and market administration techniques. Despite all the remonstrations we could indulge in here, lack of budget and limited management commitment and vision are handicaps that are there to be overcome.

One of the finest contact center consulting practices in the world makes a point of not undertaking consulting appointments if they cannot influence the strategy upstream of the customer contact process. They deem the first step to be absolutely vital to the success of any project — corporate management's understanding that the customer contact center is a part of the corporation's overall business strategy. Only then do they take the next step: identifying the processes and staff skill sets necessary to serve the business's strategic goals. The last step is identifying the technology, and then modifying existing and/or acquiring new technologies to support the customer contact center's strategic goals, processes and people to ensure a sustainable process.

Cost of sales and servicing customers are expanding in most companies as products and services get more complex and competition for market niches becomes more intense. There is a real hope that process reengineering will challenge bottlenecks in the customer satisfaction delivery process.

Despite a finite customer contact center budget, as manager, you are expected to perform to the executive management's expectations. However, if you are a customer contact center manager who is dealing with a limited corporate vision, then you must challenge the perception that the customer contact center is at worst — unimportant, or at best — a secondary channel. This can be difficult. Worse yet, overcoming limited vision on the part of corporate management can be an impossible task if you are not equipped with the necessary tools to convince key executives that more of their support is needed.

☐ PUTTING COSTS INTO PERSPECTIVE

We began this book by talking about the cost of acquiring a new customer, and the value each customer contact or call represented. Later, we will revisit a diagram we used in Chapter 1, but with more context added. From this you can develop an argument to assist you in developing rational and convincing budget request.

Although the first question usually asked concerns the center's size, since there is no such thing as an unimportant customer contact center, size is of little significance. The costs of a customer contact center are another matter — a 20-position pure customer call center cost roughly a million dollars a year. And that is only the direct costs. We have ignored the marketing investment to drive prospects or customers to call or email a request, but we have also omitted any revenue that may flow from successfully completing a transaction.

Next, you need to understand how big the project is likely to become, and therefore how big will the customer contact center have to be to serve the application. This is important from a facilities and funding point of view, but may be a difficult question to answer at the start of a project. Customer contact center sizes are based on call volume and the number of calls an agent force can effectively process in a given work period. The calculations are call length dependent and vary based on the type of transaction to be conducted. A "help desk" transaction may average 20 minutes per call as the software analyst assists the caller through a complex sequence of events. A directory assistance call to the telephone company may be over in less than 25 seconds. Understanding the average call length and the anticipated volume of calls allows calculations to be made that will provide a rough estimate of the size the center.

A small customer contact center may never anticipate growing larger than 30 telephone representatives, yet their mission may be just as critical to their small to medium enterprise as the reservation center of the giant airlines. Relate the revenue potential as a percentage of the company or business unit revenue volume. Do this by polling similar center managers, which you can find through the Telephone Chapter of Direct Marketing Association at 1120 Avenue of the Americas, New York, NY 10036-6700, Phone: 212-768-7277 or www.the-dma.org. They may also have local chapters in your town or state.

Understanding the value and the costs of the customer contact center opportunity is vital to being able to present your budget case to management. So, let's examine these customer contact center cost elements in detail.

☐ CUSTOMER CONTACT CENTER COSTS

The customer contact center will need staff, circuits and some sort of system to switch the incoming calls.

Personnel

Staffing: 30 telephone reps. 40 hour week at $12.00 at hour	$480
(This assumes a minimum wage and a commission or bonus)	
4.2 weeks per month, plus benefits and salary burden of 30%	$2,620
30 representatives, per month	$78,600
x 12 months	$943,200
Supervision: 3 Supervisors — 40 hour week at $18.00 an hour	$2,160
4.2 weeks per month, plus benefits and burden of 30%	$11,793
x 12 months	$141,523
Total	$1,084,723

In addition, there is also the necessary management and support staff beyond the basic customer contact center staff. The variables become less predictable at this level, and estimating costs are not as simple. They are both essential and are in addition to basic staffing costs, which add to center overhead.

Telephone Circuits and Service

The average cost for national 800 service is between $5 and 10.00 per hour. This can BE higher if all of the customer contact center business occurs during the business day. Given our model customer contact center, which has a complement of 30 telephone representatives, there is probably enough work to keep them employed constantly on the telephone, for at least 80% of the day, or 192 available man hours a day.

192 hours x $10.00 per WATS hour	$1,920
5 days per week	$9,600
4.2 weeks per month	$40,320
12 months per year	$483,840

Even if the carrier costs are half this estimated amount, it is still a significant budget increment.

Your customer center may not be positioned to serve a national customer base, so it may not need 800 service, therefore this part of the calculation may not apply. Nevertheless, do not omit trunk access and any usage charges associated with calls arriving on the trunks.

Email and Internet Chat or "Call Back" Charges

Contrary to perceived wisdom, the web has not displaced telephone traffic, but instead has had the overall affect of increasing total contacts, in general, and telephone calls, specifically. The average customer contact center is likely to be already staffed to handle (or is considering the addition of) email, web chat and web-generated calls ("click-to-talk" or "push to request a call back"). Many web-generated contacts can require that a telephone call be made to a conventional telephone number. This is due to not only to call-back requests, but also because of some ambiguity within a customer's or prospect's correspondence. This adds to the center's total carrier costs.

The Telephone Support System

A well equipped inbound telephone call management system, with support for 30 telephone representatives, two or three supervisor positions, the requisite trunk capacity, and reporting capability can be acquired for about $1500 a position. There are systems that cost more and some can be found that are less expensive; but, as expected, each offers greater or lesser feature sets to address the customer contact center application.

Previous published books by this author concerned buying this type of automatic call distribution equipment and argued for more features, not less. A well-equipped automatic call distributor is a vital tool. Call center performance and business results are enhanced with the right machine and handicapped by selecting the wrong one.

Since the original writing in 1987, PBX manufacturers have done a better job closing the gap in providing more user control and greater reporting via onboard or outboard call management reporting strategies. The author has been involved in the adaptation of sophisticated call management technology to some of the most popular PBX-based ACDs. The most striking discovery is the horrendous inaccuracy and cavalier vendor attitude to agent data gathering. In one case, there was no guarantee the data being gathered about one agent was indeed that agent. Given that performance reviews are based on this data, it's no wonder that many human resources department became concerned and forced vendors to address this issue. Still, some ACDs on the market and/or currently in use are based on decades old PBX systems and cannot embrace these customer control and reporting nuances. This is particularly critical now that we are adding other media channels, including the potential for a customer to switch from web to voice as a seamless transaction. Although "plumbing" and plumbing management are important, more critical is not interrupting a sales transaction by forcing the customer to end a web session and make a separate phone call. Buyer beware!

Workstations and Applications

Some sort of computer workstation will be needed for each agent. So, even though the hardware and LAN connectivity may cost less than $1500, when you add in the costs for the departmental servers, database and software systems, and the necessary training, you can easily add another $10,000 to $15,000 a year per seat.

First Year Customer Contact Center Cost Summary (20 agents)

Personnel Costs	$1,084,723
Circuit Costs	$483,840
Telephone Equipment	$105,000
Work Station Technology	$400,000
First Year Total	$2,073,563

There is no surprise here. Operating costs are ten to twenty times any hardware acquisition costs. Yet, finding, training and keeping a productive staff is vital. Just as wise and objective management is vital.

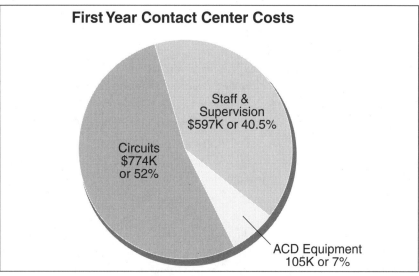

First Year Contact Center Costs

Staff & Supervision $597K or 40.5%

Circuits $774K or 52%

ACD Equipment 105K or 7%

Figure 6.1.

Since this book was first written labor costs have gone up while staff skills and availability have dropped. Hence the huge business in outsourcing, even beyond national borders.

As Figure 6.1 indicates, a start up customer contact center is not a minor investment. Of course, you should use the numbers presented here as only a guide. You can adjust your numbers around these to get a feel for your particular customer contact center costs.

Predictable Side Effects

It is important to note that deteriorating service levels cause more than disgruntled customers. Disgruntled customers complain more, then without satisfaction, go silent, at least as far as your company is concerned. However, they can be expected to tell up to 25 other prospective customers of the bad experience they have had with you.

More telephone-based complaints displace genuinely productive calls, and their resultant revenue and positive business attitude. Receipt of numerous complaints on a daily basis can affect the morale of your existing agent force, which consciously or unconsciously, is imparted to the customers, coworkers or others. This can snowball into an overall deterioration in attitude and sales.

There is a growing interest in new customer contact center discipline aimed at analyzing call and contact content to identify and short stop these types of aberrations. More about this in Chapter 15.

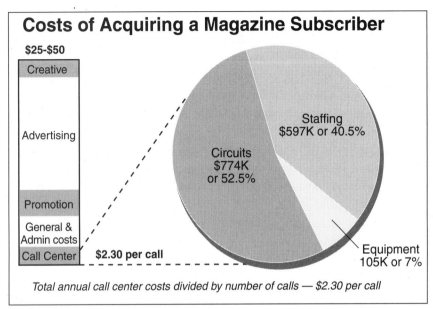

Costs of Acquiring a Magazine Subscriber

$25-$50

Creative

Advertising

Promotion

General & Admin costs

Call Center $2.30 per call

Circuits $774K or 52.5%

Staffing $597K or 40.5%

Equipment 105K or 7%

Total annual call center costs divided by number of calls — $2.30 per call

Figure 6.2. **Note:** Obviously (given the annual subscription rate), the actual subscriber, although important, is not the primary sale. The primary sale is to the advertiser — ad space rate is dependent upon the subscriber count and quality. The magazine spent $22.70 to induce potential subscribers to call, only to inadequately process the call and subvert the sale — 10% of the call handling costs were misunderstood and inadequately invested and managed, causing a ripple effect. When examined, this ripple effect causes an unexpected, though predictable, impact on the center.

Repeating an earlier statement, the value of the call and the cost of acquiring the customer are important calculations, allowing the cost of the customer contact center to be put into perspective. Lets look at the cost dynamics of serving that call.

It's apparent, after viewing Figure 6.2, that the actual call handling cost is about $2.30 a call or less than 10% of the entire cost of sale for this particular paid subscriber.

What is clear here is the amount of money that has been invested to create the demand that causes these prospects to call. With 192 hours of inbound WATS time being used per day by our sample customer contact center, there is obviously a way to back into the number of calls being received and the marketing expense being incurred to achieve this caller demand. If a call averages 4.0 minutes (which is not untypical), and 11,520 minutes (192 hours) of call traffic is offered to the customer contact center, this amounts to 2,880 calls per day.

Not all calls are sales so it is necessary to discount a certain percentage of these inquiry calls as overhead. Assuming 15% do not convert to sales, 2,448 subscriptions are sold per day at an average cost per sale of $25. The marketing investment for that day is in excess of $61,000! The revenue potential to be gained from these potential subscribers is another opportunity.

Using the magazine subscription example, where the annual subscription rate is $37.95, the 2,448 callers converted to one-year subscribers produce $92,902 in revenue. If they convert to two or three year subscriptions, the revenue potential is even larger.

To place the customer contact center cost in context suddenly takes on a whole new complexion.

Average daily marketing investment	$61,000
Average daily revenue potential	$92,902
Average daily center cost (1st year)	$5,594

Annualizing this simple example is even more graphic:

Average annual marketing investment	$16,104,000
Average annual revenue potential	$24,526,128
Average annual center cost (1st. year)	$1,476,768

This calculation assumes 12 months of operation with 22-day months.

Caution with the Numbers

When you do these types of illustrations be careful to use "shoulder" months (i.e. a month that's not the highest or lowest, a but an average month between the highest and lowest) and don't overstate the numbers. If your chosen month appears too dramatic, despite the data's validity, the numbers are harder for management to accept. There is also the veiled suggestion that this project should have been understood and begun lots earlier. Then the question that rears its ugly head becomes "should 'heads roll' due to a neglected opportunity."

What About a Pure Service Application?

When the customer contact center is a pure service application and there is no apparent revenue associated with the calls, look to the lifetime value of maintaining the caller as a customer. We discussed the cost of acquiring a new customer, but we need to be reminded that keeping that customer into the second and subsequent years of the relationship is much more profitable than the initial relationship. In the world of circulation and customer service management

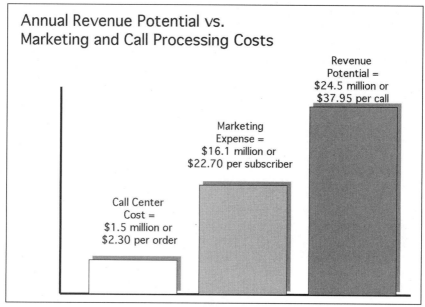

Figure 6.3.

it is generally accepted it costs five times as much to capture a new subscriber than to maintain an existing subscriber.

With the cost to acquire a new subscriber at $25, and the first year revenue being $37.75, there is only $12.75 left to fulfill the subscription over the next year. A mailed subscription renewal reminder or invoice may initiate year two of the relationship. It may be a follow-up telephone call. Whatever method is used, it is targeted at a known subscriber satisfied with the magazine, so the focused marketing cost is less than $5. That's because this particular renewal sale does not carry the cost burden of the initial marketing expense. Protecting goodwill and subscriber satisfaction is vital to holding the line on marketing expenses. This principle applies well beyond magazine publishing.

Consider for a moment that the subscriber has a problem with their subscription, needs to change their address or called to renew and was unable to reach the magazine customer contact center — the subscriber is inadequately served. This could be enough to place the renewal in jeopardy. In the context of your business is this something you wish to consider?

Do not ignore the fact that a service or response center can gather strategic market information so that the traditional sales channels and/or the customer contact center are better positioned to sell upgrades and add-ons. This is a valuable bonus positioning potential for almost all customer contact centers.

Investing Where It Counts

If the cost processing the call amounts to less than 10%, the cost of acquiring the customer and the switching equipment costs less that 10% of the ongoing call center costs. Understanding these cost dynamics and your corporate cost justification process are necessary to your success in arguing for an appropriate budget for customer contact center equipment.

This is key to establishing the perspective necessary to challenge limited thinking as customer contact center start-up budgets are drawn up. Communicating these relative costs and the investment necessary to ensure upper management expectations are met is up to customer contact center management.

Typically, expenses will be examined and those that are perceived to be more finite will be considered targets for budget restraint. Here is where equipment comes under scrutiny and a target for expenditure control.

☐ CAPITAL INVESTMENTS VERSUS OPERATING COSTS

There are two portions to any customer contact center expenditure: Capital costs and operating costs. Oil field accounting logic applies...if you can paint it, its capital goods and can be depreciated; if you can't paint it, it is an operating cost and can only be expensed. In many people's minds, tangibles are more easily subject to budget and price restrictions. Operating costs are less predictable, thus more conceptual. Things like labor and circuit usage are subject to demand and usage, therefore tougher to control, and **often fall under someone else's responsibility**.

Many times other parties to the customer contact center purchasing process, such as telecom and MIS management and/or purchasing and finance only

---

Case Study: Poor Service Costs Utility Billions

There is the story of a California telephone utility company, which in the late 70's and early 80's was besieged with customer complaints. They had gone to the Public Utility Commission (PUC) to petition for rate increases to improve service. The PUC denied the billion dollar rate request as the commissioners did not deem the utility's customer contact centers to be delivering adequate service! Simply put: the customer complaint of poor customer service gained more attention than the rate request. Until that was fixed, the PUC was unwilling to allow the utility more revenue to fix the core problem, overall service.

judge on the set up costs, not ongoing operation. They, themselves, are judged on how much they preserve a company's capital assets not the less easily defined operating costs. These parties are typically the economic and technical forces in a company. Customer contact center hardware seems easy to skimp on. "Why go with a Cadillac, when a Chevrolet will do?"

Conversely, the customer contact center manager is going to be judged over the life of the center or his tenure as manager of this center, whichever is shorter. There are also other vital accounting dynamics that also come into play.

☐ "MISSION CRITICAL" VERSUS TECHNICAL AND ADMINISTRATIVE ROLES

A customer contact center, more often than not, plays a "mission critical" role in revenue or customer satisfaction. Emotion is high. So is the ability to measure customer contact center performance.

First, the customer contact center is typically a very visible customer contact point. Often it is in either a primary sales or sales support role and therefore carries a sales revenue quota. This quota is reviewed on at least a quarterly or more frequent basis. The results and the success of the center are eminently judicable. There is great emotion surrounding the mission and making the numbers.

Second, if your customer contact center is a source of service or customer support, the measurement of customer satisfaction is measured in two ways.

1. Ensuring customer satisfaction, which actually begins way "upstream" of the customer contact center, although the customer contact center is expected to be ombudsman and "resolver of problems," most of these problems are not of the customer contact center's making. If satisfaction is not achieved, customers are lost and the customer contact center is often considered as part of the problem. Even if it was not the problem, it is perceived that the center had knowledge of the problem and did not communicate the product or service shortfall to the responsible departments so they, in turn, could remedy these. It is hard to be right.

2. The other source of judgment comes from the president's office. It comes in the form of a email or call from a disgruntled customer (maybe a stockholder) who rightly or wrongly feels they are not being served well. Down comes a dictum, "fix the problem, no matter what the cost." It comes as no surprise that lots of "emotion" is attached to the transactions processed by a customer contact center.

Customer Contact Center Mission Choke Points

A customer contact center can fail in its mission at one or more of five points in the call process:

1. An inadequate telephone network, by being under-equipped with too few and/or poor quality telephone circuits.
2. Too few local loops between the telephone company central office or long distance service and the customer contact center.
3. An inadequate or unreliable switching system.
4. Lack of understanding concerning the proper staff required to serve the inbound call volume.
5. Poorly trained and motivated telephone representatives, who have little or no stake in effectively completing the call.

An additional reason for failing can be not fulfilling the expectation created by the call, but this moves us into a region of motivation, training, marketing and selecting the appropriate product or service. These are areas that this book is not trying to address as there are already many great books on these subjects.

☐ STRATEGIC BUYING

Typically three corporate interest groups are involved in acquiring customer contact center equipment and services:

- the technical buyer,
- the economic buyer, and
- the user buyer.

This committee or group, formally or informally assembled, have different individual objectives. All ostensibly have the same overall goal: serving the profitability of the corporation by buying the most cost-effective system. But even that is relative — based on the discipline and agenda of each committee member.

The technical buyer is a telecommunications or data processing professional who has little opportunity for gaining experience in establishing a call center. This is because it is not a frequent exercise for most corporations. Understandably, technologists are typically more fascinated by the technology of the machine — what a device is, rather than what it can do for the business.

Since writing the predecessor texts, the author is happy to report that more telecom and data management are giving more credibility to user requirements. However in many companies, the battle is not over.

The economic buyer (purchasing) is looking for equipment to do the job, at the right price and under enforceable terms and conditions. He looks to the advice of the technical buyer as to the adequacy of the system. If the manufacturer describes the system as an ACD, email server etc., then that is what it must be! However, this is again a definition subject to the background of the designer, amplified by the marketing organization "spin" and broadcast by the sales organization.

The end user buyer (contact center manager), who will live with this choice on a day-to-day basis, is almost always assumed to want more than is absolutely necessary to do the job. Sadly, in some cases the end user is actually treated as having less weight than other committee members when it comes to making the final technical selection and decision. This is a challenge that the end user buyer must meet, thus the customer contact center manager needs to take on the role of the educator.

But first, this manager must understand the economics of the customer contact center in question and know intimately the tools which are necessary to achieving the corporate goals, i.e. building sales and/or delivering customer satisfaction.

Then, and only then, can the end user buyer educate those critical to the decision-making process. They need to be taught the strategic and operating issues that you, as the customer contact center manager, must deal with on a daily basis. Once this knowledge is shared with the other influential buyers, the customer contact center manager should be able to wield more influence in final equipment and service selection decisions. Thus, hopefully, the paradox of capital versus operating expense will be resolved in the customer contact center manager's favor.

Do not neglect potential allies in the sales, marketing and service delivery divisions of your company. They need the center to work so they are not burdened with additional chores that can "fall-out" of a less than effective customer contact center.

119

The ACD Conundrum

A sophisticated ACD system provides customer contact center management with a window into all the information necessary to monitor the network, local loops, business character, staff performance and achieve the goals set for the center. But, be aware: many ACD systems do not provide the depth of data gathering and reporting to understand real performance of the essential customer contact center elements. This is despite what the vendors might say! Understanding these elements and the tools necessary to manage them is vital to arguing the significance of acquiring an appropriate system.

The Small Customer Contact Center

As an end user buyer with twenty or fewer positions to staff with TSRs, you need to ask yourself:

- What are my current options?
- Is the center to occupy existing space? If so, what telephone equipment (if any) is the space served by now?
- Is the center to occupy new space? If so, will there be other company occupants and what are their telephone system plans?

The good news is that this is the hottest part of the market right now and there are more options available than ever before.

If the center is part of a larger organization and sharing space, the customer contact center can generally use a portion of the PBX that serves that space. Most modern PBX systems offer in conjunction with the administrative PBX system, an ACD of some sort. How sophisticated this system may be depends on the manufacturers understanding and commitment to the application.

Unfortunately, many PBX manufactures have embraced ACD operation in their PBX with minimal sophistication. They tend to take short cuts based on the architectural and feature decisions necessary in building a PBX. This is an area that deserves more explanation, which we will do in Chapter 9 when we compare operational compromises of PBX- and CO-based offerings against standalone ACD systems.

Once a customer contact center moves beyond twenty positions, it starts to consume space and telephone resources that are out of proportion with the number of staff and telephones this group represents.

Typically the customer contact center management also begins to request services and information beyond the capabilities of the generic PBX. At this stage there are many more options to consider.

When starting a customer contact center from scratch, with no previous switch decisions to live with, the customer contact center manager can choose what appears to be most appropriate. It may be a key system, PBX/ACD, or a standalone ACD. At twenty positions the cost dynamics of the center clearly justify specifically designed universal customer contact communications distribution equipment.

However, even below twenty positions, the options can be extremely rich.

- Most small PBX or Key systems have the capacity and rudimentary software to support a small customer contact center. You can even find automatic call sequencers as add-ons.

- A PBX system with an ACD feature set and effective reporting.
- A small standalone PC-based ACD may be an option.
- An application specific customer contact center ACD system.
- Central Office ACD services offered by your local phone company.

☐ AUTOMATIC CALL SEQUENCERS

Historically, an automatic call sequencer (ACS) was a solution added on to an existing PBX or key system. This was attached to a group of trunks or station lines. When all agents were busy, the ACS answered, played an announcement and placed the caller on hold. After the delay message, music could be played. The ACS alerted the agents to a call's presence by some lamp illumination technique (flash rate) or add-on station display. This also indicated which call had been holding longest so an agent could pick up and serve the longest holding call. Sequencers provided some rudimentary reporting capability and provided increased status monitoring capability.

The big difference between an inexpensive ACS and an ACD is the ACS is "passive." The agent must decide to initiate the answer. There is no compulsion to answer the call. The hold status is broadcast to all agents and it is up to them to pick up the call.

Most ACD systems deliver a call to an agent and they are obliged to answer. Removing the decision and discretion from the agent is considered desirable.

Both ACS and ACD systems report via a combination of call count and call length information, whether or not the callers were served and agents were answering as planned.

With up to ten positions an ACS may make sense. Selecting an ACS is a short-term solution as the missing features incur a number of operational and cost handicaps.

As ACDs, key and PBX systems have some important operational shortcomings, although they are cheaper than a true ACD. Above thirty positions, the economics show a true ACD has a clear advantage.

Today, with the availability of inexpensive standalone ACD technology and the availability of sophisticated used systems, smaller centers have access to better systems at lower prices and need not compromise operational efficiency.

121

☐ THE UNIVERSAL
CUSTOMER CONTACT CENTER PLATFORM

With the advent of the web and all that it implies in alternative communications channels, there has been a realization by a number of smaller vendors that the market is open for true multimedia customer contact management systems. The goal is to easily merge calls, emails, faxes, web chats, web initiated call requests and (potentially) video calls as common customer contact events that need to be handled by a universally equipped sales or service agent.

Many of these vendors have begun to have an affect on the traditional tele-centric-based customer contact center providers, whose response has been to offer add-on components to their existing systems in an attempt to offer an integrated solution.

Then there is the Voice over IP (VoIP) phenomena. Although this technology is in its first stage, it still adds a complication for not only traditional customer contact center equipment vendors, but also the end users. Although VoIP isn't quite ready for a leading role in the call center environment, if a contact center has less than 50 seats, look into VoIP. There are systems available today that are both traditional PBX-based and IP-based that can support VoIP. Ensure they have sound answers and delivery plans for robust call center features, workflow distribution and redundancy. With these issues addressed, VoIP calls are routed just like any other call even though they arrive under a different carriage and addressing format transparent to the agent and caller.

Nevertheless, the best advice to all customer contact center managers: Be careful when selecting a vendor to specifically address requirements for multi-channel and cross-technology integration within your center. Marketing may be ahead of reality.

☐ FINDING ACD BARGAINS —
SECONDARY MARKET ALTERNATIVES

If you need a new ACD and are not ready to make the leap to a VoIP strategy, do not ignore the option of used equipment. A customer contact center may be very handily satisfied with a machine of relatively recent manufacture that is no longer large enough for its original purchaser. Customer contact center equipment is often removed from service for reasons other than obsolescence and age. Some of the earliest installations of true ACD systems are still providing reliable daily service to their owners after 20 years. These systems have paid for themselves many times over.

Whole systems and parts for existing installations are readily available on the secondary market. The thought of a customer contact center acquiring used equipment is not overly attractive to a manufacturer. This is like buying parts for a car that is already owned. There are not the options for alternative parts sources after you have committed to one or other manufacturer by buying their system and are now a "captive customer."

Knowing this, the manufacturer prices parts and upgrades accordingly. The gross profit margin that may have eroded in the "competitive heat" of the original sale is not threatened nearly as much when it is time to upgrade. Most of the popular systems are available in the secondary market. Some of these systems have seen as little as twelve months worth of use.

The thought of acquiring a machine of recent vintage at thirty to fifty percent off the list price is very attractive to a customer contact center manager on a limited budget. The question that immediately arises though is how "old" is "old" and what is the upgradability of these systems? Will the buyer of such equipment give up important modern features?

Because most of these systems are predominantly software driven, they can be upgraded to the most recent versions of software for the cost of the software. Successful manufacturers are in the enviable position of having an established customer-base. The new systems these manufacturers introduce must maintain a degree of "backward compatibility" with systems they have already sold to existing customers. Backward compatibility is a fancy way of saying that they cannot ignore history and arbitrarily obsolete systems sold to earlier customers. End users do not like to buy systems at the tail end of their market cycle, so are generally assured that their purchases will not be arbitrarily made obsolete. The commitment to customers to protect their investment in previous product generations and a manufacturer's history in regard to "backward compatibility" and seamless upgrades are vital.

A manufacturer's history of protecting customers from obsolescence is normally a matter of pride and is relatively easy to determine. Note that vendors with a smaller customer-base are more concerned about this issue than the larger vendors, as they have more to lose by ignoring or abandoning a product and user base.

In this age, where software development and maintenance of a system is an immensely expensive proposition, PBX and ACD system manufacturers cannot afford frequently to introduce entirely new software packages. The two most well respected ACD systems have control software that has evolved over nearly two decades.

The systems subsequently introduced are typically hardware upgrades with the previous software version enhanced to reflect the new hardware and introduce some additional features. This is called "value engineering," i.e. getting the most value out of the product supposedly to the benefit of the existing customer-base and the manufacturer.

These new features may be minor, or may introduce a new strategy and direction. The latter approach is usually accompanied by a high degree of marketing effort. Take your time to understand your way through the vendors marketing position. There may be very little new being introduced. The positioning may be defensive as another competitor may be affecting the success of the manufacturer. All of this is important to the customer contact center manager when considering the secondary market.

If a company is introducing a totally new machine, which does obsolete an existing ACD model or product line, financial arrangements may be made in the way of discounts to prod existing customers into the newer machine. The replaced machines typically find their way into the secondary market.

There is also an opportunity for the customer contact center on a limited budget during a manufacturer's product transition. Offer to buy a newly superseded machine at a reduced price. Generally, the manufacturer has been making presentations under non-disclosure statements to key prospects and customers. The timeframes for installation precede the production delivery of the "new" machine, so interim arrangements are made to provide "old" versions, then a later replacement. The "old" equipment needs a home. These traded systems are often available directly from the original manufacturer at substantially reduced prices.

Buying from another user who no longer wants the machine is one alternative, as is buying from an equipment broker who specializes in this brand of system. Most of these machines have been used up to the day of de-installation. The manufacturer or its authorized maintenance representative has maintained them. Obtain assurances from the seller and ensure you are provided some limited warranty. Refurbishing may occasionally be necessary. Most manufacturers will refurbish the equipment for you at their factory, but usually only at an arbitrarily high price. Field refurbishing is often adequate at greatly reduced cost.

Negotiate payment terms and do not make final payment to the seller until the machine is installed and working to your satisfaction.

Software Title Ownership and Support

What about title to the system software? In most cases, the original manufacturer does not grant title to the software that drives the system. A license to use the software is sold with the title to the hardware. The vendor of the used equipment may have title to the hardware but legally cannot transfer the software. Inquire of the seller as to the status of the software or call the manufacturer and ask what their policy is regarding software title. Also determine that if you acquire used hardware, what position they take, if any, when it comes to support.

Most vendors have an unpublished rule; we would sooner see a customer contact center buy our equipment, even though it's used, than lose the system to a competitor. The reason for this is that successful customer contact centers will grow in size and sophistication, and the user will eventually need to do business directly with the manufacturer. Nevertheless, there are certain manufacturers who have never understood this and levy punitive software licensing fees on those individuals who buy on the secondary market. There is the added disadvantage in that most of the newer features (such as CTI and VRU links) require the latest software version, typically not available with a system available on the used equipment market.

Secondary market sellers are quick to point out that software licenses are not needed if the secondary market purchase is only for parts to upgrade an existing system. Not obtaining a software license for a new site is also an alternative, but do not expect any support from the manufacturer until you establish a relationship.

Make sure you obtain all the user documentation. Training may be available from the manufacturer. If you are a licensed user, a manufacturer will not deny you training. Negotiate terms carefully and obtain possession of all of the equipment (which you've determined to be in good working order) before the final payment is made.

The Secondary Market at the Dawn of Voice-over-IP

Today we are at a significant point in telephone evolution. Traditional "mainframe" PBX and ACD systems are under significant pressure. First, from the universal customer contact routing platforms that integrate and blend customer contact center transactions and simplify database interaction. Second, from the promise of VoIP, i.e. routing some or all customer contacts as Internet-based transactions over an IP network.

Unless it is a capacity issue or to equip a new site, buying a new PBX or ACD today that does not support VoIP is tantamount to applying an expensive "paint job" to your existing switch. Vendors are feverously working on projects to ensure their ACD can deal with emerging technologies. Many development teams are working on building state of the art VoIP ACD software capable of transferring voice and data on the same network. Once this is perfected much of the ACD's work will be achieved without using any hardware switches or PBX to transfer and route calls, leading to a purely soft switching VoIP ACD.

You should also note that all of the "mainframe" customer contact center switch manufacturers have roughly 90% of their features in common. The major deliverable difference is often only price.

If your company cannot embrace a VoIP strategy today, but still must deal with a significant customer contact center system requirement, an elegant bridging strategy may be used equipment. This is particularly attractive if you can buy an application specific ACD system, like an Aspect or Rockwell ACD, for 20 to 30% of the original cost.

Managing Telephony Workflow

Throughout the book we frequently discuss customer con-
tact centers and the equipment they use. In this chapter we focus on the telephone call component and the management of telephony workflow.

The automatic call distributor (ACD) remains the hub of any customer contact center reengineering process as this device provides the services "rationing,"
load balancing and reporting that allows a customer contact center to best deliver on its promise of service. An ACD might best be thought of as a telephony
workflow engine. Similar devices or feature sets with varying degrees of maturity, which are focused at email and web interactions are also making their way
into the customer contact center. The ultimate device is a system that synchronizes and harmonizes all of these channels. ACD systems have a stable history
and are the basis for development of multimedia or cross-channel, cross-application solutions in the universal customer contact center.

Although we offer other chapters dealing directly with the technical make-up
of the ACD, the reader may ask at this juncture, "What exactly is an ACD and
why is it different from other call handling devices such as key telephone systems (KTS) and private branch exchange (PBX) equipment?"

An ACD — automatic call distributor — is a misnomer, since the name suggests that its primary function is to "automatically" distribute a large volume of
incoming calls in a predetermined and equitable fashion to customer contact
center personnel. And, although, that is its day-to-day advantage, its primary function — and justification for why ACDs should be discrete from administrative

switching — is that it collects and stores very detailed process information. Lots of information in a way that a generalized administrative switch (PBX) does not.

The ACD and customer contact center manager can manipulate that information in myriad ways to better understand the character of a specific business and so manage the business opportunities offered by its callers. Thus, it's possible to identify the precise level of service a customer contact center delivers (to different classes of customers or transactions), the cost of delivering this service and the image it projects on behalf of a company.

Never take your eyes off the information gathering and presentation aspects of the system. Never accept less than you believe you need and never accept fitting your business goals into a vendor's often-limited vision.

This becomes particularly true as you move into the applications value offered by the new customer contact center technology platforms of CTI, intelligent desktops, and client/server technology supporting various customer relationship and customer experience tracking systems.

CTI (Computer Telephony Integration)

There has been an interesting shift since CTI became a significant call center feature. The writer was privy to a presentation by a major computer manufacturer; they pushed CTI and the relative value of mainframe host and gateway software. The manufacturer had similar products for midframe environments and what appeared to be a telephony server. Using their technology, they optimistically projected the average price to complete a full CTI application was between $100,000 and $400,000 with a payback of less than a year.

This was extremely aggressive. Few significant CTI projects ever deliver a payback within a year and require up to an average investment of $3,500 per workstation, NOT including the telephone switch (if needed) or a desktop PC of recent vintage (if needed), plus significant commercial power upgrades. The vendor countered by saying that CTI was really "just plumbing." And went on to state that the truth of the matter was that until process reengineering and workflow management was addressed, CTI offered little value beyond the time saved in automated customer identification and screen launch or "pops" (population).

The vendor's excuse for not presenting the opportunity this way is that it scares most call customer contact center managers who do not have the vision or authority to embrace or initiate projects of such breadth! From that vendor's perspective, using such methods to build a market demand was totally justified.

The goal of this book is to equip the reader with the tools and expectations to embark on realistic customer contact center process improvement, which just may involve CTI.

☐ INBOUND CUSTOMER CALL FLOW

There is a new focus on how the whole process of managing inbound customer contacts works, particularly after adding web-based contact media.

The process begins way "upstream" of the actual contact, beginning in the promotion or information broadcast that stimulates the customer to pick up the phone and call or visit a website. It is most important that the customer contact center know why this call is arriving.

The next issue is the number of calls or customer opportunities offered. This may be an order or a request for help or information. Each of these has a measurable outcome — an order or a served customer.

Each call follows a "state profile" that can be simply compared in microcosm to a sales cycle — taking the customer state from "suspect," to "prospect," to closed order or satisfied customer. Over the course of the transaction (but not ignoring the bigger relationship concept) the number of opportunities diminish. If 100 prospects call the customer center, yet only 10 result in sales, what caused the attrition and how can it be corrected? Often it has nothing to do with technology.

The new thrust in ACD reporting is to get to the reasons for and the remedies for this lopsided "conversion ratio," i.e. offered opportunities to results. This is being done in numerous ways, as discussed throughout this book — particularly when we get to customer experience mapping.

The leading edge standalone ACD systems are no longer simply "a better telephone system," but a better way of doing business. This is because of the process reengineering thrust offered by these standalone systems and the adjunct services delivered by their vendors really attack business transaction management. The PBX-based ACD vendors are still stuck pitching "phones," no matter how hard they try to abandon pure hardware.

Transaction and process management tools are what give your ACD system power. But, first we will begin by focusing on what an ACD is best known for — distributing calls to agents.

In a pure telephony environment, the basic notion of a customer contact center is the processing of large volumes of homogeneous telephone calls. If serving inbound calls is typically the primary function of the center, trunks carrying calls into the center and the telephone positions in the customer contact center are busy more often and over longer periods, than a typical trunk or telephone serving other parts of the business.

A major difference between customer contact center applications and regu-

129

lar administrative telephone use is that call centers carry a disproportionately large volume of call traffic when compared with typical administrative (that is, normal office) telephones, circuits and switches. The exception to this statement is CO (central office or the telephone company exchange) switches and network circuits. In both the ACD and CO examples, some phenomenon has occurred to concentrate and focus the unusually high volume of telephone traffic on the customer contact center or the network service.

In the case of the long distance telephone network components, physical concentration of administrative traffic has occurred at subscriber PBX systems, local and other devices. When it is clear that the calls are for the long distance network, the calls are concentrated and long distance resources allocated. Concentration of the traffic has occurred. The network is engineered for the anticipated demand. Typically this is done for the average demand, so a short delay (queue) and occasional busy announcement is acceptable.

In the case of a customer contact center, this concentration has occurred through establishment of and advertisement for the customer contact center as a focal point for customer and prospect inquiries. Instead of these inquiries being vaguely directed to the main company switchboard number or PBX, they are concentrated on a specific service group. Hence, the concentrated telephone traffic occurring in disproportionately large volumes for the number of trunks and staff present in the customer contact center. All calls to a customer contact center can be considered as having the potential to generate or protect revenue. Although a short queue is acceptable, it does need to balance against the revenue potential and the competitive landscape.

Since the customer contact center is a focal point for a volume of customer or prospect calls, there is also the necessity to ensure the callers are answered and served correctly. This introduces two additional concepts that are extremely critical in the world of customer contact centers and the reason that ACD systems came into existence in the first place: massive data gathering and user programmability. We will discuss these later.

The people that staff a customer contact center, typically called agents (from airline "sales agents"), telephone sales or service representatives (TSRs), or customer service reps (CSRs) are organized into groups or teams that are trained to serve the requests from a particular type(s) of inbound customer. These customers are routed to the agent group based on their request for service.

☐ THE UNIFORM CALL DISTRIBUTION (UCD) SYSTEM

There are various systems available to aid in this routing process. One system is the UCD, which is generally used for overall call distribution. A UCD tends to be less expensive and more rudimentary than an ACD, and although their use has diminished over the years, UCD routing principles have gained new popularity with cross channel/media applications. Yet, even then, the UCD's activities are typically restricted to overflow or backup, not primary service groups. Since the price of basic ACD software bundled in a PBX or key system has fallen so low or in some cases is provided "free," there is little or no reason to accept the operational inequities and information limitations posed by a UCD. The net result would be the requirement of a larger staff to handle the same customer call load than with an ACD system.

A UCD uses a top-down or round robin call routing. A TOP-DOWN routing or search (sometimes called "hunting") means the system begins to look for an idle station at a physical hardware or station address beginning from the top of a predetermined list of stations. To illustrate, we will name this station #1. Once the first incoming call of the day is received and is in process (at station #1) any subsequent calls follow the same search pattern. Beginning at the top of the heap and searching down the list until an idle station is encountered. However, once calls end at the low number stations and new calls arrive, they are passed to them first while stations later in the numbering plan remain idle, theoretically never receiving calls. A number of "hot seats" are created where the agent works harder than their peers at the higher numbered stations. This is considered inequitable in an overall center environment, but can be an elegant solution when calls are selectively overflowed to an email or other administrative service staff group.

In an attempt to remedy any work distribution inequity, a ROUND ROBIN distribution technique was introduced. It worked just like top-down but was distinguished from the top-down search because it began the search for an idle station at the next station after the last station that was delivered a call. This produced better distribution. Yet, it still didn't distribute calls truly equally. A UCD using either top-down or round robin distribution cannot deviate from its predetermined routing plan since these systems are neither dynamic nor easily reprogrammable, particularly by the user.

If the customer center management does not have control of their systems and its ability to respond to customer demands, they are already behind the service curve. The issue of user reprogrammability is a significant user feature that we will discuss at a later point.

Email Revives the Value of the UCD

Previously we discussed how agents separately focused at either telephone calls or email may work to support each other in call traffic peaks and valleys.

The theoretical solution is to divide your contact center staff into three skill preferences:

- The first group (group 1) is call-centric (in and outbound preferences may be an additional sort for this group).
- The second group (group 2) has better correspondence and email skills.
- While a third group (group 3) is agnostic.

Give the entire staff ACD call and email workflow features on their desktop and triage emails to group 2 first and calls to group 1 first. Group 3 is the first group that receives call overflow from group 1 when all its agents are busy and the calls are on hold beyond desired queue time. Group 3 is also the first to receive emails that fall outside the capacity and desired response criteria (target turn-around time) of group 2. Group 2 becomes the group of last resort when calls swamp groups 1 and 3.

This is where a UCD routing structure would be ideal in group 3 and group 2, so that incoming calls do not randomly interrupt every email or correspondence focused agent.

☐ THE ACD

To repeat the basic concept which distinguishes ACD systems from other systems. An ACD system will tend to equalize the workload across telephone agents or telephone representatives. This allows for higher productivity based on the knowledge that all agents are being treated equally as far as work distribution is concerned.

An ACD differs from a UCD in that it typically routes a call to the agent who has been idle longest. The ACD maintains an availability or "free" list of agents in a particular group. As an agent finishes the last call, they become "available" and are added to the bottom of this free list. As time passes and they remain on the list, they progressively rise to the top of this list until they become the target destination of the next call. They receive the next call and are removed from the free list, only to be added or recycled upon releasing the current call. To do this an ACD needs substantial processing and memory power.

As an option, a well-designed ACD system will also allow the user to employ rudimentary UCD routing with ACD routines. There are times when there is a distinct need for a top-down distribution technique in conjunction with the more equitable call distribution provided by the "longest idle" technique. An example of this occurs in a low traffic period, when there is a minor amount of call overflow to a second group. Rather than randomly interrupting everybody in the overflow group from alternative work assignments such as manual email responses, the system biases the interruptions to a few agents assigned to the stations at the "top" or beginning of the agent group. The rest remain focused at some other job — updating the mailing list, entering previous orders, outbound calling or whatever — unless the call volume really picks up.

The recent generation of ACD systems introduce a number of routing "wrinkles" such as conditional call routing, skill based call routing, knowledge based call routing, the inverted queue, and most recently intelligent call routing.

Conditional Call Routing. This uses the concept of the free list but also allows "interrupt" routing based on encountering a service, time, caller identity, call type or traffic load condition. This causes the machine to look to a table of conditional instructions, which reflect call treatment based on the desired business objective.

Take the example of a queued private bank client with a million dollar stock trade. This client insists on a live trader. When identified by a special called number (directory number, 800 etc.), inbound caller identification or voice response prompted account number, this caller is recognized and served ahead of the retail penny stock trader, even though the larger trader called seconds later.

Also, it may be that once a predetermined queue length is reached, calls are overflowed earlier to backup groups than under a normal queue scenario.

Skills Based Routing. This is a wonderful concept, but as executed, it simply allows agents to sign on in more than one group, identifying the second classification as having certain properties or skills that match specific customer demands.

For example, a customer contact center may have a requirement that it serve both Anglo and Spanish speaking customers, although its Spanish speaking customer load is not so burdensome so as to dedicate Spanish-speaking agents to that service. Thus, the Spanish-speaking agents would carrier a lighter load than the English speaking agent group.

The better strategy is to build a primary group of all agents (Group 1) and build a second shadow group (Group 2) of bilingual agents. The system then should have the ability to differentiate Spanish-speaking customers from the general caller population . This can be accomplished by publishing a distinctive telephone number for Spanish-speaking callers wishing to be served by Spanish-speaking reps or via Dialed Number Identification Service (DNIS) or a DID number. Or even by a specific caller ID being gathered from the network (ANI) or by a voice response unit (VRU) requesting an account or SSN and being compared against an existing customer database record that identifies this customer as desiring Spanish language service. The ACD system then looks at the shadow group and the shadow queue that only Spanish-speaking reps are part of.

What happens here is that the system, because of this skill "interrupt" reduces its search for an eligible agent, based on the unique skill required to serve this caller.

From a pure efficiency point of view, great care must be taken when employing call identity and load granulation. Skill-based routing is a significant feature that can be overused and dilute the effectiveness of the call center. Some of the worst examples can be found in software support. Here a caller identifies the software type, platform and version, only to be routed to such a small universe of skilled agents that the queue becomes extended, the service is diluted because the economy of scale becomes so diluted to the point of defeating the use of an ACD. Understand why this level of caller identification and granulation is desirable and balance this with business objectives — not just the ability to triage to the caller. Now, with CTI and integrated web-based self-service, the customer contact center manager can find better ways to do this.

A recent book by William Durr, *Navigating the Customer Contact Center in the 21st Century*, suggests that many owners of systems capable of skill-based routing are defeated because the skill set databases, once established on agent hiring, are seldom updated.

Knowledge-based Routing. This is another version of skill-based routing that purports to consider more than just a particular agent's skill set and eligibility to serve a particular customer.

Intelligent Routing. This is the latest iteration of trying to connect a particular caller to a particular agent and offers all the potential to turn an ACD into an inefficient PBX call processing system. The real difference with intelligent routing is that the routing process considers more than the call center universe. It also looks at caller data, such as when the customer last called, the subject of the call, to whom they last spoke and the feasibility of connecting that customer to the same agent. Of course, all the issues of service level and prompt response to a customer's request are factored into the routing decision so that callers don't go into impractical and infinite queue. Nonetheless, the whole economy of scale concept is diluted.

The Inverted Queue. Another inbound routing feature that has gained popularity in the leading standalone ACD systems is the ability to invert the call type. That is when an inbound customer call encounters a delay, the caller is given three options, to wait on the line for service, hang up and call back or leave their number via a VRU prompt, hang up and expect a call back. This latter call back feature may be driven off the ANI capture. Here the system captures the inbound caller number and because the customer hung up before service was received, inserts a callback event into the inbound queue.

The ACD system then can automatically interleave the outbound call back with inbound calls. What happens here is that the nature of an outbound call is changed to "look like" an inbound call. Most inbound agents claim to not be good at proactive calling, but fine at reactive calling, i.e. if a customer or prospect asks for something that's fine. But if they have to initiate a sales contact, they have a real objection to the process. This comes up as an issue when we discuss "call blending," as offered by some of the newer outbound predictive dialing systems.

Call or Case Tracking. This is a recent tool that has been introduced in leading customer contact centers. Call or case tracking was an early manifestation of customer relationship management. The goal is to gather topical case data so the state of the customer inquiry can be tracked. If a caller calls back and speaks to a second agent, that second agent can answer the call and customer in the context of a customer relationship. Inquiry state management is also descriptive of this process. It has myriad uses in many different customer contact centers, the most obvious of which is to equip as many possible agents with the knowledge to answer each customer call better. This begins to build the ACD back to its original goal of rationing the supply (your call center resources and agents) across as large demand (customer calls) as possible — back to economy of scale.

The Typical Switching System

This chapter deals with the hardware, architecture and relationship between the components and subsystems of a switching system, which includes the automatic call distributor (ACD) system. The discussion is limited to traditional switching systems and, for the most part, ignores Voice over IP (VoIP) technology.

There are seven groups of components or subsystems that make up any switch:

1. Common control or intelligence of the system.
2. Internal switching network of the system.
3. Ports into and resources attached to the switching system.
4. Backplane or signal distribution subsystem.
5. Internal power supply.
6. Power distribution subsystem.
7. Mechanical packaging of the system.

All these elements are packaged together in a physical housing that allows integration of these subsystems with the software that is stored in the memory of the common control or "brains" of the system.

All switches on the market today are computer controlled. At the core of the system is a specialized computer running specialized code that allows connections to be set up between two (or more) ports in the switch. Each port represents the two parties to a conversation. An inbound caller occupies a trunk and the agent that staffs a station. Unsurprisingly this path between the parties is

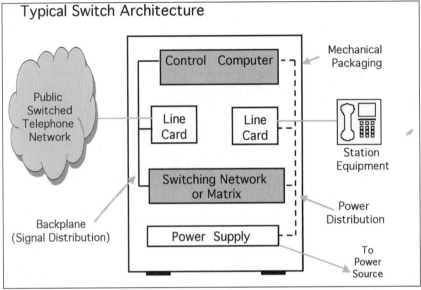

Figure 8.1.

called a "talk path." Note that the two parties to this connection may be a human and a machine, for example, a queued customer listening to music is attached to a port attached to a music source, announcement or whatever.

This computer technology has evolved from earlier processors and architectures to the more powerful 64 bit processors available today. Yet, tightly written machine level code can run very well on smaller distributed processors. Old, therefore, is not necessarily bad. Many manufacturers are usually reluctant to upgrade core processor technology for the sake of technology since often the many hundreds of man-years invested in software would be unsalvageable. There are, for example, ACD systems that have been in service for over 20 years that are still providing reliable service; although the cost of supporting some of these systems may be prohibitive. Manufacturers find that newer technologies, like VoIP, can offer greater potential for ongoing development and evolution and code development is easier and faster in these newer environments. This is one justification for earlier and faster adoption of VoIP in a "green field" application.

The common control or intelligence of a traditional switching system manages the system resources and allows call origination, completion and feature use. The system brain may be one single computer or a number of distributed computer processors that talk to each other in a parallel (shared responsibility) or on a hierarchical (master/slave) basis. How this is achieved becomes important when a system fault occurs. How the system recovers from this, so as to cause little or no disruption to service and your business, is important because calls represent revenue now or can erode goodwill and represent lost revenue later. The common control system manages a number of processes. These may include:

- The switch resource status management process.
- The switching process.
- Data gathering.
- Utility and diagnostic programs.

These programs run on an operating system that uses various program tools to execute the different functions.

Resource Status Management. This process merely keeps track of what is going on in the switch. Resources are all the ports occupied by a device. A device may be a trunk, an agent or station, an interactive voice response connection or utility device of some kind. These devices are typically in one of two states, idle or busy. To change from one condition to the other, these devices "request service" from the common control device. The computer then looks to a table that tells it what to do with the request for service. The computer may expect to receive further data from the device in the form of the dialed digits that make the request more specific.

The Switching Process. This process allows the setup of a connection between the two resources requested by the device (for example, a caller station) by following the treatment instructions residing in a table in common control memory. This then enables the common control computer to set up a connection between trunk A and agent Y. The status resource management aspect of the system monitors the connection for the duration of the transaction and "breakdown" the connection and then returns the devices and communications path to idle upon completion of the call transaction.

Data Gathering. Throughout this process, data is being gathered about the transaction. This may be minimal and only used for internal system management and diagnostic purposes or it may be more extensive. With more extensive data, many more applications can be addressed. These may be call accounting, network optimization and/or trunk and station productivity analysis. Remember, the more information you have, the more insight and control over the application. But also, the management burden is increased in tandem with the additional information gathered. When you have more information you have to do something with it. Otherwise having or not having this information is irrelevant.

Utility and Diagnostic Programs. Concurrent diagnostic programs watch that all of the components and resources remain within certain performance ranges and should anything fall out of range, alarms are triggered, time out of service clocked and reports generated.

□ THE INTERNAL SWITCHING NETWORK OF THE SYSTEM

The system switching network or "matrix" is a series of components that allow the connection of any trunk, any station or any utility device, to any other device. A utility circuit or device is a service device that is switched into the call transaction to allow use of this device in the call setup. For example, a tone generator delivers dial tone during call setup; a tone receiver "hears" and interprets the digits. In the case of an ACD, such a utility device could be a modem or fax machine. These consume ports when present, and switching capacity when in use. The network or matrix has addresses: one where the communication enters the switching system network and another where the communication leaves the

State vs. Record Machines

It is important to recognize the philosophical difference in switching system design, which particularly affects ACD operation and management. Some systems are defined as "state" machines, others record machines. This is defined by the way they gather, report and record call-processing data.

A *state machine* collects and reports every change that a device or call goes through. This data is delivered in true real-time and can be used in a number of ways. This data allows real-time screen updates on a field-by-field basis. If it is written to disk, it allows the collection of a complete audit trail of transactions via device, resource and personnel productivity.

A *record machine* must complete a transaction, typically a call, or wait for a system "snap shot" to be "taken" before the total machine (by device) status is reported. The machine never catches up with itself and is incapable of producing real-time displays for the supervisors. Most PBX-based ACD systems are record machines. Watch a supervisor system screen and if the whole screen goes blank and repaints at regular (10 seconds) intervals it is a record machine. One popular PBX-based ACD system, starts "load shedding" by shutting down reporting functions in peak hours as the switching process demands more and more central processing power! This and other systems like it were never designed to be an industrial strength ACD.

There are always many other program modules running in today's switching computers. These programs, described as utility programs, allow the system to do more utilitarian things.

system. These are called ports that are, in turn, occupied by a component such as a trunk, station, or utility device.

The majority of these systems are digital. Analog systems use a less flexible, but extremely reliable scheme, that views the network as a matrix of X and Y coordinates. There are many analog systems still in daily use, and this technology will continue to provide reliable and cost effective service for many telephone callers well into the 21st century.

The common control sets up a connection by referencing the physical X/Y switch coordinates and "cross point" switch that accomplishes the connection between the particular X port and Y port devices. The switching matrixes were originally a network of physical switches. Subsequent solid state networks did away with physical switch networks by accomplishing the same function with silicon chips, which substituted a "physical matrix" and physical switching address with a logical address that translated to a hardware (trunk or station) address. The number of simultaneous conversations the switching system can support is governed by the number of cross point or talk path connections the analog or digital switch can handle concurrently.

All digital systems use an addressing scheme that allows a device to access a communications buss or pathway. This pathway resembles a freeway with entrances and exits. Each entrance and exit represents a port. Each port offers traffic in the form of "slices" of a message. These are like cars entering a freeway. Each message has an intended address, just as the driver of the car has an intended destination or exit. Unlike the car, which represents a complete unit, a conversation is made up of potentially millions of message parts. By assigning each message with an identity "flag" and an address, the "digital parts" of a conversation can be connected between port A and B. These are the respective parts of a conversation, albeit in tiny increments. By assigning the message with (entrance and exit) addresses, a buss can connect port A to port B.

Messages are sent backwards and forwards between ports. Our freeway example assumes a one way or asynchronous transmission not a two way or bisynchronous conversation. There is a second timeslot assigned for the second port.

These systems transmit conversation samples 8,000 times a second. They divide time vertically like a sliced loaf of bread. Each slice represents one 8,000th of a second conversation sample. This sample is sent from port A to port B on a time slot. B responds and sends a message (also sampled 8,000 times a second) back to A. This occupies a second time slot. Each conversation requires two time slots. To find the total number of simultaneous conversations the machine will support, take the total time slots in a machine and divide by two.

When considered in the context of hundreds or thousands of ports, and hundreds or thousands of calls in a busy hour, the load on the common control computer managing all this activity is high, which leads to a need to have adequate system capacity or bandwidth. Note that all computers have a finite load capacity.

This same two-way messaging packet and bandwidth question applies to your local area network when VoIP is considered. With VoIP, the address packet now becomes an IP or Internet protocol address.

☐ PORTS INTO AND RESOURCES ATTACHED TO THE SYSTEM

The common control and the switching devices make up the core of the traditional switching engine in any switching application. Attached to and controlled by this switch are the devices that accomplish connection and communication to and from this engine.

The switch package is made up of a rack(s), and the shelves within the rack(s). These contain the various components or resources that make up the remaining elements of the system. These elements are the reason the switch exists. They represent the trunks, stations and service or utility circuits that need to be connected to accomplish the particular call transaction application this switch serves, meaning the customer contact center or business communication system.

Each trunk or station port is considered a "user available" port. That is a port the user can assign a resource or component to, such as trunk, station or with growing frequency, voice processing devices. Other ports in the system are occupied by utility devices, shared or "pooled" across the system, thus shared by the entire universe of user ports. These are referred to as the utility circuits.

Ports may be dedicated to a single function. They may be pre-assigned to be occupied by a trunk, station or utility device. A system with "universal" ports allows almost any switched device to occupy almost any slot within the switched device slots.

"User Available" Ports

The reason the term "user available" is used to identify the net size of a system is that vendors talk grandly about the "capacity" of a system when it's put into use by a user, although the system might not support the full device capacity presented by the vendor. Gross and net capacities differ because there is more to a switch operation than hooking up stations and trunks. For instance, utility devices and real-time limits within the computer's processing power prevent a machine from reaching its maximum advertised capacity. Also utility devices that occupy ports can "hog" limited computer processing. This means that ports cannot be occupied (even though they are physically vacant) due to lack of available processing power.

A vendor may give the user some say in the amount of utility devices in the system, although typically these devices (i.e. tone generators, receivers, conference links etc.) are engineered into the system to meet anticipated traffic load. Today, however, you find that more and more of these devices are single integrated circuits (chips) and no longer need to occupy a unique slot.

Vendors tend to be conservative in making sure the system is configured with enough discrete (but port occupying) devices, such as service observe ports, and they often provide more than are necessary. The ports these devices occupy may be freed up to support an extra couple of analog trunks, agents, whatever. Check with your vendor, you may be able to "squeak" through some tight capacity situations by displacing utility devices, but this is a secondary or transient strategy only. Most vendors discourage this as it unduly complicates system administration for vendor and user alike.

143

User Devices consist of:

Public Network Components, i.e. trunks — analog and/or digital circuits consisting of:

- Local loops.
- Foreign exchange circuits.
- Intra and Interstate WATS lines.
- Other bulk service long distance lines such as virtual private network circuits and dedicated tie-lines.

When telephone service is delivered in an analog format it can be bought a trunk at a time. Digital service generally arrives 24 lines at a time in a T-1 span.

Customer Side Resources, i.e. telephone terminal equipment or instruments, such as:

- Plain old telephones (2500) sets.
- Proprietary telephone sets.
- ACD devices, such as —
  - ACD agent instruments,
  - Integrated desktop workstations with "soft phones,"
  - Supervisor instruments,
  - Training instruments, and

- Monitoring or Service observation ports.

Utility Devices, i.e. devices that are necessary to process a call, such as:

- Tone receivers.
- Tone generators.
- Modems.

System Devices could consist of:

- Announcement devices for delay announcements.
- Music or promotion-on-hold devices for delay purposes.
- Training or emergency recording devices.
- Automated attendant devices for call screening and call prompting.
- Ports for accessing voice messaging and voice response devices.

Public Network Devices

Attached to the matrix or internal network are the public network devices. These are the circuits that allow calls to be placed to or from the switch to other points on the public network. These circuits exist for specific applications. Some are usage-based services such as WATS; others allow call or caller type discrimination, e.g. dial 800-555-1234 for Sales, 800-555-1235 for Service, etc. In the case of an ACD (although cost is a consideration) specific circuits may exist for reasons of geographic coverage. They can serve specific regions that you wish to receive customer calls from. Combinations of cost, call type and regional coverage add to the apparent complexity.

These circuits may be digital or analog. Today, most call centers of any size buy digital service (T-1 as a minimum) since it's usually more economical. The signals on analog circuits arrive in analog form and then converted to and from waves of electrical energy to sound waves by the telephone instrument headset or handset. This is contrasted with digital circuits that transmit acoustic representations in digital form.

A digital T-1 channel offers both the telephone company and the user substantial gain in circuit capacity and reductions in circuit cost. A two pair circuit (4 wires) can now carry 24 channels or separate voice conversations. Formerly a circuit pair (2 wires) was limited to one conversation or voice path. Telephone calls arrive at your switching system in digital form and are terminated on either T-1 digital trunk cards or an interface device called a channel bank that must unscramble or decode the signals, first to individual conversations then natural speech. To break up this data stream into 24 separate incoming conversations, the switch or channel bank must demultiplex (or "demux") the data stream on the T-1 span into 24 conversations. The outgoing conversations are multiplexed ("muxed") into one 24-channel T-1 span.

As a rule when planning for trunk types, trunk group organization and sizes, look to the business objectives first. Then, and only then, look to achieve these objectives in the most economical manner. By all means seek technical advice, just don't start there; strange results can occur when technical judgment is first applied to select the types, groups and quantities of trunks to serve a customer contact center.

Customer Side Resources

On the user side of the switch, there are the user resources. These may range from a simple 2500 series telephone instrument (a black phone), a modern electronic telephone set, a high call volume ACD agent instrument, or an 'integrated voice and data workstation where a pop up screen feature, known as a "softphone," may reside. These devices may address voice applications only or they may be hybrid voice and data devices.

There are two schools of thought in delivering instrument features to the marketplace. They originate in either hardware or software philosophies. Many instruments have features hard coded into them, i.e. there is a specific button for every feature. A feature change may require an additional button, ergo the hardware philosophy. Then there is the software-based feature philosophy, where the manufacturer adds features via software, thus enabling the easy addition of features without hardware changes.

A user feature control table, known as a class of service table, may control some of these features. This resides in a database in the switch. Each instrument (or user if it is based on logical sign-on identity) has features extended to them based on a table granting levels of permission. Much like a system security access. The higher the class of user, the more features are available to the user such as agent, supervisor or monitoring supervisor.

There is a great deal of importance attached to the integrated voice and data workstations. They will be discussed later in the book.

Monitoring/Recording Capacity

Unique to ACD systems, these devices resemble the category above, but are designed to supplement the supervisory, monitoring or quality assurance and training functions in a customer contact center. These devices often differ from the normal TSR or agent instrument because of unique hardware and/or software-based features that extend the instrument beyond general ACD agent functionality.

A position that is equipped for monitoring must allow a supervisor or call auditor to patch into a call as a third party without affecting the quality of the call. This means SILENT monitoring. Typically, this can mean a reduction in the volume of the call, although the monitor position should not introduce extra sound on the line when the supervisor plugs into a particular call, hence the term "silent monitoring." Even when automated quality monitoring is added to a customer contact center, some informal manual monitoring access should be maintained for those exceptional occasions where out-of-plan review or discipline is required.

Note that when an automated quality monitoring system is attached where specific monitoring or service observe ports are designated on a switch, the ability to monitor manually is usually not lost. The exception is if the system only offers 16 or 32 monitoring ports, and all are required to attach the automated call recording device for the quality monitoring function leaving no room for manual monitoring.

Also note that if the attachment of the monitoring device is by way of an ACD terminal emulation, there is limited tactical sampling capability. Even more important, there is the hidden effect of displacing more than one instrument from a traffic and internal conference port perspective. Attachment via terminal emulation is a short-term strategy.

Utility Devices

These are the tone receivers, tone generators, modems and other devices that are necessary to process a call. These may be either physical or logical devices.

The underlying rule in detailing each of these devices in a system is that each discrete device needs an address to enable it to participate in the call transaction process. Each discrete address therefore consumes communications capacity and displaces an alternative device. This principle will become increasingly clear as we move into capacity issues, number, throughput and control limitations. But first, let's look at the public network devices.

Software vs. Hardware Class of Service

Take care when you are evaluating an ACD that the hardware devices for agent, supervisor, or monitor are physically all the same. Ideally, instrument features as assigned to the staff role should be controlled by matching software-based class of service definitions. In this way you avoid stocking more than one type of instrument.

System Devices

Announcement Systems

One example of a system device is the prerecorded announcement system that provides the initial announcement to an incoming caller. Initially they were simple message playing devices such as tape or solid-state memory devices, but more often than not these are now voice response devices.

The earlier tape and solid-state technologies provide informational content and were originally designed to deliver a simple "delay" message. These have been expanded beyond the inevitable "...please hold, one of our agents will be with you eventually; this is the only boring message you will hear...." to provide information and user specific promotions. They should not be confused with the newer voice processing tools (such as automated attendant and IVR), which provide call screening and prompting for service selection and other functions that expand the possibilities of call greeting.

Simple announcement devices occur in two forms. A mechanical continuous loop recording (a tape recorder and player) or solid-state digital devices. They may be an integral part of an ACD system or a port allowing attachment to a discrete external device.

Digital announcement devices are preferable due to sound fidelity, instant reset and replay (the caller always hears the beginning of the script) and reliability issues. The disadvantages of the older tape technologies were this: a mechanical tape playback device runs at a constant speed and is continuously playing the delay message. Calls that are to receive the message are queued up in a micro queue ready to receive the message. This is so they don't barge into the middle of a nearly completed announcement. All new call arrivals must wait until the beginning. Calls do not arrive simultaneously; therefore they are all queued at different lengths, until the announcement is completed. This adds to individual call length, which can be almost the entire length of the message. This increases line occupancy and circuit hold time. A powerful and expensive nuance!

In many cases these devices use multi-track tape with multiple channels per individual tape. Two or more announcements can be recorded simultaneously on the same tape and played back to two or more separate callers. But these announcements must be of identical length. This leads to decreased flexibility. Even though the words may not take up the same space on the tape, the first, second and night announcement occupy the same tape space. A announcement will still take up the same physical tape length, and therefore time, as it occupies a channel with its time predetermined by the longest announcement on the tape loop.

Today, most systems use solid-state or IVR systems as announcement devices, which eliminate mechanical tape issues and recording degradation problems induced by wear on the moving tape and playback heads. Digital announcers have no moving parts, so they are also significantly more reliable. But, most significant is the circuit time saved. When a caller "barges out" of an announcement as an agent becomes available, the digital announcement is instantly reset to the restart point and is immediately available for the next inbound caller. No additional delay is imposed on the caller while the mechanical tape loop winds on to the beginning of the next play cycle. Though this may be seconds, seconds add up quickly when hundreds or thousands of calls are being processed.

Disk-based versus solid-state announcements. Major ACD vendors have elected to do away with separate recorded announcement devices in favor of a disk-based voice response systems and/or solid-state announcement systems. The advantage of both is almost unlimited flexibility. The main disadvantage of the disk-based system occurs in the cost of acquiring and bulletproofing such a system. There is also the conundrum of the additional time added to the call answer time since each call requires disk access; especially when compared to the time it takes to access the "flash" memory of a solid-state announcement system. There's no way around it — it takes longer, even though it's just microseconds, to access disk-based announcement data.

Nonetheless, if you have an older technology in place, using disk-based announcement technology to deliver on board automated attendant functions, ACD voice mail and rudimentary IVR capability can be a good idea. However, with the added performance that comes with adding a disk-based system to older technology, there are trade offs in reliability and administration that need to be balanced against increased customer clarity and care.

Recorded Announcement Flexibility

Adding business applications in a customer contact center demands more announcements; many of the major popular PBX-based ACD systems have significant limitations in this area. Take care to understand how many unique recorded announcements can be played simultaneously to incoming customers of the same and/or different types.

But in the same vein, in the case of simple greetings and announcements, a solid-state system is not only expensive to acquire and to operate, but is overkill. In this situation, the solution might be to introduce a disk-based system and then load the frequently required, disk-based announcements into short-term memory and play back from there. Disk access, as any PC user will attest, is quite measurable. It takes time to determine where the file is, access and play it back.

A number of large production-based customer contact center will not consider a disk-based system unless it also offers a short-term memory compromise which can handle its high frequency, routine announcements. Note that in a large customer contact center, an additional second per call average can amount to as much as $100,000 or more a year in expense impact.

There are three other issues which the contact center manager should be aware of and which vendors do not explain about disk-based system that can be critical to the customer contact center.

First, if the system is configured as a redundant system to allow non-stop operation, the "B side" of the system must duplicate the "A side" disk-based voice technology and database. If it does not, the B-side cannot duplicate operation during an outage of the A-side, resulting in no disk-based voice service redundancy. To keep the price down in a competitive bid situation, vendors often omit the duplication. They can truthfully say it is a redundant machine, but, in reality, it will not operate as a mirror image in regular operation.

Second, if the voice databases are not maintained as absolute replicas of each other, mirror operation is also impossible. More recently this has been addressed with RAID disk technology that manages this as part of the individual disk routines.

Third, often vendors heavily pitch their ability to offer integrated announcements, auto attendant, voice mail, and integrated voice response. Frequently, the IVR function is so rudimentary that should the customer contact center require sophisticated IVR scripts and data communications with a host, these ACD vendors must introduce yet another IVR from a third vendor.

This means a redundant ACD system with fully redundant voice services and sophisticated IVR requirements could require three separate disk-based voice service devices.

Music and Promotion Devices for Broadcast to Delayed Callers
The customer contact center industry knows a caller is less inclined to disconnect during a queued delay if the caller is provided some feedback that confirms they are still in queue. This feedback is provided by announcements and music

played during the delay. The implication of providing this feedback is that the caller has not been forgotten. Today, this feedback has been expanded to include promotional and educational messages and even "all news" broadcasts.

Voice Messaging and Voice Response Interfaces

These interactive announcement and tone recognition technologies allow a caller to "converse" directly with a computer system using a touch-tone pad. This can be used as a triage or "call prompting" process where the caller is able to select the service they wish to receive, (push 1 for service, push 2 for sales) ask for information and participate in other transactions, without the involvement of an agent. With the right programming, many callers satisfy their requests without ever speaking to a live agent. If these "self-help" strategies are executed wisely, they are tremendously effective from both a customer satisfaction and customer contact center resource optimization perspective. The alternative, however, can create more and longer calls and unnecessary customer ill will.

These sophisticated devices are now ubiquitous in customer contact center applications. They offer great service enhancing and labor saving potential when correctly applied.

This market has become extremely confusing for the average customer contact center buyer, as there are so many offerings, which may or may not be appropriate for use in your customer contact center. This caused problems, since there were five distinct applications that were typically supplied by four different types of machines. Each vendor type extended their devices' reach to capture more market and, in turn, reached into vaguely related applications. The result was "marketing over reach" and sales to a number of dissatisfied customers

150

WARNING: ASCAP (American Society of Composers and Publishers) aggressively tries to levy royalty fees against users of music on hold, as this is considered the unauthorized rebroadcast of copyrighted material. This has encouraged the use of promotional material and the "all news" alternatives. ASCAP even alleges this is not a way out as even commercials contain copyrighted musical material. There is no clear solution to this royalty issue. ASCAP is calling small businesses and trying to intimidate them into paying. Sadly many succumb without questioning their rights. There is no absolute answer to this, but ASCAP can be caused to back down by disconnecting the allegedly copyrighted programs.

As an alternative, there are many providers of professional announcement and music on hold programs that can be extensively customized to your business applications. These suppliers provided the license for unlimited use by the customer contact center.

who reached machine limitations well before the machine was obsolete for its originally intended purpose.

There are basically five different types of voice technology applications beginning with the humble home answering machine. This serves as a model and a basis for explaining each of the different types.

Answering Machine: These devices serve one line, have local memory and are non-interactive. A call comes in, it delivers a message and then it will record a message. These systems are typically narrow devices with no other applications, though a number of PC board systems are available for use in your PC. The answering machine in its purest form requires no other interface other than connection to the telephone line. The user interface is based on recording the original outgoing greeting message, then recovering and clearing each of the incoming messages. This is primarily a consumer technology.

Automated Attendant: This serves more than one trunk, delivers a message to an incoming caller, asking the caller to select their destination. This may be a function (Press 1 for Sales) or an extension or even an alphabetic directory of names. The system expects entry of touch-tone (DTMF) digits to select the address or name. It interrogates a local table that dials the appropriate extension identified as sales or the desired extension. The automated attendant is connected to a key system or PBX. In its original application, local memory and storage were limited and it had but one role, simple call triage. It did not generally interface with any other devices such as computers.

Voice Mail: This is an application designed to allow interactive non-simultaneous messaging. Voice mail is an ideal description. The system is served by a number of trunks or connections, serves a large universe of callers, and is highly interactive with extensive local memory, computing power and message storage. It can work with an automated attendant connected to a key system or PBX. In its original application of exchanging voice messages between users, it did not generally interface with any other devices, such as computers.

Audiotext: This is a machine that manages a huge library of audio-based information for playback to a caller at the caller's demand. Each piece of information represents a volume in the library that's directed at answering one specific request. The audiotext machine stores these volumes and allows the caller to access the appropriate information message via a directory of services that are available. The caller must find and select the information they want via touch-tone selections. These systems are served by many trunks and have significant local processing power and vast local storage. Based on the limited interactivity of the recorded data, there is typically no communications with an external computer or database.

Integrated Voice Response (IVR) or Voice Response Units (VRU): These systems connect to multiple trunks (or extensions behind the call center switch) that have significant local memory and processing power. They differ in two significant ways from voice mail devices. They are expected to follow sophisticated user programmable scripts and interface and use external data from another computer.

These are among the most sophisticated applications of voice technologies because they go beyond simple messaging into actual data gathering, interfacing to external databases that require a separate computer look up. Then they can articulate the data delivered by the external computer (text-to-speech conversion).

Voice Technology Connection

There are a number of ways these devices can be connected to the system, and a number of ways they can be implemented. Connectivity can be accomplished in two ways: via analog (2500 set emulation) or digital links to device ports. This latter strategy may require both a voice path (station port attachment) and control via a CTI link.

Analog links are less efficient and occupy more switch real estate when compared to digital service. This has significant implications where the specific services on the switch are used at the expense of traffic capacity. We will discuss this more extensively when we cover quality monitoring and customer experience mapping.

People vs. Technology

Be careful. Care needs to be used to effectively implement these voice response devices and processes. Every channel between the switch and the voice response device equals another occupied port. So the device can be a drain on the ACD capacity. Be aware that with interactive voice response and automated attendants, technology is not the prime issue. Unambiguous scripting and caller acceptance are the keys to making these work. Great attention must be paid to the design of the human interface.

Callers can still be "turned off" when greeted by a recording. A well conceived application and script can accelerate acceptance when IVR and automated scripts are used. There are many tricks to ensure your use of voice technologies is successful with your callers and fulfills the promise of service improvement and cost savings.

Recording Devices

Frequently the occasion arises to record a particular call. This can be for any number of reasons, such as:

- security, tracing the "crank call,"
- maintaining an audit trail in a high value financial transaction,
- risk in a public safety call (911),
- sales verification,
- training, or
- monitoring or quality assurance and counseling purposes.

Originally these devices were simple tape recorders or commercial grade multi-track tape records, or perhaps a recorder using a VCR format, which attached at the trunk or agent station termination blocks and usually did not require a dedicated port. Although this is not the case for a quality monitoring system that records specific agents on schedule for quality applications. These systems attach to the ports designated as service observation ports. The application determines the termination. We will talk more about this in Chapter 15 where we discuss quality monitoring and customer experience mapping.

Where all calls are being recorded for audit, training or sales verification, a voice activated multi-channel tape works better when terminated concurrently with the agent stations. Most ACD systems include a system-wide emergency

Voice Technology

Key Lessons: Make the recording short, the choices simple, and use a brisk script with a thrust toward solving the caller's problem. Always provide an apparent incentive of better service because there is no wait for a live agent. With email and websites as an alternative, do not forego reminding customers there is another, often better, alternative for an answer to a complex question.

Script Lessons: These are simple and bear repeating — dial 1 for Yes and 3 for No are still powerful alternatives to a basic question. People have trouble remembering and or deciding on more than three alternatives offered by a voice response device. After option four, remembering whether to dial 1 or 2 for sales is an issue for 80% of your callers? Five alternatives and you have lost them!

Choose Your Audience Carefully: Then you must educate them extensively. Take as much time with internal customers (co-workers) as you do with external customers. Do not surprise your audience. Otherwise confusion can turn to anger and a feeling of no longer being important to your business.

recorder for crank calls, which needs only one recorder port, assuming your center does not receive more than one crank call at a time. However, there are a number of device considerations when you move beyond the simple application.

There are a number of recording devices in the market that have application in the customer contact center. However, just like the early days of voice response, each application and the technology solution is very specific and typically cannot perform a function it was not designed for.

There are four segments in the recording device market and it's necessary to understand the applications segments and the devices used in each:

Logging or record "all" calls. These systems are also known as voice loggers. This type of device received stellar coverage for its role in the Bankers Trust scandal of the 1990s where it dutifully logged all of the misdeeds of the FX traders. Voice loggers record all calls on the switch or that an agent takes or makes on their phone.

The system connects to either the incoming lines or to each agent's phone. This connectivity option means that each monitored line or phone has a dedicated port on the recorder. For large centers needing hundreds of recording paths, this can run into hundreds of thousands of dollars.

These systems are considered mission critical and religiously take every call. Initially, they wrote an audio record to tape in the first generation systems, but more recently these systems use disk technology, which enables the identity of each call by channel (trunk or agent station) time and date. This time and date "stamp" and channel number are the index points for each call record. Since ALL calls are recorded, to locate a specific call, the only items you use to "find" the call is the channel (agent or trunk) and the date and time of the call.

The problem with any playback is actually finding the call you need to playback. The disk-based systems all play back from a dedicated PC that has access to the recorded voice record via the index. Voice recorded on tape typically means manual review with or without the help of indexes. For non-digitized tape-based systems, voice access is sequential, i.e., you have to start from some point and fast-forward or rewind to the requested part on the tape that you think has your target recording. On VHS tape systems that means going through the parallel channels up to 32 times or until you find the target call.

Until recently, the big flaw with logging or call recording systems is that digital data tags were seldom used. It is necessary to know the physical location of the recording on a tape to find it, or if disk-based, time or port information. The recording was not accessible as a digital computer file and therefore could not be tied back to any call transaction data generated by the ACD system.

The second generation of call logging devices have enriched the sequential format of call recording by adding an additional index layer that contains CTI gathered data indexes and meta tags that allow consolidation of various records so they can be more easily collected, sorted and presented for replay and reporting. However, these systems still basically follow a sequential tape index structure even though the records are stored on disk. These second generation platforms or media storage systems are predominantly proprietary in nature with closed databases.

Scheduled recording of agent voice calls for quality assurance. More recent entries into the customer contact center management systems are quality assurance recording devices. These offer a significant breakthrough in reducing effort and increasing the flexibility of monitoring and quality assurance.

An interactive monitoring system allows a supervisor to "schedule" automated call recording instead of real-time supervisor monitoring. The system records the agent(s) based on that schedule. These systems typically connect to the switch like a "supervisor phone," i.e. via the ports on the switch designated for service observation. This connectivity option allows the customer to use a smaller number of channels to monitor a larger number of agents. There are distinct limits to this architecture, as the volume or applications can grow beyond simple quality sampling.

Play back is done from any standard touch-tone phone, much like voice mail. This type of device is primarily software-based but requires a non-proprietary server containing voice connectivity cards. In simple quality applications, this approach is generally more flexible than the proprietary hardware and database of logger-based systems. Alternatively, call review may require the supervisor to play back from a dedicated PC or from one equipped with additional monitoring hardware. From the interactive monitoring system the supervisor can access voice files by group, subgroup, or by agent ID and since they are digital in nature, the system can go directly to the requested voice files. These files should be able to be tied back to any digital data generated by the ACD for reporting purposes. There is an important and critical difference between this and basic logging or record-on-demand when no CTI connectivity is included in the recording structure.

Recording a call on demand. These record-on-demand systems are used to record an agent's conversation based on an agent or supervisor request. The typical applications are "Sales Verification" and "Over the Phone Transactions," where active customer consent can be used to execute a contract. In these applications, the agent needs to record the customer saying, "Yes, I want to buy your product."

Initially the recordings were manually invoked, but more recently they are connected via a CTI control link to the switch. They receive a request to start recording an agent, then receive a request to stop. The most recent innovation is to embed a macro command in the data entry field associated with the positive order that automatically launches the demand recording

Replay access is usually done from an analog link to the tape recording or disk-based record. There are also the emergency recorder applications to catch crank calls and bomb threats that are inherent to an ACD. These are typically not robust enough for verification application. There have recently been some interesting developments in verification as it relates to paperless order confirmation, which we will discuss in Chapter 15.

Screen capture or data recording. With the advent of automated voice monitoring came the need to concurrently view the data screen being used by the agent. This was so the reviewer could see the agent was correctly representing the information displayed on their screen to the customer. These are data applications and more of a LAN capacity issue than that of an ACD function.

There are two ways of doing this: the older and hardware intense screen duplication or the software approach (via software-driven port replication on a mainframe or software screen image capture) that allows a mirror of the screen. The latter is better and more flexible and can be displayed by the reviewer who is equipped with a similar display device.

The software-based system also records what's happening on the desktop. It can record either host-based sessions or all of the desktop activities. This becomes important when a company moves from real-time monitoring (expensive and inconvenient) to monitoring past recordings.

This is desirable as the reviewer does not need to be on the phone concurrently with the agent waiting for an appropriate customer call. A major concern in implementing screen capture is the impact that recording this data will have on local area network bandwidth.

These systems usually have a logical link through the network or reside on the host computer itself. Access is usually done from a dedicated PC or from any PC connected to the host.

Recent work has been done to connect the voice and data recording monitor systems to provide an even more complete view of the customer/agent transaction to be reviewed.

We will discuss the uses and benefits of these systems when we discuss customer contact center management style later in the book.

Ports

All these devices are attached to physical ports in the switching system. These ports are finite and when they are in use they occupy a path through the switching network consuming network resources, thus reducing the switch's available capacity.

☐ BACKPLANE AND SIGNAL DISTRIBUTION SUBSYSTEM

A backplane or signal distribution system may be made up of a single printed circuit board in a single shelf system, or a complex of backplanes interconnected by ribbon cable or other mechanical path for the passage of electronic signals. A single shelf backplane is called a "motherboard." These devices have finite physical capacities on a shelf, rack and system basis. They may be simplex or alternatively, duplex to ensure redundancy.

Internal Power Supply

As expected, electric power is the lifeblood of any computer controlled switch or network. Lose power, lose the computer, the switch or the network. Along with receiving and providing a degree of power conditioning, your system's internal power supply reduces the voltage of the incoming power to the voltage required by the individual components in the system. The internal power supply may be just one device (simplex) or multiple devices in serial or multiple devices arranged in parallel (duplex). These power systems may be redundant. Check, just because there are multiple power supplies in the preferred system, they may not be redundant. Also check for that single point of failure, such as a single fuse or buss.

Power supplies generate heat. This heat needs to be dissipated. Always leave room around a power supply. The bigger the power supply the greater the heat. Find out:

- How is it cooled — convection or mechanical fans?
- Where is the power supply — at the bottom of the device (heat rises!) or at the top.
- In a system where cooling is mechanically assisted, what happens if a fan (or fans) fail?

Avoid storing supplies or equipment on top of a switch or computer rack as you may be interrupting the airflow necessary to cooling the system.

The Power Distribution Subsystem

The power distribution subsystem is the power distribution network in the machine that allows electric power to be distributed at the right voltage to each of the components around the system. This can be the weakest part of the system — especially when it is overloaded. Supply (the front end) or demand (the back end) can overload it. That is the power being fed to it or drawn from it. Power variations are a common source of problems for all computer-controlled devices.

What options do you have regarding electric power? AC (alternating current, or the power coming off the electric company's grid and through the standard wall socket) or DC (direct current provided by filtering commercially available power through an inverter and batteries which "steps down" the power to the switch). By using a DC power source you have a higher degree of power regulation and reliability.

The more the computer companies got into the switch business, the less attention was given to delivering conditioned (DC) power, which was traditionally a telephony standard. However, providing AC power and the batteries and inverters necessary to power a switch added additional installation and maintenance cost, and in a competitive sales environment, was considered undesir-

---

Case Study: **Shared Electric Power can be a Problem**

An ACD was installed in a Las Vegas, 40 agent, 24 hour a day, hotel reservation center that handled 90% of reservations for a billion dollar hotel casino. Need I state that in this 1000+ room hotel, the guest reservation center was critical to their revenue and competitive position in the market. Every morning between 8:30 and 9:30 and every afternoon between about 4:00 and 6:00, the system would "crash" in a big way. The service organization did not find the pattern or frequency of the failures until they put power-monitoring equipment on the power outlets. But this wasn't done until after they had changed out the computer control and power supplies. Only when the power monitoring equipment was in place did they discover that the ACD shared a circuit with a large three-story hydraulic elevator to the executive offices! This elevator drew massive power requirements, during the "rush hour" crunch, as the pump fed the elevator ram. The sag in power was more than the AC powered ACD could stand.

able. The cheaper alternative is standard AC. With any desire to increase reliability, the cost of acquisition and ownership is higher. Consider this against the mission critical nature of your customer contact center.

The more critical your customer contact center is, the more critical it is that you pay attention to power regulation and backup. Our humble opinion: a full-blown DC battery backup system powering a DC machine makes most sense. Although an AC system with an external power regulator typically works fine and is less expensive when used in small applications.

Now, with complex applications available from CTI and customer contact center process reengineering, the need for redundant servers, LAN connectivity and robust desktop machines becomes even more critical. Yet, most system integrators and available hardware are short on power system answers that have been considered standard operating procedure in the telephone industry for eons.

With the advent of VoIP, even greater concern needs to be paid to ensuring non-stop operation for the customer contact center, as now the points of vulnerability to power outage are multiplied innumerably.

159

The Mechanical Housing of the System

All of this technology must now be packaged up in an effective unit that meets regulatory (safety), manufacturing, marketing and cost objectives. The package consists of rack or cabinet mounting for shelves, which contain slots for printed circuit boards and subassemblies. These represent the various components. These are connected to the backplane at right angles from where they receive power and communicate with the system at large.

The Switching Engine

In calling out the various components and subsystems of a switch, it becomes clear that a device, a component or a system has finite capacity, which occurs at all system levels and exist in three dimensions:

- The number of devices that can be physically accommodated in the mechanical housing.
- The number of simultaneous communications paths supported by a single stage switching network.
- The raw call transaction processing power of the common control computer(s). The questions that should be asked in any switch purchase are:

1. *Physical Capacity*: How many devices can it physically support and are there any tradeoffs if analog versus digital circuits are used? This is a switch "real

estate" question, i.e. how much room is in the switch to accommodate all of the required devices? Digital devices (such as a T-1 span of 24 channels) may be one card, whereas analog trunk ports may be four, eight, twelve or even 16 to a card. The impact in an analog system is more cards may be required for the same number of physical connections.

- What is the maximum number of devices per shelf?
- What is the maximum number of shelves per rack?
- What is the maximum number of racks per system?
- What is the maximum port capacity of a single stage switch?
- If it is possible to connect single stage switches into a larger switching complex, then how is this done and what is the trade off?
- Are existing device ports displaced to accomplish this intermachine connection?

At this stage we are only looking for points that may introduce physical system blockage. That is a caller in queue unable to reach an available agent because there is not an available talk path. We are not yet considering the impact of software being unable to logically support a physically nonblocking configuration.

2. *Switching Capacity*: How many devices can be supported and simultaneous conversations carried by the switch and what is the actual traffic capacity of a single stage switch?

3. *Common Control Computer Capacity*: What is the largest referenceable installed customer?

What we are looking for is the ability of the switch to handle calls. You must watch the language the various vendors use. Some refer to "BHA" or busy hour (call) attempts, which means what it says — attempts only, NOT completions; whereas other, more honest, vendors talk about "BHC" or busy hour call completions. A switch can deal with more attempts, if it does not have to route, connect and monitor call status, then break down and "housekeep" the data following a completed call.

The Advantages of A
Purpose Built ACD
System

Processing an automatic call distributor (ACD) call is
from FIVE to SEVEN TIMES more complex than an average administrative
call processed by a PBX, and therein lies a key issue as to why a generalized
switch such as a PBX or CO does not make the most desirable ACD platform.

It consumes more resources, it is switched more times and is more intensely
managed by the computer controlling the switching process. Ideally more data
is gathered and reported about the call status, status changes and call disposition. And this only gets more complex as we add IVR, and CTI applications
such as call tracking or customer inquiry state management in an attempt to
reduce customer contact center staffing.

Because of the revenue and/or goodwill implications, the call carries a great deal
more importance and emotion than the average administrative PBX call. Simply
put, the customer contact center is considered a "mission critical" application in
most businesses and is a whole lot busier than the regular PBX phone system.

Manufacturers have recognized this to different degrees and offer differing levels of system features and robustness to fit every type of customer contact center.

The features and values of an ACD system can be categorized into five
broad areas:

1. The architecture and capacity of the switch.
2. The ability to provide service to callers in a system outage.
3. The ability to provide a high degree of information about all facets of the system, its operation and customer contact center resources.

4. Ease of use at every level so effort in using the system does not impair the primary caller transaction.

5. An open system control architecture so that the ACD system can interface with other customer contact center resources.

☐ ARCHITECTURE AND CAPACITIES

The first question to ask a switch provider offering to sell you an ACD concerns the history of the switch. What did this system start out as? A key system, a PBX or a purpose built ACD? A CO, PBX or key system based ACD, did not begin life as an ACD, and is consequently encumbered with certain design decisions made for the originally intended application (administrative phone support) and market (size).

When a system (any system) is designed certain engineering compromises are made to meet marketing and manufacturing cost objectives. This expedience often caused the last generation of switches to become generalized PBX solutions.

With generalization comes a lack of clear engineering focus. "We are designing it to be a multi-purpose switching engine capable of serving any voice application." Yet, competition, market reality and engineering design considerations are diluting factors. An ACD system is a very particular application with the need to accommodate the many human factors and management issues of a real-time cultural and business environment. Many of the needs of customer contact center management and ACD design contradict good PBX design.

Generalized switches, with their intended market application, often become the manufacturer's basic platform for the application or market that is identified as the next revenue opportunity. The system may provide modifiable call routing, can be programmed by the user, can produce call statistics, and maybe can be sold as an automatic call distributor. But is it truly an automatic call distributor?

Manufacturers generally do not lie, but what if they do not understand the application well? Compromises are made and if the manufacturer doesn't understand why specific features exist, it may "shoehorn" the application into a less than ideal machine with constrained computing power. The user of such a generalized system subtly loses opportunities to save money and take control of their business destiny. Nobody lied, but neither party to the transaction may realize what they are giving up from a business perspective. And the lower initial acquisition price may appear to favor the PBX or hybrid platform.

A generalized approach compromises three aspects of customer contact center management:

- direct control of operating costs,
- ease and convenience of gathering and analyzing real data for analysis, and
- the business insight and business advantage that comes from such data.

The various opportunities to control operating costs (trunk and staff) over the life of the system, far outweigh the increased acquisition cost of a standalone ACD over a PBX-based ACD. In spite of this, many buyers balk at the increased system cost of a standalone ACD when compared to a PBX-based system. It is important to understand the critical mission of the customer contact center and the opportunity to save operating costs. If you can live with the compromises offered by a PBX-based ACD system and your business loses little or no advantage, then a PBX-based ACD may be just fine. However, once a business understands the compromises brought by a PBX-based solution, they usually sing a different tune.

A purpose built ACD system has very specific user focused properties, for instance, the system must:

163

- Be both physically and logically non-blocking so it can connect any two ports (resources) attached to it, at any time.

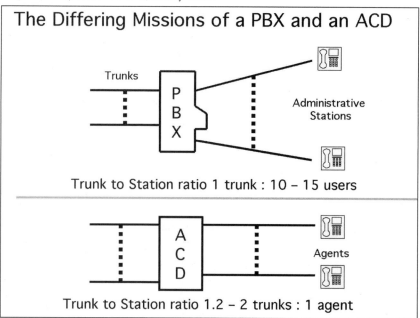

The Differing Missions of a PBX and an ACD

Figure 9.1.

- Allow a high degree of flexibility to the user in order to allow trunk, agent, routing and reporting arrangements to reflect user requirements. For instance, the user should be able to easily manipulate the call routing tables at any time, so as not to abdicate customer contact center flexibility to vendor service availability or cost. Ease of user "programmability" is a significant issue. "User programmability" does not mean writing system code, rather it means the changing of system tables to reflect your business objectives — on demand, easily and quickly.

- Offer a high degree of reliability. Because of the critical revenue winning or revenue protection of a customer contact center, great importance is placed on uninterrupted operation. Typically some degree of resilience is built into the machine to retard failure.

This is a world of subtle, but expensive nuances. If you don't understand the issues, you don't ask the right questions and you are never aware of the implications. Blissful ignorance!

Physically and Logically Non-blocking

It is important that all calls be answered and served promptly, because of the revenue and image maintenance issues associated with the call. There are few exceptions to this rule.

Physical blockage is the inability to connect a caller desiring to reach another person or device to that person or device due to a lack of talk path capacity in a switching machine.

In the case of an ACD, a talk path is the communication pathway that is established between any component and another, for example, a trunk and an agent position so the conversation can occur between a caller and an agent. This conversation may result in an order or request for help. This is a physical path in an analog switch, or adequate timeslot capacity in a digital switch. Most switch manufacturers claim to provide non-blocking switches.

A little known (and unadvertised) fact is that many physically non-blocking switches can introduce blockage. This occurs when the software that queues and routes the calls is written so that it doesn't take advantage of the non-blocking switch capacity. For example, when an inbound call has waited for an available agent in the primary answering group (Group 1) beyond a certain predetermined threshold, the call is intraflowed (overflowed) to a second group of agents. If an agent in the second group (Group 2) is not immediately available, the call continues to wait in queue.

In some systems, if an agent becomes available in Group 1 (the group primarily responsible for serving the call) while the caller is on hold for Group 2, that agent cannot answer the call, as the software does not allow for "look-back" capability. This is an example of "logical blockage" occurring even though the physical talk path exists to serve the connection.

This has the effect of increasing the number of agents in Group 2 to meet the increased or "stimulated" traffic load and service level they now must meet from prospective understaffing in Group 1.

In outbound call routing, "look back" capability in queuing tables is known to contribute approximately a 15% increase in trunk group efficiency when the trunks involved exceed thirty. Although no research or writing is known to have been done on look back routing in customer contact center efficiency, the lack of multiple level simultaneous queuing is believed to incur a similar penalty in efficiency. The sizing of the secondary and subsequent agent groups must now allow for the extra load that is offered to them due to the lack of the look back feature. The logically blocking routing scheme stimulates the traffic load they now must serve, and more staff is required to continue to meet the desired service level. This type of problem is typically encountered in the popular PBX-based ACD systems.

With the addition of conditional and skills based routing, there is an increase in the ability of these systems to focus the call on an increasingly definitive agent resource based on a specific skill set. This, however, has two differing effects: it greatly increases the complexity of setting up optimal and efficient routing tables and queues, and if done badly or employed too frequently, it can significantly reduce the efficiency of the ACD. Thus, the reason for the customer contact center (to serve many by a few) now has the potential to become fragmented to the point of denying the reasoning for the center in the first place, i.e. to take advantage of economy of scale. The effect on increased operating costs becomes significant.

New strategies are being developed to deliver the agent significantly more tools to allow almost any agent to serve the caller. Although the notion of a truly "universal agent" is probably illusory, there are many things that can be done to expand the skill set of your agents so as to expand the pool of agents available to serve your callers. That is back to being a true, purpose built ACD. We will deal more with this throughout the book.

Flexibility

An ACD system must allow a high degree of flexibility. Though most ACD vendors allow a great deal of user access to the system for reprogramming of the routing tables and parameters, a little examination of the philosophy governing this feature is necessary.

First, it is absolutely unnecessary in this day of microcomputers to deny a user control of their own destiny. We now have the power of a 1970s mainframe in a chip, and the capacity of an online DASD farm storage in a battery driven lap top PC. It is absolutely unacceptable for a customer contact center to have to build its business and call handling processes around the limitations of a vendor's shortsighted decisions or machine design. Yet, in the world of ACD systems a number of vendors expect customer contact center managers to do just that.

The background that has led to this is understandable when you examine the software philosophies that were popular at the time these systems were initially designed. ACD systems that are available today reflect four attitudes to user programmability and their effect are typically experienced by a user when making call processing changes to the system via an administrative computer console or CRT screen.

1. *The "system build."* This first generation computer controlled switch allows significant user flexibility when originally ordering the system, but does not allow for extensive subsequent changes on site by the user. Changes are possible, but are normally conducted with the assistance of the manufacturer. There are costs, inconvenience and time delays associated with any change. A very technical user could make changes to the system, but not without extensive training and skill. Personal computer programs have been introduced to "front end" the change process and make it less difficult for a user to make a change.

2. *The "table driven" system.* The next generation of switches allows the user to access control software via building and changing user accessible tables. These tables can be changed by the user, at will, to reflect routing steps, agent groups, agent properties and other system aspects. These systems use plain language in their command structure, and simple command syntax: noun, verb, and optional modifier. An example of this is "Revise Route (number) 01." The objective of this strategy is to reduce the skill and labor content necessary to manage the system. System operation still requires considerable training.

Explanation: Although the phrase "user programmability" is used, no literal programming skills are necessary to manage a modern ACD. However, there is an advantage in giving the system management responsibility to people not in awe of technology.

However, the resultant independence from the manufacturer and customer flexibility is of substantial value.

3. *The "menu driven" systems.* This is an advance in third generation ACD user interface techniques. All of the options are offered on the administrative CRT screen as the system manager steps through the system revision. The manager merely selects an alternative by moving a cursor and pushing enter. The change is made. A variation of this is the introduction of Excel-like header screens with cursor/enter selection sequences.

4. *Object-oriented user programming*: Here icons or symbols are used to represent actions, objectives or sequences. The user merely takes the object and places it in the order they require it to be executed.

Today, most systems have updated their interface standards to make the ACDs easier to use. The penalty a manufacturer incurs for not being easy to use is confinement to markets that can accept rigidity or can not afford to invest in the staff necessary to use a more sophisticated system.

Take time to understand how hard or easy it is for you to make changes to a system. What staff commitment and training levels will be needed for support? Also understand the scope of the changes, both system and component level. Understand to what level in the machine rule changes operate.

What is the extent of your ability to change the system? Down to what level of detail? Command granularity — systemwide, trunk group, trunk or call-by-call? A good example of control over your ACD (command granularity) is answer control. Here a call-answering rule is applied by the system. If a call arrives on a trunk and there are no agents immediately available to answer the call, it may make sense to allow it to ring. The logic behind this is that not all telephone calls are answered immediately, so it is quite reasonable to expect the caller to accept two or more rings, until an agent becomes available. Here is the key. If the call is not answered, billing does not begin. This is a major advantage to an inbound WATS subscriber, because the alternative is to answer the call, and while the caller waits to speak to an agent, you play what has become very expensive "elevator" music. It's not the music that's expensive but the unnecessary phone line costs involved to play it.

The question of command granularity is this: When the delay parameter (or any other like parameter) is set up, does it effect the entire system, or just the trunk group, or perhaps only an individual trunk? Or is it dynamically assigned on a call-by-call basis?

Ideally you should be able to control the smallest logical unit your ACD affects — the individual customer call. If you are restricted to constant treatment at a trunk level or worse, trunk group wide, or worse still, system wide, you become increasingly unable to adapt to instant business conditions.

When is a Telephone Call a Call?

The next issue is the optimum delay dynamically adjusted to reflect agent availability. If there are agents available, does the system provide the minimum one ring then immediate "cut through" to the agent, or does it introduce a flexible caller delay period that can be longer or shorter based on agent availability?

A little telephony explanation is required here. This is just like setting up a computer-to-computer communication. There is the hardware connection — the physical path. Next there is the signal requesting a response — essentially, "is anybody out there?" Then there is the answer and handshake between systems. A CO, once it has a call destined for your pilot ACD number, selects your physical trunk (access) and sends a supervision in — ringing voltage on an analog circuit or the digital equivalent on a digital connection (ISDN or DSL). Your ACD recognizes the inbound call service request and responds with a digital answer signal or "answer supervision" in an analog world. The call is connected and billing begins.

The following comments are directed at network expense, which fortunately is less of a cost today, but remains an important and costly consideration. There is a commensurate or alternative labor expense also.

The ultimate question for the vendor is "do you offer dynamic call-by-call answer (supervision) delay? Unfortunately, they will all answer, "yes." Typically because they do not understand the nuances of the answer.

But don't stop there. The next question should be, "at what point does your system return an answer (supervision) signal?" Again, this is the signal to the telephone company to connect the call and begin billing. The answer is when the ACD answers the call.

Then ask: "can you delay answer no agents (IVR ports, whatever) are available to serve the caller?" Again, the answer will be, a yes. Now the question and their answers become more critical. Find out what governs the return of the answering signal (supervision) — the trunk card settings or an interrogation of the routing tables and agent availability?

Here is what all the PBX-based systems do: On a trunk card by trunk card basis, they will allow a fixed ring delay to be set. That is X seconds. The trunk card or trunk group can be set to deliver a 10 or so second delay. NO MATTER WHAT! What these systems don't do is detect the inbound call and ask the central computer to check the agent database to determine if an appropriate agent is available before they return answer (supervision). The reason they do not do this is the basic architecture of a PBX/ACD versus a true ACD.

When any computer-based system is designed, the system designer is tasked with building the system as inexpensively as possible. One of the easy ways to reduce cost in switching design is to provide less computing power and storage. (It is as though these folks have never visited a PC store to see how cheap computer horsepower, memory and disk storage have become!) One of the ways to reduce computer processor demand is to reduce the number of repetitive tasks the system processor needs to be involved in. One such task in a phone system is answering a call. The logical next step, if a process is always the same, is to relegate that task to a lower process than the central processor.

In a PBX system, one such task is answering an inbound request for service. This means that any call to a trunk card is answered with no further questions asked of the central processor. When a system originally designed as a PBX is pressed into service as an ACD, there are certain things it can never do without dramatic redesign. Coupling a inbound request for service, with the dynamic interrogation of a routing table by the main processor when there was no physical and logical design linkage, is next to impossible.

This little shortcoming will cost a call center bunches. Here's how to measure the impact on your customer contact center. First, how many calls does your center queue a year? Of all the calls, how many exceed 6 seconds in queue, before being connected to an agent. How many exceed 12 seconds, 18 seconds, etc. A system that can retard answer supervision on a call-by-call basis can save you at least six seconds of billed INWATS time per queued call, or 12, 18 or more respectively. One recent customer contact center visited by the author received 75,000 calls a day of which 54% were queued per day for an added cost of $29,000 a month just because the system answered calls with no heed to staff availability.

A manufacturer may have you believe this is an insignificant feature, but American business sends millions of needless dollars a year to long distance carriers because their systems cannot search for an available agent before the answer signal is returned on queued calls. Back when AT&T made PBX and ACD systems this made great sense to them — there's something reminiscent about "the fox watching the hen house!"

Just because computing power has gotten less expensive, it does not mean that software redesign has caught up with the available power. Check. The absence of this feature may mean thousands of dollars a year in higher INWATS expenses to your customer contact center. Subsets of this feature are:

- "When is a inbound customer call recognized as a call?" "Supervision in or supervision out?" (System answer?) As we have just explained, the reason this is important is first, for the issue of dynamic answer and recognizing the presence of a call so that the routing database can be questioned. If the database cannot be interrogated as an unlinked question, the system will answer all calls blindly and add unnecessary call length and billing to your customer call operating expenses.

- "When is the call counted as data for reporting purposes?" Supervision in or supervision out?" (System answer?) If a call is not measured before answer, a significant amount of trunk performance and potential network problem identification data such as "ghost calls" is lost to management.

- Is the presence of a call measured before the answer signal (or supervision) is returned? Again, for the same reason as mentioned immediately above. In the case of an ACD that allows CO ringing to be factored in as part of the queue cycle (much to the chagrin of the long distance carriers), it is important to determine how long a caller is in this state, so caller tolerance to delay can be measured. A ring cycle is six seconds in duration; two of ringing and four of silence. If no one (or nothing) is available to answer the call, two rings saves 12 seconds of billable WATS time per call and so on. It is important to understand the length in the ring pre-queue for true service level and queue management purposes. In the case of one call center, a second added to the length of the average call, adds $1,000,000 a year in expense to the teleservices budget. Move this off your "books," back to the carrier, and although you will require available trunk capacity, it is cheaper than billable trunk capacity, carrying expensive music.

The ability to drive this command and management detail is first a question of the manufacturer understanding the issues in a customer contact center, then providing the features by providing adequate computer power and software.

When an ACD began life as a generalized switching device this type of detail was next to impossible to deliver to the user.

"Dynamic Barge Out"

Finally, there is dynamic barge out. Now if the system provides a fixed delay, such as 10 seconds or so, what happens if an agent becomes available during that time? The desired answer is that the system will detect the available agent, answer the call and deliver the caller to the available agent. This is called "barge out." Because the delay decision is limited to the realm of the trunk card or trunk group and no other issues, if an agent becoming available in the target group, there is no way to barge out!

Available agents, queued callers, this system is logically blocking for a moment! Available agents twiddling their thumbs while callers wait for service. Again these are the subtle limitations of a PBX-based ACD system. As subtle as they may be these shortcomings conspire to cost a call center owner thousands of dollars in inefficient operating practices.

☐ SYSTEM RELIABILITY AND FAILURE RESISTANCE

A true, purpose built ACD system is also built with the knowledge the system will fail at some time in its service life. (An outage may also occur as a planned event.) This knowledge of the system not being available is a critical design consideration. There are ways to minimize the impact of an outage. These range from totally redundant components to no redundancy at all.

Understanding the levels of progressive system degradation available from your chosen vendor can equip you to minimize the impact of a failure. You can then be intelligent on how you spend money to reduce the impact of a failure. You may not need the diesel generator on the roof. However, if the MIS department already owns one, hook it up to your battery recharge system. You are a mission critical application.

ACD systems can be taken down intentionally for a number of quite logical reasons — hardware or software upgrade or physical move of the equipment. If the center operates 24 hours a day there is no ideal time for this to occur, so an ability to continue to serve the incoming callers is desirable.

Understanding how system reliability and failure resistance is engineered into your preferred system is important.

Power

A system can fail due to the unavailability of power. To minimize the impact of power failure, three strategies can be pursued. These may be used individually or concurrently.

Conditioned power: An ACD can be driven with AC or DC power. AC or alternating current is normal commercial power coming out of the nearest power socket, delivered by the power company off the local grid. It is subject to all the vagaries of surges, spikes and drops in voltage that disturb computers.

When the office lights flicker, it's a power disturbance. If these are not considered of significant consequence, wall power can be quite satisfactory.

To mitigate minor transient power problems, a voltage regulator may be used. This is a device that's plugged into the socket power source. It contains voltage buffering or regulation electronics designed to catch and isolate momentary power transients. This isolates the system from very short-term power problems, but does not supply power in an outage. Pay attention California.

Uninterruptible power: This is the other extreme of power protection. This is a completely isolated system with it's own or alternative power source. In this case the ACD is completely isolated from commercial power. The ACD is typically DC powered. The power is provided by a DC or direct current system. This DC system "floats" the ACD away from the commercial power. There is a battery, which is constantly charged by the commercial power supply grid. The ACD is supplied by the battery source, which has a life of an hour or more. The ACD's power is supplied by the battery. An inverter maintains the battery charge. When power fails, the batteries supply the system until they are used up, or the power returns. Statistically most power failures are less than 60 minutes in duration and within the planned battery capacity.

Backup generator: The building housing the customer contact center may have an AC power generator that "kicks in" at the instant of a power failure. The question becomes how fast will the power be available and will it be transparent to the ACD. If it takes a minute or two to come on line the ACD will die, unless the power system (typically a bank of DC batteries) is engineered to accommodate this. The MIS department cannot tolerate this either, so the building management should understand the problem.

If you have access to such back-up power, make sure your company includes your customer contact center as a candidate for back-up power support. Having the order entry terminals running on the telephone reps desks and no phone calls is a little ludicrous. Now the opposite is equally undesirable. Get the ACD on the same system. Once again, selling your management on the mission critical nature of the application is necessary.

If you have an external generator, make sure you have diesel in the generator's fuel tank! Don't laugh, we know of a huge credit card company which got caught running on empty.

These strategies only ensure power is available to the system. Now it is necessary to consider building a redundant system.

A Redundant ACD System

Here we make the assumption that the failure is caused by a failure within the actual system. A processor fails, a disk crashes, a controller or switching network hiccups. Any number of system glitches can cause the system to stop processing calls. The major element is the brain or control processor.

In the area of common control technology, the system can have redundant computer systems. That is a second computer system that backs up the first in the event of its failure. It is important to understand what redundancy really means in the case of your chosen vendor. Is it the entire system or just large parts of the system deemed to be critical? The control computer, the internal system power supplies, power distribution, switching technology and the individual line, station and utility components are all indispensable elements. Are these redundant? Typically this is not the case.

A new generation of technology introduces more redundancy than simply duplexing the common control. Then it allows the "hot swap" of failed parts. Check which components your vendor means when they say "redundant."

173

Common Control

Common control can have two types of redundancy:

Mirror image or "nonstop" computing or uninterrupted operation. This first method uses mirror image processing. Two computers "mirror" each other until the point of final execution and one completes the last step or process after comparing the consistency of the separate results to this stage. The last step is completed by one of the processors and everything begins again. This type of redundancy in the control system should insure consistent uninterrupted operation. To implement a system of this nature is more expensive in the short run, but is cheap insurance in the long run, because the business you protect is short-term revenue and long-term goodwill.

"Hot standby". This is the second and more popular method. A second processor kicks into operation when the first fails, i.e. a duplicate or B side system stands ready to kick into operation should the A side system "die." The problem with this technique is that calls in process can be "dropped" during the system swap over. Note that newer systems do not drop calls and look more like the mirror or nonstop application mentioned first. Again, initially expensive but cheaper long term business insurance, but this redundancy only affects the system intelligence. If other major parts fail, the system still dies.

Typically a vendor will identify the maximum number of elements that can be affected by any one outage. A system outage will affect no more than xx lines or agents...a shelf...a cabinet. Consider this as a percentage of the customer contact center and it can take on a whole new meaning. If "only 48 ports (trunks, agents, etc.) are affected by any outage" and your customer contact center only has 96 ports, you just lost 50% of your customer service capacity. If they are just your trunks or just your agents, you may be dead in the water.

Some systems, especially those using onboard voice processing subsystems, require a complete duplication of the voice processing subsystem on the B side to perform identically to the A side in an A side failure. As we mentioned earlier this may not be configured the same, especially if the vendor's searches of "wiggle" room to lower costs in a competitive deal. It is absolutely critical that not only must the B side mirror the A side, but the database must be synchronized so that everything works identically.

Software Redundancy

This is important if the reliability "bug" resides in the software. Is the mirror software an exact replica of the first, bugs and all. Well designed redundancy typically runs a second and unique copy of the software that does not inherit the errors occurring in system A over a lifetime of use.

Power Fail Transfer

Another type of redundancy, which is really a fallback strategy, is to build onboard bypass into the system so that with any failure (power or system) telephone calls on analog trunks can still be received, since they are continually supported by CO provided power. The system must be equipped with sufficient power fail transfer trunks to allow the system to operate. A typical PBX ratio of 5 to 10% of the trunks being equipped with power fail bypass trunks is totally inadequate in a customer contact center.

Power fail transfer (PFT) trunks are telephone lines from the central office that are hardwired to a specific phone instrument during a switch outage. The PFT feature must be present in the line card for it to work. These individual phones become single line instruments much like a single residence line. They are not switched by the PBX or ACD during the outage since it's dead, but remain directly connected to the phone company exchange and can receive and originate one call at a time.

PFT trunk cards are also more expensive than regular telephone line cards and they take up more space in the switch. As an example, 8 regular lines may occupy a single slot but when PFT line cards are specified they only allow 4 or even 2 per slot. One vendor trick used to keep the proposal price of a PBX/ACD low in a competitive battle against a standalone ACD is to omit PFT trunks or keep them to a minimum. When bypass features are incorporated in a PBX/ACD quote, the price gets a lot closer to the more fully featured system.

Alas, with the greater use of bulk digital telephone services, which rely on reliable power, the notion of power fail transfer or bypass to central office power is less and less available to call centers. VoIP accentuates this vulnerability further as all connection is over a computer network that depends on a pristine and reliable power supply to operate.

Key power fail bypass questions:

- Do you provide power fail bypass? Response should be. "yes."
- How many trunk-to-agent power fail connections are included in this price proposal? Response should be either: "some," "5," "10," or, even perhaps, "all."

What you need and what the system provides can dramatically change the PBX/ACD system price and bring it closer to a standalone system with inherent bypass.

175

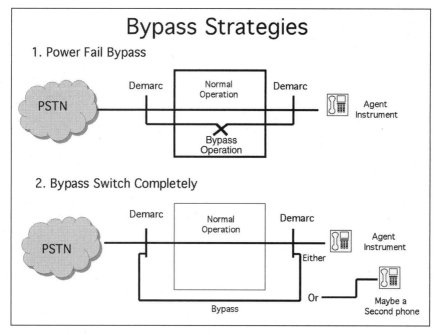

Figure 9.2.

Power Failure Operation. Upon system failure and fall back to power fail transfer or bypass, two methods are used. The first method is where the first trunk is now directly connected to the first agent position. The second method is to pre-assign each bypass trunk to a specific station. The latter technique is more flexible as you do not have to staff quite as rigidly. The only rule to be aware of — you cannot simultaneously process more calls on bypassed lines than there are manned positions.

As the system fails, before the power fail trunks start carrying the calls, all calls in process are interrupted. The actual switch to the bypass mode may be manual or automatic. Restoration of the switch to full automated operation may also occur in either method. The vendor should take great pains to explain how this occurs. Take time to understand how it works. There are some expensive and inelegant solutions.

New features and technology such as VoIP will increase the complexity and cost of redundancy.

Since a number of the standalone ACD and now PBX-based ACD system vendors think that basic announcements and music can be better delivered by a disk-based system, they must also configure a duplicate disk-based voice response system in cabinet B (the back up system). This adds significant costs. In a competitive bid, if a buyer is not watching carefully, the vendor will sell redundancy, and leave out this feature. This means in a switchover, the routing tables of system A are going to go and look for announcements, music and the like, that are not present in system B. This means that the process and subsequent databases are out of synchronization and the caller will encounter unintended interruption. The vendor acts surprised, as if they didn't understand what you meant when you asked for redundancy. It also takes a significant time to resynchronize the database and restore operation. The solution is a mirrored voice response system in system B. The user has little choice.

Note that any T-1 service coming into the system will not operate as the system needs power to maintain and decode any calls carried on a digital circuit. Also be aware of analog trunks on a digital switch. Calls arriving on analog trunks, attached to a digital switch, will not work in a system or power fail condition unless there are separate arrangements to wire these around the switch. VoIP has exactly the same problem.

The Calamity Switch

A simple and inexpensive way around this is to wire the customer contact center with separate analog telephones, and install manual bypass directly to any analog trunk. This technique and bypass block has the quaint name of "calamity switch". They work fine and constitute a simple and elegant solution provided the carrier can impromptually redirect the calls to these analog circuits or even another operational call center.

☐ HIGH VISIBILITY INTO THE SYSTEM

The ACD system must provide a high degree of visibility into the system, its operation, and the resources attached to or working with the ACD system.

There are "choke points" in a customer contact center, which can negatively affect call processing. These are:

- Inadequate network capacity.
- Too few trunks (or local loops) to carry the offered inbound call load.
- Inadequate switching capacity.
- Misunderstanding the demand to staffing ratio, so extended queue delays inflate caller abandonment.
- Poor training and motivation of the agent staff.
  Also added to these choke points can be issues, such as:
- Inadequate VRU capacity.
- Poorly designed and ambiguous voice response scripts.

A well thought out, thorough ACD system will provide an extensive array of tools to allow a great deal of insight into the flow of your business, with both real-time and after-the-fact reports. Because of the intensity of the customer contact center application, the staff with the responsibility of managing this center on a day-to-day basis are best qualified to judge what management reporting tools are necessary.

The customer contact center business has a great preference for the capture of the lowest level of call detail at all times, that is every keystroke (alpha or numeric character) generated by an agent, and trunk, delay and process detail by call.

☐ EASE OF USE

"If it is hard to use, it won't be!" The ACD system must be easy to use at every level. A customer contact center is a demanding environment to begin with, without adding a whole new level of manual data gathering discipline on the agents, the supervision, the administration and management.

The effort involved in using the system must be almost transparent to the user. The agent instrument or workstation should reduce keystrokes and confu-

177

sion so the agent can focus on the caller and the business at hand. Use of the ACD should not impair the reason for the primary caller transaction.

The staff, supervisors and management should have accurate and unambiguous reports. Plain language is essential and jargon should be kept to a minimum. Check the report column headings. Do you need a Ph.D. in abbreviations to understand them?

Administration of the system should be simple and not require layers of specialized staff. Large centers do need skilled people to manage these systems and the telephone network interfaces. Small to medium centers do not need dedicated staff, although access to two or more people trained in the use of the system and trained to deal with the system vendors and telephone network are essential. This, however, need not be their primary function. A complex ACD system increases management insight. It should not unduly increase management effort.

Today, your corporate communications department is probably completely overworked trying to buy the most reliable and least expensive wide area telephone and data networks, maintain the PBX and local area networking systems, move the phones and do other administrative tasks. The last thing they need or want is a bunch of demanding managers in the customer contact center constantly asking for ACD changes, then changes to the changes. What they do not understand is that a customer contact center is involved in the opportunistic business of sales or service as opposed to a relatively predictable task like accounting. Marketing, sales, service and other "customer facing" tasks demand taking advantage of change and opportunities.

If the ACD system administration is difficult and requires outside help from your telecommunications department, or worse still, the vendor, business opportunities will be often ignored because of the degree of difficulty introduced by a hard to use system. The command language and ease of user administration is a key differentiator for true ACD systems.

☐ INTERFACE WITH OTHER CUSTOMER CONTACT CENTER RESOURCES

Increasingly, sophisticated systems and business methods have arrived in the customer contact center, for example:

- Voice response.
- Database and list management interfaces.

- Workflow management.
- Work force planning.
- Autodialing systems.
- Customer experience recording systems.

These systems are commonly attached using Computer Telephony Integration or Client/Server Computer Telephony connections.

Connection with host and customer database applications for delivering screen "pop" information concurrently with the call is becoming increasingly popular. Computer telephony integration is a middleware or protocol translation technology whose existence is necessary when closed telephone operating and database systems are encountered. Applying CTI connectivity and techniques speeds up the call process, has the potential to re-personalize a homogeneous transaction, reduce the workload and allow the customer contact center to do more with fewer resources.

If the ACD system does not easily interface with these typically external systems and processes, considerable inefficiencies can arise. Complexities increase and response times suffer to the point that the business goal of the call is impeded. These complexities have severely hampered the deployment of third party computer to telephone links due to the complexity, time and cost of making these connections and applications run as advertised.

The arrival of open VoIP makes it easier as IP telephony and open architecture promise common operating systems and IP address architectures. If this promise is delivered as envisioned much of this section will be rendered progressively obsolete in a very short time.

☐ A LESSON TO BE LEARNED

One very expensive lesson that few ACD vendors have yet to learn is this: the call switching process has to become subservient to the customer application and database. The switch must become yet another input/output device. This is particularly true when an ACD is used in conjunction with outbound applications.

Today, there are still remnants of the battle for ideological supremacy between the switch and the computer vendors. The ideal solution is to let the customer win and in doing so remember why all this technology exists.

VoIP and packet switching renders this argument completely redundant. Closed proprietary circuit switched telephony is on its way out. And with it much of the closed thinking of the telecentrics.

179

ACD Basics

From the beginning we have talked of customer contact center system sizes and varying levels of system sophistication. The author acknowledges that there is no easy way to categorize ACD systems. However, the best way to begin is to look at the way that they fit specific applications and contact center sizes. While realizing that this can cause offense among some users and vendors who have chosen to install ACD systems within centers that are outside the scope of their system's typical capacity and sophistication, the author still notes that the right fit is important when considering an ACD solution.

So as to not appear schizophrenic while swapping back and forth between references to analog and digital systems, TDM or VoIP packet switching, the following rule has been established: If you understand the business objective and capacity, any architecture is subjugated to serving these requirements.

☐ CUSTOMER CONTACT CENTER SIZE

Under 20 Positions

In the small customer contact center the cost of ownership of an automatic call distribution system is usually less than 10% of its operating costs. For a customer contact center of this size, there are also a number of options for help with automatic call distribution:

- Key telephone system (KTS).
- Hybrid key/PBX system (provide rudimentary ACD functionality).

- PBX-based ACD system (many provide ACD features of varying sophistication).
- PC-based small ACD systems (these can provide the features of larger low-end standalone systems).
- Integrated PC-based systems (some provide switches and/or VoIP-based platforms).
- Standalone ACD systems (note that vendors primarily pursue markets above this size).

The primary reason that a manager of an under 20 position customer contact center would consider one of the more powerful systems (such as a standalone ACD system) is to accommodate planned growth or a smaller office being part of a larger customer focused network. Or perhaps, the manager or company has experienced the advantages of a powerful ACD system in other operations, and want to export this advantage to smaller domestic applications or overseas operations. Once VoIP becomes a standard medium, this market will see the most rapid change as large system features can be extended to any size enterprise and even the networked at-home agent.

One Fortune 100 Company regularly installs $100,000 ACD systems in its customer contact centers with as few as fifteen inbound/outbound telephone marketing representatives. They have found that the improved productivity, the increased sales and lower costs for a center equipped with a true standalone ACD far outweigh the capital cost of the system. When its US domestic ACD systems were made redundant for any reason, they were moved to their overseas markets with the same sales growth and cost containment effect. It is no surprise that this company is the leader in its market and pioneered the deployment of desktop productivity tools and customer contact center workflow management.

20 to 50 Positions

When a center reaches the 20 to 50 position size, the right ACD system can allow significant control of a center's line and staff costs. Typically the annual cost of ownership of an ACD system of any type is less than 10% of its total operating expense.

PBX-based ACD systems are most frequently encountered in this segment. Just note that many of the most popular of these systems provide ACD features of varying sophistication. CO-based ACD systems are also beginning to make headway in this market segment and you can find VoIP beginning to be discussed for new sites in this market segment, although many companies and center managers still have reservations about this technology. If you are a manager in such a situation, you might want to consider one of the integrated PC-based systems.

Due to the financial dynamics of customer contact centers above 30 positions, standalone ACD systems start showing up frequently in this segment. The reasons for this are:

- The recognition that a customer contact center of 30 agents is costing the owner a million and a half dollars or more a year and that there is a vital need for optimal management. Today, the features reporting facilities of most ACD systems that find their origins in PBX architecture materially compromises the management ability to precisely manage this sized center.

- A customer contact center of 30 or more agents takes a disproportionate amount of switching and system resources compared to thirty administrative users. From the switch capacity point of view these thirty agents are the functional equivalent of between 150 and 300 administrative (regular telephone) users. Therefore, a standalone ACD outside the company PBX keeps a potentially disruptive application off the main company switch and maintains the status quo.

183

When you combine an ACD and a PBX together you bet neither system will grow unchecked. We have already discussed that the growth dynamics of a customer contact center reflect business success. New campaigns and new market offerings that require phone support come and go and can cause havoc in traditional telecom engineering. When the business is successful often the customer contact center sees growth first, immediately followed by an increase in administrative staff to support the new business. As both the ACD and the administrative PBX requirements begin to grow they "step-on" each other.

The disadvantage cited by the PBX-based ACD proponents is system disintegration. Advantages cited for including the ACD on the PBX are ease of dialing (from a common station numbering plan) and ease of transfer between the ACD and the administrative portions of the company. The standalone ACD proponents counter with the arguments that a good standalone ACD installation reflects as closely as possible the PBX dialing plan and transfer protocols and can accomplish a transfer with a minimum of fuss.

Neither is totally correct. There is little comfort in the seamless integration between the PBX and ACD functions as other disadvantages occur. Typically each agent position is now served by a minimum of two PBX hardware ports. There is one physical "station" number for the ACD group number (this is how the ACD portion of the PBX allocates calls) and a separate physical number for the agent station (so the supervisor and others can call the agent directly). A standalone ACD doesn't usually do this as an agent can be dialed by station (software defined) or by agent identification number. Calls from a PBX station to an agent on a standalone ACD can also use the look-up table and translation capability of the ACD dialing logic.

Because of the functionally separate hardware ports (agent and station numbers), management reporting of agent ACD and administrative activity on a PBX-based ACD is often disintegrated. The ACD data is reported in the ACD MIS system. The administrative activities, such as outbound calling or intra switch calls are reported on the administrative (SMDR) reports. This call detail affects the overall agent productivity, yet it occurs as a separate report category. These fragmented reports defeat objectivity in reporting employee performance.

These are the main disadvantages of trying to "stuff" both applications into one machine. Take the car/truck analogy: although the combined vehicle can do both tasks, it is not great at either. The potential compromise with a PBX-based ACD may be your business's effectiveness.

A parallel to this lesson exists in mainframe computing. When transaction processing got too intense for the mainframes of the day, intelligent IT managers off-loaded these applications to outboard application-specific computers called mini-computers and mainframes happily continued on with their batch functions. Mainframes remained vital and grew in market share and importance doing the job they were intended to. The same is true of PBX and standalone ACD systems. A similar non-parallel evolution can be expected in VoIP adoption for customer contact center specific applications, unless customer contact center specific software evolves as fast as VoIP deployment. This is much of the motive behind the big network companies' applications and alliance initiatives versus the traditional switching vendors.

With the ACD function on an application specific switching system external to the PBX, everybody gets more business done more quickly and at a lower transaction price. Nevertheless, often the technical staff gets upset because the user population has not adhered to a corporate PBX (assuming it does call distribution well) standard.

50 to 150 Positions

At this size, the cost of ownership of an ACD system amounts to less than 4% of the total operating costs of the customer contact center. When looking for the right ACD system for a center with more than 50 positions, circuit and staff management features become more and more important, particularly when they address daily operating expenses, such as appropriate trunking capacity to serve active customer campaigns and the correct staff count to serve them.

In this market, less sophisticated systems, such as the PBX-based ACD have lower market share due to the increased appreciation within this market segment of the issues and the solutions offered by the standalone ACD system vendors.

This market size is the early target for integrated VoIP contact center offerings primarily because low capacity requirements is perceived as a safe place to launch new products. This philosophy indirectly diminishes the importance of the revenue contribution this contact center makes to your business. Before becoming your IT department's or vendor's "crash test dummy," make sure they are invested equally in your risk.

Vendors in this mid market segment are duplicated again, although on the lower end of the offering spectrum, the less sophisticated offerings depend solely on price to compete. Giving up features for a lower price is a decidedly short-sighted strategy.

185

150 to 500 Positions

When the customer contact center reaches 150 or more positions, ownership of an ACD system is fundamental to managing most of the expense of this center. A feature rich application specific system can positively affect operating expenses with a 10 to 30% reduction.

At this size, the ranks of the PBX-based ACD vendors thin, despite what their specification sheets and marketing may claim, although a number of PBX-based ACD systems and CO-ACD systems compete, there are many feature shortfalls and engineering compromises, which make these options less attractive. For instance, CO-based ACD systems have immense capacity but are short on control and MIS features. In the 1996 edition of this book, we forecast this to change. Well, the more things change, the more they stay the same!

There are two standalone vendors effectively competing in the large customer contact center market — Aspect and Rockwell.

VoIP has not begun to penetrate this market other than being the basis of an interesting discussion.

One really interesting wrinkle that has occurred is the replacement of solid state announcements with VRU based announcements. The merit of this more system intense approach arises when complex routing requires prompting and customer touch-tone feedback. This approach greatly enhances flexibility to change and dynamically adapt to new business conditions and marketing offers.

More than 500 Positions

The logistical issues and options for an owner of a customer contact center in excess of 500 positions have little to do with ACD system selection. For those who consider this an attractive customer contact center size, the choices are limited. While you can find ACD vendors that compete in this market, again only two ACD vendors (Aspect and Rockwell) can provide nonblocking switching systems over the 1000 agent position range. The other vendors can only begin to portray this capacity by linking switching modules together. However, when they begin linking switching modules, the database management, physical and logical blockage, all conspire to complicate an already complex project. And that's just the beginning. For instance, you must deal with other potential headaches, such as facilities, potential staffing shortages, disaster recovery, etc. that seriously challenge the wisdom of such huge customer contact centers.

☐ THE ACD — FEATURES AND DEFINITIONS

An ACD system has key features setting it apart from other call switching applications:

- The physical and traffic capacities of the switch.
- The trunks and agents capacities and tradeoffs.
- The call processing capabilities.

Capacity

Typically an ACD system size is expressed on a "trunks to agents" ratio, e.g. 256 trunks x 200 agents. Although many vendors talk about the "trunk side" and the "agent side," ideally a switch should have universal porting. That is every slot in the system can carry any device, whether it is a trunk, agent or utility component.

Analog trunks generally take up more "real estate" than digital trunks. Talk with the vendor and determine the necessary compromises; this may push you to T-1 spans (digital) early in your planning. It is now increasingly cost effective to make the switch to digital since even the local telephone company has begun lowering prices in its push to sell digital service in order to extend its plant capacity.

The internal network of the switch may limit the physical capacity — that is the number of talk paths or simultaneous conversations that can be connected by the switch. This, in turn, may bring into question the port capacity of the switch and processing capacity of the common control computer.

The test is "what will the vendor guarantee in busy hour call completions or BHCCs?" We talk more about this below. This is an acid test that equally applies to VoIP. Do not be surprised if VoIP vendors "dance hard" or stumble with this question. Do let your business be anyone's test bed!

A VoIP environment begins to complicate things dramatically. For instance, you have to consider the number of simultaneous conversations (timeslot and packet capacity) that the LAN can simultaneously deliver. But, you can't stop there, you also have to consider the quality of the voice as it's delivered from the Internet or digital trunks and terminated on a network router at the customer contact center, plus all the associated data applications.

This begins to get difficult, as data specialists generally do not have the same respect for a transaction as do voice specialists. This reverts back to an early comment about customer contact center traffic; it's not just calls, it consists of people with emotions, perceptions and a willingness to buy and you must treat them accordingly. Hopefully your network administrators and data specialist either understand this or can be educated so they do grasp the implications.

Next, you need to look at the traffic capacity. This can be illustrated by comparing the switch and common control to an intersection with a number of converging streets, managed by a single traffic control officer. There may be sufficient lanes to carry the traffic (physical size), yet not enough time or space for all the traffic fed from the lanes to cross the intersection safely (communications pathways), even when all are under the control of a single traffic control officer (insufficient common control capacity).

Analog: In an analog switching system the physical switch capacity is limited to the number of talk paths. A talk path includes two ports, one for each participant in the conversation (the caller and the agent, music or voice processing port) and the actual communication link through the switch. The maximum number of calls, which can be connected simultaneously, is the same as the talk path total.

Digital: In a digital switching system similar resources are consumed. Two ports and a path through the switch link these ports. As this path is an electronic buss, typically using time division multiplex switching technique, the conversation capacity is measured in timeslots: one for each conversation/transaction participant. A call, at minimum, needs two simultaneous timeslots. The maximum number of conversations, which can be connected simultaneously, is exactly half the number of timeslots the system supports.

Other measurements used in measuring the traffic capacity of a switch are CCSs and erlangs. A CCS is one hundred called seconds. There are 36 CCSs per hour (3600 seconds). An erlang is an engineering term for an "hour of traffic." A manufacturer will rate each switch port at 36 CCS per hour. However, while such talk path capacity may exist, the switching computer is incapable of matching that capacity with processing power when all the ports are attempting to make and receive calls simultaneously. Nonetheless, this type of rationing is common in PBX engineering and causes real trouble when a busy ACD function is added to the same switch.

The traditional measurement of the common control or the computer controlling the switch has been "busy hour call *attempts*" (BHCAs). This means the number of call attempts the switch can recognize and *attempt* to process is rated the theoretical capacity of the switching computer.

A more reliable measurement is the number of "busy hour call *completions*" (BHCCs). This means:

- the number of calls the switch control can recognize,
- process under the call processing rules,
- connect to the correct resource(s),
- maintain the connection for the duration of the call,
- end the call,
- break-down the connection returning all the resources to idle status,
- track and report the call progress,
- write a final transaction record to disk, and
- now the call is considered a *completion*.

Both BHCAs and BHCCs are theoretical; however, there is a significant difference between *attempts* and *completions*.

VoIP: The same constraint exists in establishing the logical connectivity between the source IP address (caller) and the destination IP address (agent). But now it is compounded 8000 times or "n" (whatever rate the voice is sampled, coded, compressed and decompressed times two parties to the conversation) per second and by the bandwidth of the network.

Unlike a solid state switch where a fixed or dedicated path is set up between port X and port Y, or a time division switch that establishes a virtual connection for the call duration, VoIP packetizes the conversation and sends each logical packet (X000 per second) over different routes to arrive at the destination IP address in the correct order and at a constant rate provided the bandwidth and real-time processing are present. To do this requires Quality of Service (QoS) technology, which has been one of the major barriers to sending quality sensitive voice and video over the Internet.

The human ear and eye are both very sensitive to unnatural effects created by interruption to the smooth transmission of sound (voice) and pictures (video). This has been solved in low traffic environments and is well on the way to being demonstrated in the volume approaching a customer contact center's requirement. But this architecture needs to be unquestionably reliable and constant if it is to be used in the demanding environment of a customer contact center. Yet, it will happen faster than any other switch migration we have ever seen due to the compelling applications accompanying the migration.

The real measurement of any call processing platform, analog, digital or VoIP, is the "sustained call processing rate." This is the number of calls the system continues to process over a sustained period of time, such as a shift or a day, without degradation in the call processing switching or report preparation and presentation (display or print). One major PBX vendor's strategy is to momentarily suspend or "shed" report gathering, supervisor screen update and printout tasks during high traffic periods.

□ STRATEGIC CONSIDERATIONS

Growing pains, i.e. outgrowing a current incoming phone system remains the single biggest problem facing today's incoming customer contact center manager. It is difficult — if not impossible — to explain to top management the likely growth of an inbound center. Nonetheless, the consequences of buying too small are high in terms of annoyed and disenfranchised customers, especially if they have become accustomed to a good previous level of customer service.

There are four steps to avoid being trapped into an inadequate system:

1. Have as clear an understanding of your plans and size potential as humanly possible. As difficult as this may seem, take the time and assume growth and success will occur. (It will!)

2. Scrutinize the official manufacturer documentation carefully, watching for small manufacturer warnings to distributors and users to avoid this configuration and that application. Insist on getting everything you can lay your hands on. Get it in writing! Availability, amount and quality of this material will tell you a lot about the manufacturer and vendor. If they aren't helpful during the sales process, be warned that few vendors behave better after they have your money.

3. Have the vendor/manufacturer sign up to spend time understanding the underlying growth trends driving your business. Ensure the vendor stands by the recommended configuration from both a traffic capacity and a "no surprises" agreement, a term that tries to assure that there will be no additional cost, effort or abnormal inconvenience placed upon the user due to the need to work around some undisclosed system limitation.

Remember that a vendor is more likely to grant price, support or contractual concessions upon the initial order than on subsequent upgrades. Once all payments have passed to the vendor, *caveat emptor*.

4. If the vendor offers price concessions to win your initial business, attempt to have these apply over a reasonable period from the date of installation. Reasonable to a buyer means as long as possible, though not so it is punitive to the vendor. Ask for 36 months and settle for 18.

Growth Strategies

A vendor of an analog or digital switch uses two strategies to grow a system:
• First, fill the system cabinets to their maximum designed capacity.
• Second, when growth occurs beyond the single initial system, tie two or more together as transparently as possible.

This is done with intermachine links or tie lines and can be accomplished relatively effectively. The key to the effectiveness of this strategy is the transparency of the implementation. Does it add to the effort of managing the resulting system, and does it work as one system?

The VoIP network vendor is likely to say, "we have encountered no practical traffic limits," which means we have not grown a system to a size that we block traffic. Be most skeptical and ask to speak to a current user of a system significantly larger than you anticipate installing and growing to. Do not let your business be the first!

For any implementation, the answer to these traffic experiences can only come from a user who has really implemented this solution in a similar configuration (size and application). The "no problem" response from the vendor, although well meaning and self serving, does not mean that much to the new user who has more systems than he bargained for. Make sure the explanations provided prior to contract are adequate, in writing and understood. Additional protection from over representation by a vendor is to make all correspondence and proposal material part of the contract. Tell the vendor this in any proposal solicitation.

The switching modules may interconnect quite elegantly and the call control software may balance the load equitably, but what are the database management and administrative issues, if any? If you update call processing on one module, does it apply to all? How is this database synchronization accomplished? Is the all report data consolidated on one report system? Similar questions apply to separate VoIP LAN configurations and central server management.

Alternatively, the implementation in a switched environment may be a simple virtual private network (tieline?) with no intelligence associated with the routing of an interflowed call, yet the system produces elegantly integrated report formats and data.

The concern is that two machines with adequate cumulative port count may be placed in service in the same application. Yet, they will act independently of one another, merely passing calls off system A to system B, with no heed to equitably balancing the call load between machines, agent groups or agents.

The large standalone ACD vendors understand the implications of this and have built big, single stage switching machines or implemented interswitch load balancing in response to objections to this strategy. Take the time to understand the process and the tradeoffs that accompany such an engineering solution.

Companies like Cisco and their external telephony server based ICM products (formerly GeoTel) add a network-based intermachine control layer between disparate switches (brands, locations and even countries). This is a brilliant strategy to position Cisco for widespread VoIP deployment in customer contact centers as they are learning the control layer disciplines necessary for high volume voice events in an IP customer contact center implementation.

The Impact of Poor Implementation

The problems that can arise in poorly implemented solutions (discreet analog or digital switches) manifest themselves in one or all of the following ways:

- Inability to manage as one system, thus additional effort at the supervisor level, i.e., managing two systems.
- Increased staff to accommodate for the inability to look-back from switch A to switch B when an agent becomes available at the entry point/switch.
- The management and synchronization of two or more system databases.

A clever solution to this inequitable call distribution over one or more systems is to provide a second stage switch that acts as a system-wide traffic hub. This works to broadcast a request for service across the second and subsequent switch modules, without the entry switch system or module (the system at which the call enters the system) relinquishing call control. This allows for more equitable call distribution to occur.

As we just discussed, external telephony server-based network control layer can smooth these load imbalances in a more sophisticated real-time way. It also forecasts how the data network companies will allow large virtual private network users to manage VoIP-based customer contact center networks. This is a truly compelling vision for VoIP that can be sampled now in a multi-site traditional voice customer contact center network.

In a manual environment, physical allocation of the traffic loads and the call-processing tasks associated with trunk and agent splits must be studied and allocated by business type. This is generally satisfied by paying attention to the communities of interest within the center. The only issue that remains is the consolidation of the databases and the real-time and printed reporting processes. Vendors pursue varying strategies to do satisfy this.

☐ ANALOG OR DIGITAL

Traditionally all voice calls trunks arrive at the switch in one of two forms: analog or digital. In the old days, most trunks were analog and arrived at "your doorstep" on one pair of wires. Almost all business trunks are going digital as digital technology has dropped prices and increased capacity on existing facilities. Still you may need to deal with local trunks or local loops that are in analog format.

A digital switching system must provide "reverse" channel bank technology, albeit single circuit and a great deal more integrated and efficient than traditional channel banks. This technology is necessary to digitize the analog circuit for transmission through the digital system. Or in the case of a "reverse channel bank," reduce a bulk digital trunk down to a single analog circuit for transmission through an analog switch.

Today in North America, a T-1 span represents 24 multiplexed voice channels arriving on two pairs of copper wires. Due to telco plant preservation strategies and lower tariffs, these are now as commonplace as analog circuits. In Europe, the standard is 32 channels. A circuit is either one pair or one analog channel or two pair and potentially 24 or more channels.

When an analog phone pair arrives at your doorstep, you can attach a $20 black phone directly to it (in telephony nomenclature, a 2500 set), dial numbers and speak and be heard. Nothing could be simpler. When a T-1 link arrives on your doorstep, life is dramatically more complex. The only thing that can be attached is a digital switch with line cards that can "untangle and decode" digital messages into 24 separate conversations.

This is called a "channel bank." A "channel bank" is an electronic device that the splits up the T-1 stream of 1.54 million bits per second into 24 discrete analog circuits that can be terminated on a standard analog telephone switch — PBX or ACD. More recently most PBX or ACD systems allow for direct termination of T-1 trunk as they have this technology directly built onto the trunk cards.

Trunk Termination

One of the big advantages of digital technologies is the downsizing in componentry, space (switch real estate) and power consumption. Because of this the circuit cards required to terminate digital and analog circuits are typically of different technical and physical properties. The size issue is most significant at this stage. Digital circuits can be supported on considerably denser cards, typically 4, 8 or 16 discrete devices per card, whereas an analog card may only support 4 or as few as one discrete analog circuit. This means the system is physically larger and the shelf that may support 64 digital circuits can support a lesser number of analog circuits.

Watch for the analog to digital tradeoff issue and the physical capacity impact on your intended system configuration.

With VoIP, the space must be allowed in the routers, racks and card slots (real estate) for the appropriate amount of Internet trunk terminations (traffic sensitive) and routers and servers to support the number of agents and supervisors expected to serve customers in this contact center.

193

ACD Addressing Criteria

With either an analog or digital switch, the trunks are physically attached or terminated on an ACD either singly or in trunk groups. In the case of a T-1 circuit, 24 or more based on the span capacity are terminated on a card that "demuxes" the 24 channels into discrete circuits that are in turn considered individual trunks or ports. The port is considered a physical address. It is important to understand this, as this is one of the ways an ACD figures out how to handle an incoming call. These methods are described as either physical or logical.

These addressing criteria have become significantly richer with the addition of network derived identification (ANI, DNIS etc.) and computer telephony added meta tags from eternal databases (ACD and departmental systems, CRM, help desk, etc.).

Like a voice call with specific network or CRM tagging, a VoIP solution will use the logical packet address and associated instructions or meta tags as the routing instructions.

The switch expects to route the call by looking at the database table that applies to this physical port. This is the default route unless instructions based on the digits or meta tags are received. Once these instructions are recognized, the physical routing is ignored, and the logical table addressed. These are interpreted by the ACD system and the call is routed to the intended agent, or group of agents based on the caller or nature of the call based on the digits received. These digits may be "direct inward dialed," "automatic number identification" or "dialed number identification service" digits. This is logical call-by-call identification.

Physical trunk groups generally refer to the telephone trunk type, whereas a gate or a split refers to the call or business type that arrives at that gate or split.

VoIP envisions call-by-call routing because of the much richer meta tag accompanying each IP packet.

☐ CALL ANSWERING PROCESS

A call can be presorted by virtue of dialing a particular number (published as the number to call for satisfying this particular request); thus, this particular call type is presorted into the intended ACD gate. As a call arrives on a given trunk, in a predetermined split, a number of things happen. The inbound call supervision request from the carrier's central office is detected on the trunk and the system reacts.

An important and crucial difference in how calls are handled begins at this point. The true standalone ACD and the PBX/ACD begin to separate into application specific and generic groupings. The resulting differences, although in many cases potentially written off as mere subtleties, make major differences in the manner in which the customer contact center is managed and at what cost.

An ACD system can react in one of two ways: answer the call, or upon call detection of "supervision in" or the digital equivalent, look to a call processing control table for step-by-step instructions. PBX design and architecture dictates

Physical. Historically, this is derived from the where the trunk is attached. This physical trunk is assigned to a split or gate in the database, which indicates the type of call or caller, and therefore the type of treatment the caller will receive. When a call shows up on this port, the switch looks to the database, decides which route applies and routes the call following a predetermined software table. The call will be routed to the intended agent (answering) group based on the rules applying at the time of arrival.

Logical. Alternatively, despite the physical trunk group, split or gate the call may arrive on, the call will be preceded by 4 or more significant digits sent by the network (DID, ANI or DNIS). This may now also include the external database meta tags.

the system answers the call now (unless there is a preprogrammed delay), whether there is an available agent or not. A true standalone ACD asks the routing control instructions what to do before anything else happens, then proceeds through a list of steps and conditions that apply to that call.

Traditionally most telephone systems (analog or digital) IMMEDIATELY answer the inbound call upon receiving the request for service. Next, the system looks at the call treatment table that indicates the call should be served by a particular station number or group designated by this station or directory number. The call is routed to that directory number (extension number). In the case of a PBX this may be the operator station or pilot number in a hunt group. If this station or group is configured as part of a PBX-based ACD group and the number

When is a Call a Call?

A telephone call is considered completed when the caller is connected through to the intended party or caller. Under standard telephony signaling this is accomplished when the last central office sends ringing voltage (analog circuit) or a "supervision in" (digital circuit) message to the intended party telephone or system. The called party or system responds by answering the call, and sending an answer message (analog "reverse" supervision voltage or digital "answer supervision" signal) back to the CO and then back up the circuit to the calling party CO to establish the connection and begin the billing cycle.

In a digital world all these voltage changes and supervisory signaling is achieved with digital "start," "stop," "set-up" and timing messages. In an ISDN world these travel on the out-of-band "D" channel, or in-band as unique digital messages signaling the switches to begin, set-up or end a call.

In the analog sphere, completed calls are not the only events occurring on the telephone line. There are often spurious voltages that look like "supervision in" requests. An ACD will react by answering and treating these events as calls if they are of sufficient strength and duration. A true ACD system will begin measuring these "calls" from the time that they appear on the line, before or after answer. These events show up as lost calls, if the ACD is sensitive to reporting these events.

In the VoIP universe an infinite number of different control signals attempt to duplicate the behavior of the telephone, for better or worse. Unfortunately, some of the protocols used on analog telephony networks are compromises or workarounds that digital engineers copied for standards sake. VoIP is slavishly following some of these older rules and standards (even when often these were plainly modest standards) to accommodate backwards compatibility, rather than pioneering a new format at the expense of the installed base.

or phone stations represented by this directory number is busy, the caller should then be routed to an announcement, a subsequent group, and so on. In the case of a customer contact center, slavishly answering a call and sending it to busy phone(s) intended to serve the call is not smart.

This indicates the designer tended not to understand the nature of the ACD application in a customer contact center particularly well, because automatically answering a call when the agents intended to serve the caller are busy, is not optimal. When this happens, the owner of the ACD typically gets to annoy callers with inappropriate instructions or expensive "elevator music"

Answer Delay

A smart ACD might let the call ring ("no answer") for up to 24 seconds or even longer. While this happens the carrier is not billing the inbound WATS user. Also, callers do not perceive ringing to be a traditional delay or "queue," provided it's not extended unreasonably. Three to four rings "buys" 24 seconds when there is no one to serve the caller. Ideally this happens dynamically on a call-by-call basis.

This more desirable process is used by all of the standalone ACD systems and integrates the answer process as an optional step into the call processing rules. By doing so, the ringing cycle can become part of the queuing process. This feature is known as "ring delay," but this is a misnomer. It really means, "answer delay."

By adding a fixed or a flexible number of ring cycles to the answering process, the actual answer or "supervision out" can be delayed longer than expected. Supervision out is the phone system "return handshake" to the CO saying it is ready to accept the call. A traditional ring cycle in North America is two seconds of ringing, four seconds of silence, for a total of six seconds. European standards vary by jurisdiction but tend to be slightly shorter.

When a call comes into a central office or switch, the caller may not enter the ring cycle at the beginning. The reason for this is that ringing is a continuous cycle being generated and switched to incoming calls as needed.

There are key human factor issues associated with intelligent use of the ring cycle that work for the customer contact center manager. First, a caller when

---

WARNING: When you ask your prospective ACD vendor if they provide "dynamic answer" or answer delay, make sure they explain if this is a fixed, system-wide delay rule, a trunk group, trunk or call-by-call rule. The latter is most desirable as it reflects the traffic condition at the time the individual call arrives.

calling a business will typically accept ringing from 2 to 7 times or 12 to 42 seconds. An intelligent ACD system can use this time to the customer contact center's advantage by actively including it as part of the delay.

Good reporting of delay and abandoned call statistics will allow a customer contact center manager to really determine the actual "threshold of pain" callers are willing to accept both prior to, and after, the answer point. The logic being that it's cheaper to let the phone ring than to answer it and put the caller on hold on your finite and costly 800 lines.

Second, there is an apparent temptation to reduce call length by immediately cutting-through an inbound call when there is an agent immediately available. This would bypass the ring cycle completely, and eliminate part of the six seconds associated with a normal ring cycle. Experience indicates this does not save any time per call. This seems counter-intuitive. Our phone use conditioning causes us to expect some amount of ringing when making a telephone call. By eliminating all ringing, the inbound caller becomes disoriented and uses up more than the six seconds to figure out what happened by discussing this with the receiving party. Any saving is thus offset.

A standard trunk delay means there could be agents available but the caller must go through the fixed system, trunk group or trunk delay cycle. Both the caller and the agents are "twiddling their thumbs" while the system blindly follows the inflexible delay rules. Causing more staff and more trunk time to be needed.

Today, VoIP is considered non-usage sensitive so that much of the notion of answer delay seems moot. If we can predict telephone company or carrier behavior by the last 100 odd years, expect a usage sensitive cost to ultimately apply to VoIP calls at some point. There is no way the traditional carriers are going to give up long distance revenues without a fight.

Call Routing Tables

Although the process of handling inbound call routing appears similar, it's filled with nuances and subtleties that are critical to a customer contact center. This process is the core of customer contact center management and is a process that must be understood clearly before selection of an ACD is made. Operational costs are directly and dramatically impacted by how these tables operate. Switch (analog or digital) manufacturers call this process by different names, i.e Call Control Table (CCT), ACD call routing, Call Vectoring or routing tables.

A call-processing table is a set of preprogrammed steps that reside in system memory. These may a standard set of steps, preprogrammed into the machine prior to installation. Or, more often today, these are tables that may be changed by the user to address call processing demands as they occur. The simpler these changes are, the more flexible and, therefore, more functional the ACD becomes. If these changes are only possible with effort or factory support, the less likely (and the more expensive) that the ACD will be adapted to meet the customer contact center needs.

The Anatomy of an Incoming Call

An incoming call, arriving on an ACD, can be segmented into four elements:
- Call setup and ringing.
- Any call queuing necessary.
- Talk time (agent or VRU).
- Disconnect.

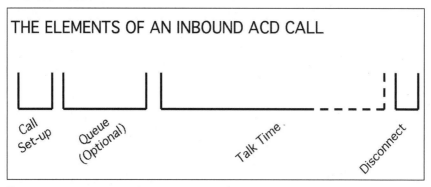

THE ELEMENTS OF AN INBOUND ACD CALL

Call Set-up Queue (Optional) Talk Time Disconnect

Figure 10.1.

Early UCD and ACD designers assumed the only portion of call length actively manageable by a user occurred once the call was answered, namely the queue and the talk time of the individual agent. The modern ACD systems designers changed this through their greater understanding of a basic rule of telephony: shorter calls cost less. They set about extending greater control of the call, and call length, to the user.

If a switch knows when a call arrives, and if it has no agents available to serve the call, then why not delay returning the supervision signal to the central

Simple Rule: Flexibility means you can control your destiny more, though with this control comes a greater need to understand your objectives and the ACD tools available to help you meet these objectives.

office? Don't answer the call unless there is an agent available to serve the caller. A true ACD switch can "watch" a ringing line and agent availability simultaneously. If an agent becomes available and there is a ringing line it immediately answers and quickly sends the call to the available agent. If an agent is not available, it intelligently uses the ring cycle as a delay until no longer courteous, then it answers and plays a greeting and delay message. Should an agent become available at any time in this process, a "barge out" step is activated so the delay can be interrupted and the caller served.

While the major standalone ACD manufacturers have already implemented these steps, the major PBX-based ACD vendors still build pale replicas that can cause a call center owner to incur additional operational costs. This applies in particular to "command granularity."

Earlier mention was made of the how deep a command definition could be forced and the effect of doing so. Ideally it should be down to the smallest common denominator in a system — a call. Data gathering should be granular down to the agent keystroke, trunk number and clocked second for every call and call state change, internal or external (in and out), complete or incomplete. This is particularly critical now that a new generation of call tracking applications is beginning to emerge.

Called Party Disconnect

In Chapter 4 we discussed the nature of call setup and breakdown control residing with the caller. This does not change with an 800 call, even though the responsibility for paying for the call shifts to the call center owner.

Control still resides with the caller. If the caller is slow to hang up after completing the call, or is served by an older, slower central office (CO), there may be an appreciable time before the CO senses the circuit has been returned to idle and breaks down the local loop, making it available for the next incoming caller. Two things occur, the call length may be extended by a few seconds and the trunk is effectively busy so a second inbound call will be unable to access that trunk. The overall load is increased and more trunks are required to carry the same load to compensate for this few seconds of disconnect overhead.

Modern ACD systems recognize the instant an agent releases or hangs-up on a call, relays this to the local CO and requests immediate disconnect. This frees the circuit for the next call. Meanwhile the agent has gone on to serve another caller arriving and possibly queued on another trunk. The breakdown signal can now arrive from the calling party end of the call, immediately, late or never at all, and this has minimal effect on trunk occupancy and required trunk capacity. This feature eliminates the impact of any delay.

How does your proposed vendor handle this? Most PBX-based ACD systems are passive and wait for distant end disconnect on an inbound call or a trunk "time out," whichever comes first.

Measuring the Duration of a Call

Before leaving the switch-based call answering process, there is one other issue of importance from a data-gathering point of view. When does a call begin to be measured?

- When it is answered by an agent?
- When the ACD system answers?
- When the request for service occurs as the CO raises "supervision in"?
  From a reporting point of view, this is important for three reasons:
- Measuring actual trunk occupancy.
- Measuring the actual point of abandonment by a caller.
- Discriminating between real calls and spurious events or "phantom" calls that could be mistaken for real calls, thus distorting reports.
  Optimally, the ACD system should:
- recognize and measure a call from the instant "supervision in" occurs,
- record the time of system answer, time duration in queue before system answer and after,
- follow the route steps taken by this call, and
- record time of any IVR participation, any re-queue and time of agent answer.
  VoIP has not yet demonstrated significant sophistication in this data gathering and reporting space. Fortunately, we have watched these sorts of initial management and reporting omissions be answered in subsequent releases of preceding generational technology upgrades. VoIP is not expected to be different.

Note: One aspect of the ACD call from a staffing and event duration perspective may include a state known as "wrap up" or after call work. If this is required on a call-by-call basis, it adds time per call and affects the number of calls an agent can handle and thus the overall capacity of the system. Now much can be done with computer telephony and call management to gather the call type and disposition data on the fly so minimal wrap-up time is necessary.

☐ WORKFLOW

On the user side of an analog or digital ACD system there are agents who are organized in teams, groups and splits based on function. The reason for the segmentation can be a number of things. A supervisor or lead agent is assigned a team of agents or TSRs, up to the maximum the supervisor or lead can effectively manage. Typically, the more complex the transaction or business type, the higher the required lead/supervisor to agent ratio. The more complex the business, the fewer agents a supervisor can manage.

Often the customer contact center is broken up into groups matching a responsibility. The individual groups are then pitted against each other to encourage better performance.

One of the most innovative uses of this strategy, maximizing network and ACD technology has been using DNIS, and more recently ANI, to obtain regional origin of inbound calls made to one centrally published 800 number. The calls are then routed to the responsible group who competes with their peers to satisfy the caller with prompt service, and request satisfaction.

201

Skills-based Routing

When a call is routed to a group of agents trained to serve the caller's request, it's referred to as application separation, skills-based routing or business rules-based routing. The system adds a database that categorizes agents in two ways: the physical group they are assigned to and a logical or skills based category. This means the agents are assigned or trained to serve a particular type of caller or caller problem(s). An example would be English and Spanish speaking agents who are assigned to serve the 90% who are English-speaking callers and the 10% that prefer to speak Spanish. A skills-based routing system will override the physical group organization (that is the majority of English speaking target agents) and search for an agent with a specific skill (an agent who is bilingual). The call center organization is confused to some degree when it applies a responsibility matrix versus a hierarchical structure to serving callers. By becoming too sophisticated in assigning skill-based logic, or primary and secondary responsibility assignment, a call center begins to dilute economy of scale. The customer service, market and business issues in your center may make this a trivial consideration, but then again it may add unnecessary cost.

All vendors and users interchange the terms: team, split, subgroup and group, to suit their needs. Be aware of the number of hierarchical levels or cross-group responsibilities in your customer contact organization and whether the voice ACD or email triage you choose will allow you to reflect and report on this structure.

The more hierarchical levels the voice ACD or email workflow vendor provides the better, as this allows greater flexibility and a higher probability of reflecting the way you do business and wish to manage your customer contact center. Three is acceptable, more is better.

Make very sure that the ACD system you choose tracks all calls as they are queued AND overflowed (interflowed) to each progressive group step in your customer contact center. This allows you to understand work load by group and any artificial increase in Group 2, based on understaffing or other deficiency in Group 1 and so on. Once the call is in the system, load management and understanding the destination and service dynamics are no less important to providing responsive service while still maintaining costs. This becomes even more important if an interactive voice response device is part of the call answering process.

☐ SPECIAL CALL PROCESSING TREATMENTS

Along with user programmable routing tables found in modern ACD systems, there are other control tables that allow greater flexibility and freedom from the drudgery of reprogramming as time of day, day of week and day of year changes occur. Vendors offer innumerable command sequences that can be built to match business conditions. Some are preprogrammed system housekeeping sequences such as report generation and database indexing and backup. Others are user start-up, reconfiguration or shutdown procedures. These are written and may be triggered by a clock in the system. For example, at midnight, the system will close the daily accounting files, clear all the daily statistics files, accumulate totals, allow reports to be generated and files sent to other applications servers, etc. This is all done automatically. Other system functions can be executed as batch files. At other times of the day the user may setup a set of procedures to be executed, files to download or uploaded on schedule.

Primary, Secondary and Tertiary Gate Assignment

How your customer contact center is structured is critical to your ACD system selection. For example if you have small groups of specialized agents attending to specific tasks as their primary responsibility, that do not have adequate workloads, you may wish to assign them to secondary groups.

These sequences are rather like AUTOEXEC.BAT start-up procedure files found in a PC. However, the contact center sequences are triggered on a time schedule. These time schedules are setup to follow a daily pattern. Although these clock-driven sequences may differ on different days of the week, and again, differently on statutory holidays that fall out of sequence throughout the year. To accommodate this your ACD system should ideally have a day-of-week and day-of-year table to allow exception schedules to be followed. This means a system can be left knowing it won't print reports devoid of data over a weekend, or won't turn off the "night closed" announcement and open the unmanned customer contact center over Thanksgiving. This increases flexibility and leaves management to focus on running the business, not a machine.

The ACD As a Customer Workflow Manager

The whole notion of an automatic call distributor (analog, digital or VoIP) is to ration caller demand across as few possible company resources as possible without compromising sales or customer satisfaction. The resources may be live agents (inbound, outbound and/or web call back requests) or other channels such as interactive voice response ports. The goal is to apply these scarce resource across as large a caller demand as possible, and do so in a manner as transparent to the customer requesting service as possible, while never compromising the business goal.

It's about balancing good customer service, good employee working conditions, against the lowest expense to revenue ratio as possible. An ACD goes a long way toward achieving this and is the most advanced model for most real-time customer requests available. Use of a good workload forecasting and planning package can dramatically assist in matching this customer demand to company resources, live and automated.

The ACD provides call distribution by imposing some delay in the service received by the inbound telephone customers or callers. This delay or queue is perceived to be tolerable and customary in the call center marketplace. To achieve this, maintain a happy customer base and make a profit is nothing short of an endless real-time juggling act with the core tool being an ACD. The ACD can expediently and intelligently assign service resources automatically and in a virtual transparent manner through the use of the automatic customer call routing methodology. Coincidentally, this is the one feature that is given little analysis and attention by most ACD designers, vendors and buyers alike, although it's the routing and queuing structure for voice calls. This model can also neatly extend to cover email and other less than absolute real-time requests, like web callbacks.

Call routing or workflow distribution tables are workflow distribution rules. Across the breadth of vendors, the processes appear similar although they are filled with nuances and subtleties critical to a customer contact center. This process is the core of customer contact center management and it must be understood clearly before selection of an ACD is made. Operating costs are directly and dramatically impacted by how these routing tables operate.

Call routing, workflow distribution and call-processing tables are sets of preprogrammed steps that reside in system memory, and may be a standard set of steps programmed into the ACD machine prior to installation. Today, these are tables that can be changed by the user to address the particular customer contact center's call processing demands.

The simpler it is to make these changes, the more flexible and more functional the ACD becomes. If these changes are only possible with effort or vendor support, the less likely the ACD will be adapted to meet the center's needs. Surprisingly, some systems are still quite rigid.

We broached the subject of call processing or routing tables in the previous chapter, as far as the call answering process is concerned. We did not however approach the steps beyond the point of answer. These steps include:

- The management of the status of all trunk and agent resources.
- The answer step.
- The connect step to available primary agents/group.
- Announcements.
- Other service devices like voice response units.
- The delay steps.
- The overflow trigger points or steps.
- The selection of secondary or subsequent groups.
- The use of voice messaging and other load shedding strategies.
- The overflow off this system to second and subsequent systems.

Although there is increasing appreciation on the part of the systems designers for integration of all these decision points into one decision process, today, few systems do so. Trunk level components detect a service request and answer the call, independent of knowing the status of the target agent group. This is typical of PBX-based ACD systems. Even retarding answer does nothing to alleviate occupancy, call or billing duration, if the "hardware answer" decision is not coupled with agent availability status and call aging.

The key in all of this customer call processing is to view the system and resources at any point in time, as a whole. This increases both call processing speed, economy of scale and system operating costs, by addressing the largest possible universe of resources. The more agents or applicable voice response ports available to serve an offered call (web call back request and/or email), the better the service received by the calling customer. Economy of scale rules this application.

With a functional ACD system and active and involved management, not only will the callers receive good service, but at the best cost to service ratio possible. How is this possible? Shorter calls still cost less, so it is critical that the routing process shave a second here a second or more there. These add up quickly to large amounts of time. A large national bank estimates that every second of increase in the average customer call length costs them an additional $1,000,000 a year in overhead expense.

It is simple to make this calculation. Take the total cost of your customer contact center operations and divide it by the number of calls received. Now determine the average call duration in seconds. Divide the total call center operational cost by the number of seconds in a typical call. You'll be surprised and will have a powerful tool in your "financial justification kit."

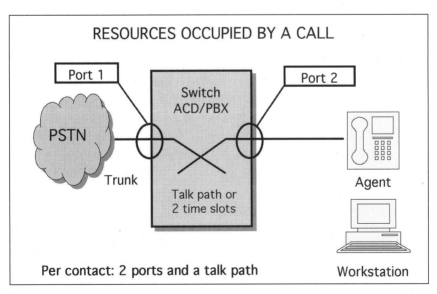

Figure 11.1.

An incoming call uses a number of costly resources during the process:
- A trunk.
- The switch, at least two ports and a talk path.
- An agent and a portion of their day.
- Optionally, a voice response device and the computer system it is attached to.
- Any supporting computer system.

All these elements are used sequentially, and in some cases concurrently, as the incoming call process proceeds. Thus shorter calls can save more than just talk time on trunks.

□ CALL PROCESSING — THE BASIC STEPS

A telephone call is considered to be an ACD call when it is connected to the ACD system. This occurs when the receiving central office (CO) sends a "supervision in" message (analog or digital). This is where the ACD gets involved.

Answer

Once the ACD receives a "supervision in" message, one of three scenarios can be followed:

1. *Immediate answer* and the triggering of a request to the call processing control table in the switch common control for inbound call routing instructions. This tends to be a "reflex" reaction that is built into a trunk component, as this is the desired reaction when a trunk requests service in a PBX application.

This is considered a "hardware" level answer decision and the term used is "hardware answer." The logic for this technique is sound in a PBX application, though not necessarily in a customer contact center ACD. In a PBX, there is no need to continually ask the common control computer what to do when a new call arrives. The same answer always applies: answer the call and send it to the operator or specific extension. Responsibility for the answer decision is delegated to the trunk card, using a "distributed processing" philosophy. This keeps the processing load off the PBX common control and frees it for more important PBX feature control.

2. *Retarding answer for a fixed duration*, then answer and interrogate the common control tables. The delay time duration can be established at one of three levels. It can be a system-wide, trunk group or individual trunk level parameter. Upon exhausting this delay, the system behaves in the same manner as the first scenario. This is also termed "hardware answer," as even the fixed delay is built into the component (in this case, a trunk card).

The question here is whether or not the system is built to "barge out" of the fixed delay period, should an agent become available. Don't take it for granted all systems do this. The big standalone systems do, but it is not an absolute rule on systems that began lives as PBXs. The implication is this: agents may be available but the ACD system insists on delaying the inbound caller. The ACD blindly follows the fixed delay parameter, oblivious to the state of the available agents. Labor is not used optimally.

3. *The final method is termed "dynamic answer".* Here the CO sends the "supervision in" signal requesting return answer supervision. The switch reacts by recognizing the presence of an inbound call, and looking to a routing table that instructs the machine to determine if anyone in the target group(s) of agents is available to serve the call. If no one is available, it does nothing other than maintain data recording the first indication of the call appearance. Meantime, the system scans the intended agent group looking for an agent to become available. The system lets the call continue to ring — this analog ringing (voltage) or digital signal is provided by the CO. Since a VoIP call appears the same to a caller, so it's assumed that VoIP ACD routing applications software will behave in a similar manner.

Concurrently the call is now being timed and aged against the route parameters. Two things will end the delay:

- an agent (or optionally an IVR system) becomes available, or
- the preprogrammed delay period is exhausted and the routing table commands the system to answer the call and, optionally, provide an announcement.

Extending the answer up to the preprogrammed delay point, coupled with immediate answer and barge out to a now available agent, are the components of dynamic answer. Assigned by the route table, this feature can save six or more seconds of billable time per delayed call. Given the fact every second counts, this can be a very large number. As software controls this answer decision it is termed a "software answer."

Connection to the First Intended Resource

This resource may be an agent position, a delay announcement or with increasing frequency, a voice response or voice mail machine. Again, there are basically two methods this connection can occur:

1. Upon *"hardware answer"* the call is connected to an available agent, an announcement device and/or music on hold. The agent begins the transaction or the delay announcement is played. This assumes a desired target resource is

available. If not, all queued calls go to music or a delay announcement, thus queuing the call at the call center owner's expense.

2. The second method is the route table driven or *"software answer."* Here, no agent has become available so the system, realizing the call has aged beyond a reasonable ring delay, answers the call. Depending on the manufacturer, this may be a discrete step in the inbound call routing table, or be an inherent subset of the announcement step.

In both cases, two issues need to be carefully understood on the part of the buyer — the flexibility of the announcement technology, and the ability to "barge out" of the delay or announcement to an agent who may become available during the hardware or software managed delay or announcement. At this stage, if an agent, under any of the described scenarios answers the call, no further discussion need be had. The call is connected and is subsequently processed by the primary or intended group.

Selection of Second or Subsequent Agent Groups

After the routing process has looked to the first agent group for service and found no agent immediately available, the system can react according to one of two parameters.

1. If the number of calls (call count) in queue exceeds a predetermined number, immediately look to the next group, abandoning the primary group as a potential resource to serve this call, or

2. Look to the time parameter in the routing table. Once this has run, the system then looks to the second or subsequent step.

Upon encountering either decision parameter, the system looks to the secondary agent group. The immediate question becomes — does this overflow step prevent the call from being served by an agent in the first and most ideal group, should one become available while the call is queue for the secondary group?

When we started this chapter, the warning was given about examining nuances. Subtle? Yes. Worthy of ignoring? No.

The issue is maintaining the largest pool of resources to serve a service request. Scale produces economy only if it is intelligently used.

Simultaneous Queuing of Multiple Groups

Vendors of ACD systems are not quick to provide clear definitions of how things really happen in their systems. This is generally due to the fact there are no accepted frames of reference or standards to judge which is a better way of doing things. Consider this scenario: three different ACD systems manned with three agent groups, the primary, secondary and tertiary groups. How would the vendor's system handle the calls?

Trunk Vectoring. This popular technique instructs the system to receive a call and look at the first group for a predetermined time. Once this time is exhausted and no agent has become available, it looks to the second agent group and so on.

The important issue here is the number of agents (or resources) being simultaneously scanned for an available agent to provide service. Typically the pattern is group one, then two, then three, so that the largest number of agents being simultaneously scanned is only equal to the largest agent group.

In most PBX-based ACD systems, the original designers pressed the "call forwarding" feature into service to implement overflow, hence the inability to do simultaneous look back when the call forwarding features are used.

Overflow Call Count. Like trunk vectoring, once a time or call waiting count parameter is exceeded, the call is passed to the secondary or tertiary group for service. In this case, the overflow decision is typically triggered by the number of calls already queued for the particular target group. If the queue call count limit is exceeded, a queued inbound call will completely ignore that group and immediately "jump" or overflow to the second or third group.

Multistage Queuing. In this case all inbound calls target a call processing control table and are handled as targets of this logical resource or table. The system then looks to match the caller with the most desirable agent/group and only when time parameters are exceeded does it "add" the secondary and tertiary agent groups as additional answering resources to the call queuing process. The word "add" needs further explanation.

A subtle, but important, intellectual shift occurs here. A process becomes the target for a call, not a physical device or person. This is a small demonstration of the evolution away from classical telephony thinking where a hardware address (station) or device is the primary call target. In all three of our scenarios we have almost identical systems with small variations in the call processing schemes. The system with multistage queuing will have a smaller and more productive staff complement because the queuing table takes advantage of the economy scale. Also it employs simultaneous "look-back" queuing.

Look-Back Queuing

Look-back queuing is much used, yet it's a little understood phrase. Being able to "look back" at a previously questioned resource is technically look-back queuing, but does it take advantage of the total available agent complement at that point in time?

It may be just a "snap shot" process. The system looks at group 1, then group 2, then group 3, then back to group 1 and cycles accordingly until the call is answered.

The look-back difference occurs when coupled with multistage queuing. Two processes occur simultaneously. First, the ACD with a call in queue, following the call routing table, looks at the first group waiting for an agent to become available. When a certain amount of time expires, the system adds a second group and so on. The second process that comes into play is the scanning of the primary agent group, then the secondary group as the additional agents are added to the logical prospective pool of answering agents. Typically the system has a preference for the first and most desirable group. In this way the largest pool of agents is built up to serve the caller.

Economy of scale becomes meaningful at about twenty-five to thirty agents and begins to improve from that point. Statistically, the greater the calls/trunks/agents the more efficient the system becomes. Trunk vectoring and overflow, without simultaneous look-back capability, deny this building of scale.

RELATIVE AGENT UNIVERSE WITH MULTISTAGE QUEUING

	Number of agents in each group	Multistage	Queuing
		Without	With
Group 1	40 Agents	40	40
Group 2	60 Agents	60	100
Group 3	20 Agent	20	120
Total at route end		20	120

No multistage queuing means the maximum number of agents = current group size vs. cumulative total of agent groups to which calls are queued.

Figure 11.2.

Overflow, Intraflow and Interflow

These terms essentially mean the same thing. A decision point is reached in a call processing sequence and the caller "overflows" to the next intended agent group. Overflow is the general description of this step. Intraflow is overflow internally to another group in a single ACD. Interflow is overflow to another group in a second or subsequent ACD or other system.

Selecting an Agent within a Group

Once an agent group has been selected to serve the call, and assuming for the moment there are a number of available agents to answer the call — how is an individual agent selected? There are three common ways: "top-down," "round-robin" or "longest available."

Top-down. In this case the system uses a technique common to uniform call distributors (UCD) and early ACD systems. Here the call is directed to a target directory number in the system that amounts to a group pilot number. The call begins hunting for an available agent at the top of the list of extension numbers these agents occupy. The call hunts through the list from the top until an available agent is encountered. The call is then connected to that agent.

When a new call arrives it begins at the top again and rotates through the same hunt sequence. If all extensions in the group are busy, the call "returns" to the top of the extension list and cycles through the list again, looking for an available agent. The effect of this hunting scheme is that it always places an inequitable workload on the agents early in the hunting sequence. Agents refer to those seats (extensions) as the "hot seat(s)."

In some limited applications this may be a desirable. Take the example of a back-up agent group primarily assigned to do non-phone work unless service levels deteriorate. With this top-down hunting method, all the agents are not randomly interrupted with overflow calls unless the incoming call load is high, then they are all pressed into service as intended.

Round Robin. This technique is less frequently encountered, but like the top-down method, it begins with an agent extension and hunts in a rotary fashion looking for an available agent. The difference here is the hunt sequence begins at the last agent extension that answered the call, plus one. Although, less of a "hot seat" effect occurs, the workload is still not equitably or logically distributed.

Longest Available. Here the ACD maintains a table known as an "available" or "free list," where each agent is assigned upon going into the available (for a call) state. As each agent joins this list, they are assigned an arrival priority. That is when they became "free" to join the list. The "oldest member" of this list is the next agent to receive a call.

Again, there is a subtlety in the shift to logical assignments of values, in this case it's the time the agent has been in the available state, that allows a substantially greater degree of flexibility and the ability to more appropriately reflect a customer contact center structure. Traditional telephony uses hardware addresses and does not allow this interaction between hardware and software tables.

Secondary and Tertiary Group Assignments

Most modern standalone ACD systems also allow more specificity in targeting certain individual agents in the customer contact center based on the skills or specific services these agents can provide. The typical example cited is the use of foreign language agents in a predominantly English speaking center. The problem that arises is this: the call volume to this particular skill split is low enough not to keep them constantly busy. As a result the total call load is not balanced across the agent group. Some agents will work harder than others, this results in inequity of work distribution, which has many implications beginning with perception and going as far as compensation.

To solve this problem, the standalone vendors provide for agents signing onto multiple groups, based on their skill. Then the system manages the equitable distribution of calls, preferring those agents with a given skill, when such a caller calls. While ensuring everyone shares an equal a work load as possible. This provides not only an equal workload, but also plays the economy of scale game as closely as possible. Reports are delivered based on the primary, secondary and tertiary splits, assignment and ultimate routes.

☐ OTHER CALL PROCESSING STEPS

Other Agent Groups. Modern ACD systems allow connection to as many agent groups as desired, under a single call processing control table. The limits here are the number of allowable steps in the table or the total number of agent groups in a single system. This does include interflow to a second or subsequent remote site.

Delay or Hold. This is a delay step, which allows for a delay of a predetermined time. Some ACD systems allow this to be inserted prior to the trunk answer point. If this is the case, ensure that "barge out" is inherent in this preanswer step. All delay or hold steps after the trunk answer also should allow barge out to a newly available agent. Do not take barge out of a pre-answer delay for granted. Check with the vendor. Watch for the fixed "forced delay" step (in lieu of dynamic answer) that many PBX-based ACD vendors sell. Also some systems don't have barge out, which means you can have available agents, queued callers and no ability to connect the two for five or ten seconds, depending on the fixed delay. Five seconds a call, for all calls that are answered immediately, quickly adds up to significant additional trunk and labor costs.

Announcements. These are optional resources that allow informational messages to be played to the incoming caller. How the system is instructed to whom and when to play these announcements is generally a function of the inbound routing table. Some older ACD and UCD systems assign this function in hardware. Ensure that barge out to a newly available agent is a standard option.

Interflow. This is an overflow step that can be found in larger more sophisticated ACD systems. The intent of this technique is to allow multiple, geographically dispersed ACDs to be networked together. It allows even larger "pools" of agents and resources to serve an inbound calling population. Large airlines and catalogers make extensive use of this technique to great effect — to shift workload, optimize staff and cut overtime expense.

Other Resources. Interactive voice response, voice mail, etc. could be used. The discussion of whether to use them or not is over — they're proven. Customers have accepted them in appropriate applications and they can dramatically improve a company's service delivery at relatively minor incremental cost.

Companies use interactive voice announcement and response technologies to supplement their agent forces. Introducing voice response technologies into your service options present many intriguing possibilities to be discussed later. The most critical issue at this stage is the perceived delay they introduce to the caller service level. Typically these voice or audio response systems are inserted as a call prompting or overflow step in the call processing sequence. They can be used to ask the caller to enter critical identification information for coupling with your account databases and so on.

Call Prompting. This term and technique was first used by AT&T in Enhanced 800 Service. Now this feature can be found as part of any voice processing scenario from the cheapest auto attendant to the most expensive IVR available.

It inserts a step into the call process as the caller arrives at their intended destination. This step uses interactive voice technology to ask the caller to identify with more specificity the department or function that they are trying to reach. The greeting script may go like this: "Thank you for calling Acme Corporation.

To better serve you please enter the number of the department you wish to speak to, 1 for Sales, 2 for Service, etc." The caller enters the number on their touch-tone phone and is queued directly to the functional group they desire. An advantage occurs because functionally dedicated trunk groups are no longer required, i.e. publishing one number for Sales, another for Service, etc. For example, the marketing department can use this feature to reduce the number of 800 numbers they use to advertise the company's sales and service numbers. "One number service" becomes viable as an alternative to multiple circuits for different customer services.

With call prompting, greater flexibility occurs, but now secondary queuing becomes a concern.

Call Overflow. Here the same technology is used, but at a different stage in the call. The voice response device is used to backup the live agents during a busy period when service to a minimum of callers may be unreasonable. Instead of "eternity queue," the voice response device takes a call back message or provides a service alternative. In this case though, successful implementation depends on how quickly and conscientiously the call back occurs.

Voice Messaging as Overflow. Adding voice messaging as an overflow feature is a two edged sword. Unless there is a compelling reason for an agent to pick up the phone and return calls, this is not going to work.

First, most ACD systems are marginally staffed. The ACD was bought to aid in carefully matching supply (agents) to demand (callers). If this is the case, the agents have little or no time left to return calls left in the messaging system.

Second, it is a well known belief that inbound (reactive) agents are not good at making outbound (proactive) calls. That is the perception and many agents act this out. The last thing they wish to do after a heavy inbound call session is to begin making outbound calls.

Third, the calls have to be retrieved from the voice mail system. This can occur three ways:

- Have the agents call the voice mail and physically "transcribe" the message, caller name and caller phone number and then make the call.
- Assign an agent to this task and distribute the call back calls to agents. Neither of these is particularly seamless or transparent.
- Integrate the voice messaging system with the ACD common control. One ACD vendor has done this to great effect. This system offers callers in an extended queue the option of leaving a request for a call back and hanging

up. Part of the message requests the caller dial their call back number into the system. The system records the voice message and the touch-tone digits. The system then monitors the customer contact center service level. As the service level improves it "inserts" these call back requests into the inbound call flow. The system "plays" the voice message to the agent and automatically out dials the phone number.

A newer strategy is now available when callers enter their account identification using a social security, account, phone or some other number via IVR, the computer system builds a call back list and presents these call backs to available agents while allowing passive or active preview dialing. This is particularly elegant in that it removes the call back decision from the agent, occurs almost seamlessly and makes the outbound call "look like" an inbound call. After all, "the caller wanted to talk to the customer contact center."

Using automatic number identification or ANI is another way of capturing the inbound caller's number. The ACD system must be equipped to gather this data and do something with it. This offers a secondary challenge in that you do not capture the individual identity of the caller. This can be done in real-time or at month end when the 800 billing is rendered. The lead will atrophy quickly, so "after the fact" direct mailings from the abandoned caller list to a caller denied service will have little effect unless they are particularly innovative and compelling.

217

☐ VoIP ACD FEATURE MATURITY

This is an interesting point in time as a new "plumbing" technology begins to take hold and show up in new or replacement opportunities.

Normally when a telephone vendor adopts a new technology, they try to preserve the investment in the work they have invested in common control, feature and application programming. They do this by grafting the control software onto the new switching paradigm, in this case packet-switch Internet voice with some new translation layer (from addressing a circuit switch to managing packet-switched voice addresses). After a vendor has made the core plumbing change, hopefully as transparently as possible to users and their customers, they will then address the control software.

Vendors are typically reluctant to change anything because of the cost. If it is not broken, don't fix it. The two exceptions to this are competitive pressures and value engineering. The hot breath of competition can be credited with many technology advances. The other is euphemistically called "value engineering."

This is simply finding a way to build something less expensively, hopefully enriching the profit margins, unless competitive and market pressures dictate a price reduction.

The maturity of traditional circuit-switched ACD feature software is deep. The question becomes how well will this translated into VoIP customer call center server software. The answer is very dependent on the vendor background in true ACD software. It they are telecentric, their solution should bias the user. If they are data centric, it will bias your technology department. Respectfully, many of these folks have neither spoken to nor managed significant volumes of customer calls.

When should you migrate to VoIP? When it can be demonstrated that your customer contact center will receive the same mature ACD workflow features you enjoy now. The technology cost savings of VoIP are trivial compared to the business that is at risk from a "premature upgrade." This is not to say leaders in the business aren't promising this marriage now, the question is delivery.

Bullet-Proofing the Customer Contact Center

There's nothing as dead as a dead customer contact center! And that means your business.

One simple question: How much revenue and profit will your company lose if your customer contact center crashes for one hour? For a day? For a week? For a month? And if you don't think it can happen to you think of the hapless companies who sat at the end of 35,000 Illinois Bell phone lines after the infamous Hinsdale central office (CO) fire in 1988. Many lost their phone service for as long as a whole month. Four weeks without calls, without revenues, with no profits from the customer contact center driven portion of your business. Lost business opportunities and erosion of business credibility. Forget the money, think of the company anguish and the possibility of furloughing or laying off employees. Think of your career. How it could have been enhanced if your preventive plans had saved the day. Think how far it would be put back if you do not buy and build a customer contact center which can resist almost the worst possible disaster, including flood, fire, earthquakes or power outage.

It takes money — though surprisingly little more in terms of the capital investment than you planned to invest in the contact center originally. This is particularly true when measured against the revenues and profits at risk should an interruption occur.

As the world moves from traditional analog and digital telephone and computer systems to completely integrated voice over IP networks, bulletproofing a customer contact center is more critical than ever.

☐ DISASTER RECOVERY
AND BUSINESS RESUMPTION PLANS

The notion of disaster recovery is a flawed notion to begin with. Unless you are a very small business with only one business location, putting all your customer contact center "eggs in one basket" is a risky strategy. A minor disaster, such as a telephone pole being hit by a car can take you out of business completely. Customers contacting your business with their individual problems have little interest in, or sympathy for, your problems.

Often the fall back is to develop a business resumption plan, but again this language assumes the business has experienced a hard interruption.

This recalls a few years back when a major mutual fund manager lost power to their ACD as a contractor inadvertently severed an electric cable to the building. That day the company had launched a major new fund and placed full-page ads in numerous financial and city newspapers to the tune of many hundreds of thousands of dollars. They could not answer one call! Because the installed PBX/ACD architecture would not allow full power fail bypass and because of digital instruments, which required 100% power, they lost millions. VoIP systems work the same way. Every element of the customer contact center must have fallback, redundancy or power fail bypass.

For a small business, putting together a back up strategy with a telephone service bureau is a partial and effective intermediate strategy. North America is moving into a period of energy uncertainty and anticipating regional rolling electricity blackouts are now a common part of business planning. Be careful when choosing your outside redundancy service because a regional disaster could swamp a regional provider with all clients calling on it simultaneously.

A true TDM circuit-switched ACD should normally be configured to allow for fallback, redundancy or power fail bypass. A VoIP packet-switched ACD package presents a number of critical points of failure with a power loss that need to be anticipated and built around. Even with history as the teacher, the vendors of VoIP networks have treated these hard won lessons from the analog and digital world lightly. Expect users to demand a more robust answer as VoIP features mature to match traditional ACD switch functionality.

☐ A BUSINESS CONTINUITY STRATEGY

A far preferable and realistic strategy, once your business grows to a size that justifies it, is to balance business across geographically separate sites. This may mean working with other company divisions or establishing a relationship with a organization that outsources back-up services as a business.

These centers, yours or that of a supplier, must be equipped and staffed to handle business operations on a day-to-day basis. Some of the staff may include your temporarily relocated staff. Here is why this makes sense. Looking at it from a business continuity perspective means having staff and equipment available at a second site. But it becomes problematic if you treat this as a switch from site A (that has experienced the disaster) to site B (the back up site in times of emergency), which is only a disaster recovery mode. You will find that staff and operations processes don't "translate" as easily as hardware and software. But when you have a continuously operating second site with experienced staff as an alternative site for overflow, this means that what other companies characterize as disasters, are really lesser events for you.

Hardware

Maybe you will wish to over-configure the system shells and populate them with extra components (gleaned from site A if accessible or rented for the duration of the event from your vendor) in the event of a major call traffic redirection. This avoids your company carrying expensive inventory just for a once in a decade event.

During the January 1994 Northridge (Los Angeles) earthquake, staff could not enter condemned buildings. Even though all of the telephone equipment continued operating, freeways were jammed, streets closed and no one wanted to leave their homes and families out of fear of subsequent aftershock damage. Many local COs were affected, making calls in and out of Southern California almost impossible. Emergency agencies were telling people to stay home so the thought of moving staff to a contractor's back-up center was illusory. Flying them out of state, impossible. Callers from unaffected regions kept calling, and as there was no business continuity plan, those calls went unanswered. Once burned, some of these companies learned their lessons and established continuity plans which permanently established interflow and load balancing between national customer contact centers.

Details

"Bulletproofing" the customer contact center is not limited to the reliability and failure resistance of the ACD system. The more you protect — the inbound circuits, the power source, the automatic call distributor, the networks and computer support system, building lighting and air conditioning — the better off you will be.

Planning an "off site" business resumption strategy is the ideal. The author has observed a number of businesses during natural disasters and their customer call center recovery strategies. The basic rule: the longer you take to recover full customer contact center operation — minutes, hours, days — the longer the residual effect.

The other uncanny corollary is how quickly a company can forget the effect of a catastrophe when they discover the cost of a backup plan! There are a number of major businesses in the Los Angeles Basin with single customer call centers, who were mercifully out of business for brief periods of time during the 1994 quake. In the cases the author directly experienced, it was not the systems that died, but building closures until inspection and granting of structural safety clearance, then the availability of staff, that kept call centers off-line. No amount of single site redundancy can protect against this, but a service bureau or sister center can mitigate this impact.

Circuits

Circuits have five elements involved that need to be examined from a reliability point of view. Without adequate circuits and access to the public network, the customer contact center is effectively out of service — no calls, no email, no web access. Tracing a circuit from the public telephone network to the customer contact center's switch or network servers requires identifying the separate elements:

- The long distance network (or VPN) provider "point of presence (POP)."
- The link between the network provider POP and local CO.
- The local serving company Central Office.
- The outside plant or cable route between the CO and the customer contact center.
- The building entry and termination point(s).

It is also necessary to duplicate this same network for your customer contact center's data and web access with just as much attention paid to each potential failure point.

The Carrier POP office, CLEC or ISP Central Office

According to modern mythology, telephone company central offices never die. Although generally unbelievably reliable, they have been known to occasionally breakdown. Sometimes they burn. There is little a customer contact center manager can do to guard against this, unless there is a disaster contingency plan.

The more valuable your incoming calls and contacts (in terms of revenues), the more compelling the argument is for a disaster protection and contingency plan. It is not very comforting blaming the local telephone company for your lost business, despite reality. Tariffs protect the local telco and the long distance carrier from direct and indirect damages due to nonperformance of their service. The most you can generally recover from the telco and/or carrier is the abatement of the telephone bill for the duration of the outage, provided it exceeds 24 continuous hours.

With the advent of the Internet as a growing customer access and carriage medium for web browsing, email and orders, the same back-up rules need to apply to your Internet Service Provider's local point of presence and connections to your customer contact center. As we have discussed, with any technology evolution the technical features associated with the mundane, such as redundancy and housekeeping are always last to be built out to meet the maturity of the generation being replaced.

Local Service Access Plan. A strategy rarely followed, because of the cost, is the termination of circuits at two separate central offices. Now, on notice, the long distance carrier can alternatively route inbound calls through the second central office and back-up cable path to the customer contact center. Obviously, a physically separate route for the outside cable is recommended. This also can be requested for your Internet access. Ensure enough bandwidth is available to carry at least two thirds of your total demand on each alternate path.

Long Distance Carrier Plan. Most major long distance carriers offer an arrangement with 800 service where they will guarantee no calls will be lost after you have notified them to redirect your inbound 800 telephone calls to a second site or other sites. Take a look at this as it is a great solution. Understand EXACTLY how it works and WHAT IS REQUIRED of your company to ensure a fast "switch" to a second site. Test it with real traffic before you "take it live." There is more to making this work than just the technical redirection of your 800 service. The target customer contact center(s) must be available to serve the volume and class of customer contacts they will now be presented with.

Staffing Plan. As discussed earlier, setting up a relationship with a company that provides disaster recovery or back up services is a great idea. There is only one flaw — they too have limited resources. There is an assumption that only a limited number of customers will demand their services at any one time. In the case of the 1994 LA quake, these companies quickly ran out of capacity to support their subscribers. This argues for an exclusive company-based business resumption plan. Remember your advertising campaigns are not typically affected by disasters and they take more time to "turn down" than you customer contact center. The volume of service requests is affected in a positive way in that call volume increases, as does the chance to disappoint.

Outside Plant

This refers to the POP or ISP to your CLEC CO and the CO to your premise. Every customer contact center is linked to the public network via a local CO. Between the CO and the customer contact center is a connection called the local loop. A similar high volume connection exists between your CO and your long distance provider's POP and ISP. These circuits are typically digital circuits, although there may be a cable or microwave radio link somewhere in the mix. All are subject to interruption by way of physical cutting of cable or loss of power. In the case of microwave, loss of line of sight connection is an issue also. Somebody builds a building or one of the antennae blows down. To guard against this, a second alternate local loop is often used. On the face of it, this is not an inexpensive measure, but when considered in the context of the business and the revenue at risk, back-up circuit routes can be cheap insurance. One major domestic airline books $36,000 per minute. Loss of any portion of the customer contact center or reservation system is tremendously costly. If this option is pursued, the routing of these local loops should be absolutely separate even down to the building entry point.

Building Termination Point

The local loops or trunk elements may arrive in predominantly a digital (T-1) or, less frequently, analog format. At the point of termination, prior to literal attachment to the ACD system, there will be connecting blocks allowing the individual trunks (bulk or single) to be broken out into their individual wire elements. In either an analog or a digital environment, total power loss ends all call processing unless uninterruptible power is supplied to the switch and, in the case of T-1 spans, terminated on channel banks.

There are five different scenarios:

- A fully digital environment, digital trunks and digital ACD.
- VoIP.
- Mixed analog/digital trunks on a digital ACD.
- Mixed analog/digital trunks on an analog ACD.
- A fully analog environment with both analog loops and ACD.

We have discussed the issue of digital trunks, digital switches or VoIP networks requiring uninterrupted power for operation. If channel banks are necessary for terminating digital trunks (T-1) on an analog ACD these must also be supported with uninterruptible power if they are to receive and decode 24 channel T-1 spans into a single channel and ultimately an analog conversation during an outage.

Power

In the case of digital ACD systems and VoIP networks, DC is the required power source at the device(s) due to its predictability and stability. This requires battery, inverter and generator back up to maintain uninterrupted operation. Note that pure AC systems are likely to require more power conditioning and be subject to potentially higher maintenance requirements due to the less precise power source.

Pure Analog Systems

With those few pure analog ACD system that may still be in use, a fall back position is often taken to avoid back-up battery or DC power systems. Power fail cut-through or bypass capability is included in the system. This allows the ACD to receive calls on those particular trunks, connected to predetermined instruments. Power from the CO is used and the instruments acts the same as with a single line home phone. All other system features are lost during the outage. If you are still using a pure analog system, check that the number of cut-through trunks configured are satisfactory to meet a substantial portion of the total trunking. Ten percent is inadequate, while one hundred percent is unrealistic unless this is a standard feature. Remember that each power fail position must be manned to take a call. Also remember that any front office computer system that is required to process caller requests must still be operational or be backed by a well-rehearsed interim manual process.

Non-Analog Systems

With a digital ACD or VoIP system there are two strategies. Complete DC power support, either from a battery source or UPS (uninterruptible power supply). The latter is more expensive and batteries need to be recharged or they too will run out of power. Most battery back-up configurations bet that the power outage will not be for an extended period of time, so battery is typically provided in minute or hourly increments. The longer the battery support time, the higher the cost. Years ago Bell Labs did a study and found that over 90% of all power outages lasted fewer than five minutes. Generator support is required for outages beyond an hour or two.

"Graceful" Degradation

One innovative approach taken by a third generation entry into the ACD arena was to provide staged power degradation.

First stage: they provide full time DC power and full battery support for the system.

Second stage: should power be lost to the inverter which charges the batteries, the batteries would support the entire system, disk drives included, until the batteries become substantially depleted.

Third stage: the system bypasses the switch to digital instruments still supported by the remaining battery to allow simple call receipt and origination.

In this way the system can still operate over an extended outage. More recently even this type of solution has become impractical due to the increased use of cross-channel cross-application customer contacts that all require clean uninterrupted electrical power.

Overlooked in this discussion is the fact that any time power is lost, other customer contact center facilities are also lost. Air conditioning, lighting and any supporting computer systems and networks. People are unable to work in the dark, so natural lighting is often designed into the center. Not all regions are dependent on air conditioning year round so there may be provision for natural circulation.

Simple call transactions may be able to be completed with pencil and paper or messages taken for callbacks.

Don't forget to include your long distance carrier(s), ISPs, and suppliers of your 800 trunks in your disaster planning. With some prior planning, they can move all or part of your incoming 800 and web calls from one customer contact center where a disaster has occurred to another operational center in a another part of the country. Just make sure the target center has the physical and staffing capacity to accept the redirected call load.

Redundancy

Telephone systems redundancy was discussed in detail in Chapter 9.

Computer Systems

There are few customer contact centers left that do not use computer systems for "mission critical" applications. Many of these systems are considered so essential that fault tolerant systems are the rule. Initially these were more expensive than their non-fault resistant counterparts and like all other systems needed clean power 100% of the time to operate. The prices of these "fault resistant" or "nonstop" systems have dropped, while the options have increased dramatically. These systems are increasingly finding their way into mission critical customer contact center applications such as credit authorization, customer service and order entry.

With more customer contact center systems following standards like NT, there is an assumption that inexpensive PC technology will satisfy platform requirements for server technology. This is a valid assumption on two counts: there are a number of viable providers of competitive systems, but more importantly, there are many cost effective industrial strength redundant PC systems designed for high availability applications.

These systems are now being deployed with hot-swappable power, fan and disk units, which are the most typical failure points, short of complete power loss.

Computer Telephony Applications

There are two architectures that are promoted for CTI applications. These are:

- The traditional host/terminal "star" configuration. Most likely to be encountered in a legacy environment as most front office sales and marketing applications (CRM) are more recent, and follow a client/server architecture. The terminal or workstation power in this "mainframe" scenario data is provided by a local "wall power" source. Lost of commercial power at the workstation means loss of the support system.

- Desktop PC and client/server configurations. If the host, PC, LAN, server or CTI link is lost, so are the applications they support. As you integrate CTI applications it is critical that a fallback strategy or redundancy be built into the system and deployed at every point in a client server network. As expected, this is simpler with a centralized computer ("host") system, as you only have to "bulletproof" that system. You can still lose terminals and connectivity to the ACD switch.

Client/server CTI environments are proving to be increasingly popular in CTI applications due to lower cost, faster deployment and increased flexibility. However, they offer more failure points than tradition terminal systems.

To build an integrated nonstop customer contact center with linked computers and switches you need to consider bulletproofing the following components:

The ACD Switch. The switch needs to have redundant common control and

---

Rule: When your company's MIS department insists their application needs uninterruptible power, do not allow the call switching process to be overlooked as a candidate for the same treatment.

MIS processors. This is typical of most standalone ACD systems and some PBX based ACD systems.

Any IVR System. Having more than one IVR system and balancing trunks across these can mitigate this. Redundant IVR systems are available but are still costly for the utility they provide.

MIS Links Between the Switch and the Host. The voice and data signal coordination typically comes from the ACD MIS. If there is a redundant MIS system there must be two parallel links to the computer system holding the application and the database. Watch how the vendors position this. In some cases they deliver more than one link (for load distribution purposes only), but the message traffic is distributed across both links. They do not exist for redundancy purposes! Two redundant CTI links are ideal.

The Front Office Database System. Building nonstop computer systems is an established art, although the cost of buying a hot standby computer system is expensive. In a recent experience, the cost to equip a call center with a hot standby system added 33% to the price of the CTI project, versus 18% to deliver a warm standby or fail over database server. The difference was about 50 seconds to accomplish the application switchover, with no loss of calls, just the "screen pop" and related functions during that 50 seconds.

Client/Server. In the case of a client/server system, which is served by a local area network — this is a relatively stable technology that can be bought in redundant configurations. But again, is the additional cost justifiable? LAN problems are readily diagnosed and remedied. If VoIP is your chosen direction, there is no choice.

Desktop PC/Soft Phones. The Desktop PC is important with "soft phones," where the desktop telephone is a "window" on the PC screen. If the PC is lost, so typically are the telephone functions, unless the manufacturer provides an external (to the PC and its power source) device that allows manual bypass operation. Again, VoIP architecture removes any option not to deliver uninterruptible power to the workstation and supporting network.

These same comments apply to each component in a LAN system carrying VoIP as the voice packets now travel across the same gateway, LAN, wiring, servers and client workstation configurations.

Telephone Terminals and Workstations

This book began life as *The Incoming Call Center* in 1987. The seismic change brought about by personal computers and the WinTel alliance has placed even the most complex communication application within the reach of the smallest call center. Back then we made the observation that the arrival of the PC in any business remade the economic landscape like a swarm of locusts remake a lush landscape. We had no idea!

We first should take a look at the instruments used in the traditional call center environment.

☐ PLAIN OLD TELEPHONE OR 2500 SET

Most telephone instruments evolved from the plain old single line instrument, most commonly found in residential applications. Plain Old Telephone Service or POTS has birthed a thousand innovations and the evolution of the plain black "phone" or 2500 set in American telephone company parlance is the modern original. If you have ever used a phone, it's probably been a 2500 set. It is a marvel of industrial strength consumer electronics. A single line desk set is called a 500 set. The "2" in front means it's touch-tone.

The rotary phone evolved to the touch-tone phone with its 12 keypad, 0 through 9 and its two special keys. Only recently have these special keys come

Note: In this chapter the word "terminal" refers to the any terminal device or agent instrument, and is not limited to a computer display or CRT device.

into use with voice response and station user programmable PBX and CO feature sets.

The touch-tone telephone has 12 keys that generate distinctive dual tone multi-frequency (DTMF) tones and a switch-hook. The switch-hook amounts to a control signal. As the station user goes "off hook", or depresses it momentarily, the controlling system is alerted and the following signals are system command instructions, that is the beginning or the end of a call. The placement of the switch-hook is traditionally in the handset cradle, away from inadvertent activation, because along with serving as a control signal it terminates the call. A momentary switch-hook depression or "flash" became a standard telephone system station user originated alert or command signal. Using the switch-hook to send other command messages to the switch was somewhat successful but led to great confusion. "OK, I'll transfer you but if I lose you, here is the correct number to call!!!"

The layout of the touch-tone pad is an inverted numeric accounting keypad minus the decimal point key. Consistent with a numeric pad, the "0" key is at the bottom of the layout, though centrally positioned as a single key, versus the double wide "0" key of a computer or calculator numeric pad. This is flanked by the special # and * keys (octothorpe and asterisk). There are many stories as to the logic of inverting the numeric pad. The most popular is the impact the inverted keypad had upon slowing the number entry rate so the central office technology of the day could keep up with the tone identification and disposition tasks.

The POTS phone remains a basic instrument and the telephone terminal of choice for very generic telephone applications. The TAP or Flash phone is a minor variant introduced to do away with much of the confusion which arose out of using the switch-hook as both a command alert signal and as the call termination "switch", where replacing the handset or holding down the switch-hook ended the call. After extended conditioning of users to end calls in this manner, the telephone industry introduced feature activation via depressing the switch-hook and dialing additional digits. Residential call waiting is a good example of a simple switch-hook routine. The accompanying comment became "...if I lose you (during this process) the number to call back on is 1234". User hostile technology!

To reduce this confusion and allow increased feature use by station users, a number of telephone system manufacturers introduced a 13th key on the instrument, known as the (switch-hook) flash key. This reduced confusion and restored some confidence on the part of station users that use of the flash key did not threaten to end the call inadvertently. Others committed to extensive use of proprietary telephone instruments, but with this strategy came increased costs.

A number of early ACD vendors used these Flash or Tap key phones as ACD instruments. The main advantage to using "tap" key instruments was the low cost and universal availability of these generic instruments. The disadvantages occurred in the relative complexity of feature use, and consequent impact on productivity. To mitigate this confusion, graphic templates were regularly attached to the face of the instrument so an agent could "join-the-dots" to complete a transaction. To further increase usability, optional display units were provided with the instrument. These are typically a separate alphanumeric display of showing call type, instrument status, time in this status and in some cases allowed supervisor to agent messaging.

There is a clear limit to the number of features that can be accommodated on a 2500 set. Very quickly a point of confusion occurs in the sheer number of steps needed to activate a feature. Training became increasingly arduous as the number of steps required to activate features was expanded. Fortunately, most ACD vendors avoided the 2500 or tap phone in all but their earliest implementations.

Most companies dropped the generic instrument in favor of proprietary devices, rich in features, yet simple to operate. This added cost to a traditional ACD purchase. They validated the experience of the standalone ACD manufacturers who chose to develop proprietary "industrial strength" high volume agent instruments in the interest of operator speed. To be absolutely effective these instruments must be supportive and transparent to the role the system performs to support the agent in the business function they serve.

☐ THE PROPRIETARY PBX INSTRUMENT

The PBX/ACD manufacturers have long used their proprietary feature telephone sets as the ACD instruments of choice. This allowed them to use existing devices, parts and proprietary signaling plans. The majority of ACD systems use instruments that began life as PBX instruments. Most PBX\ACD makers have pursued this course. The advantage accrued because manufacturing and inventory costs were only incrementally more, as opposed to the manufacture of a completely proprietary ACD instrument.

The disadvantages are relatively minor unless the instrument began as a poorly designed device. Every one of us has sat at a telephone instrument that offended us. It was cheap, the keys and buttons were illogically laid out, not large enough, crowded onto a small space, cheaply housed, offered poor tactile or auditory feedback. An early proprietary instrument, which won many industrial design awards due to its aesthetic appeal and manufacturing ease, became a hated ACD and operator console due to the design of the keys. These were perfect cubic protrusions, butting up to one another in a visually pleasing and design award worthy manner, yet an operator, particularly one with long finger nails found the "splash effect" when hurried operation caused the contiguous key to be struck inadvertently. This led to all sorts of unintended results. No serious professional keyboard uses anything but chamfered keys to minimize this problem. Yet to this day, many telephone instrument manufacturers still seem to ignore this reality.

Another issue with proprietary instruments is the labeling of the keys. Typically high use keys should be clearly and unambiguously labeled. The labeling should be applied in such a way so as not to wear off with use. Molded or gas etched keys are ideal. Keys with removable cuboid clear plastic covers are generally too delicate for high volume use. These covers are removable so the key function can be assigned dynamically. It also certainly indicates the instrument began life as a generic instrument.

☐ ACD INSTRUMENTS

The leading standalone ACD system manufacturers chose to build their own transaction intense ACD instruments. These rather neatly reflect the era in which the instruments were designed and understanding the evolution in ergonomics over manufacturing cost is educational.

One vendor provided a sturdy, well-designed instrument (1975) as an agent instrument or supervisor version. They have as few keys as was felt necessary when they were designed yet the vendor has continued to add features through use of a shift key mode. The increase in features and changes complicates the use of the instrument, and the agent training cycle increasingly points up the need for a more comprehensive key field. This company added an external display to allow increased status information to be provided to the agent.

In 1975 another vendor pioneered the use of an expanded agent specific key field with increased use of status lamps and special key fields. In 1982 this company introduced a display instrument that showed queue status, replaced many of the binary status lights with descriptive status fields and provided alphanumeric descriptions of call type, city of origin and supervisor to agent messaging. There was an oversized three line by about 32 alphanumeric character low power LCD display. Although to be seen easily, the LCD display must be at the exactly correct angle, lighting and proximity to the agent; an oversized display attempted to bypass this problem.

In 1987, came the most innovative ACD vendor and introduction of all. Along with providing hard key feature definitions, their first instrument adopted many of the context control and soft key lessons of PC software development. They used advanced call context control and dynamically labeled soft keys, which followed the call state and could be programmed to follow a "data script" thus forcing the agent to follow a certain wrap-up sequence and enter certain data. Dynamically labeled soft keys allowed the dynamic reassignment of key functions based on call progress and the logical flow the transaction is following. The system did not allow illogical deviations from the call flow. It displayed the redefined key labels on the LCD display above the associated function keys. Using software definable keys increases the flexibility of the manufacturer and increases the life of the instrument tooling. When coupled with call context control this solution can also decrease confusion and speed training on the part of the agent. This vendor also used an audio online instrument tutorial via the system voice-messaging feature.

The PBX/ACD vendors also introduced some of these display features into their instruments, but again they have continued to do this almost in a generic competitively reactive fashion.

☐ TERMINAL DISPLAYS

ACD and PBX manufacturers added displays using two methods. First they added them as external devices. The first and most pervasive was the PBX instrument. Early instruments used light emitting diode (LED) technology, but this proved too power consumptive and often required an external transformer and power source typical of a desktop calculator. Later models switched to lower liquid crystal display technology, typically found in wristwatches. This display needs abundant light and careful positioning to be effectively read by the agent. The viewing angle is absolutely critical. The size, content and use of these displays varied with the manufacturer.

The next step in the evolution of the ACD instrument was the integration of ACD instrument functionality into on single desktop display device. The long awaited integrated voice/data terminal. Early examples of these devices are seen as display components of some data processing system, rather than a fully functional CRT/volume call handling devices.

Two strategies have been followed in traditional telephony and a third is evolving with VoIP.

First, was a single proprietary applications terminal with concurrent data entry and instrument functionality. Examples of this are found in plug-compatible replacement terminals incorporating basic PBX station functionality in the device. To use these devices meant replacing the entire CRT, and in the case of one vendor, the terminal communications controller technology. This was seen as a necessity by most data processing departments. Early examples could not demonstrate the superiority of such a combined device short of common wiring that is a benefit of VoIP in a new installation. Good examples of this strategy however are being used in outbound dialing machines that never made it to the ACD business before the PC became ubiquitous as a desktop data entry and inquiry device.

The second strategy allowed for the preservation of the PC with phone functionality added as an external (desktop or knee well) box. This brought functionality and applicable station instrument capability to a PC device. The only caveat was whether the existing keyboard can accommodate the voice instrument key functions.

The value of this integration means a closer operational linkage between the applications software, data gathering and telephone function control. Also only one footprint occupies the desk thus saving valuable desktop real estate. Early integrated voice and data terminal (IVDT) vendors erroneously argued fewer cables with lower installation costs and less confusion were the benefit. The missing link was compelling applications. Early terminals had little local computing power to run macros or functions locally, so they basically failed.

The most important value occurred in providing the telephone rep with a single integrated device to interact with: less confusion, less refocusing at different display devices and less station to instrument hand movement and most desirable, less training. It was not until the arrival and maturity of robust versions of Microsoft Windows and client server architecture that the integrated desktop really flourished.

Now there's the third generation, which occurred with the adoption of Windows-based (98 or later) PC application shells, which can incorporate a usable screen-based telephone. By using Windows as the standard operating environment, any number of standards compliant PC applications can be acquired to run along with all the telephony functions in a "window." This is essentially "first party" computer telephony. Provided the applications are all DDE or ELA-HAPPI compliant, the dialog windows can swap data and telephone commands based on links that are established by the programmer building the desktop environment. When these are integrated with telephone and computer functions, and the customer applications that may run on existing host or "legacy systems," a great deal happens to improve agent productivity and automated data gathering. All this happens without complex mainframe code rewrites.

☐ THE PC TELEPHONE CONNECTION

At this point it is necessary to discuss computer telephony architecture. There are two ways of building this structure in a traditional telephone environment by either employing first party CTI or third party CTI. Once again, VoIP promises a better idea.

Third party CTI. This was the first iteration of CTI. This involved a computer running a CTI compliant application, a telephone switch and a link between the two, often on a dedicated server, and popularly called CTI middleware. The first generation was proprietary to the vendors; e.g. IBM's CallPath, Transaction Link (Rockwell), Meridian Link (Nortel) and CallBridge (Siemens). These are all early examples of the switch side link. More recently companies like Dialogic and Genesys have produced discrete packages that have become almost standards. These are CT Connect and Tserver, respectively.

These middleware applications called "third party" as the device requiring the "action" must speak to the device required to "deliver" the action via the "third party" or middleware. Typically this is a gateway or emulation system euphemistically known as a "telephony server." It really is a gateway or translation device that adds an additional layer (and latency) to the process.

First party CTI. This works in almost the opposite manner. The device, which requires the action, "speaks" directly to the device whose resources are required at the desktop. For example, a PC running a number of local client applications talks directly to the server housing the account database for screen pop purposes, then through PC terminal emulation software it asks the mainframe for the master account file data. This requires no change to the host application to become CTI compliant as it still thinks the "party" it is requesting data is a standard "green screen" (3270 or VT100/220 terminal) due to off the shelf emulation software. The desktop operating system shell works the magic through a complex of dynamic links (DLL) that are constructed by the workstation environment developer. For example, when the final field is entered on the order entry screen and completed and the entry key is struck, it automatically sends a "hang-up" message via the soft phone program to the telephone extension that is being served by the telephone switch. There are layers upon layers of code to achieve this, but no one server with the middleware.

We will talk more about the detail of this operation as we move into operation and applications.

VoIP changes the fundamental packaging of voice signals so that voice now looks like sequentially addressed data packets. There is a ton of magic going on behind the scenes to make sure these all arrive in the right order and timing (IP quality of service considerations). Because these are addressed packets, there is no reason linking of data instructions cannot be done at the "carrier layer," eliminating much of the messaging and network latency introduced by the connection of traditional telephone and data applications.

The "plumbing" is present and the market is evolving some promising applications. Once this is proven in large "industrial strength" call volumes, the floodgate will open to new applications and wholesale adoption as here will be the compelling business reason and return on investment to dump traditional switches for improved business process.

☐ THE PC REVOLUTION

The PC first began to remake the customer contact center with its use as a replacement for the computer terminal or dumb tube attached to a host of some sort. Now, robust PC multi-session systems (like recent versions of Windows) have combined voice and data terminals almost 20 years after they were first mooted as Integrated Voice and Data Terminals (IVDTs) of the early 1980s.

PC-based Telephony

There are five considerations when looking at PC-based telephony:
- The PC chosen (size and speed).
- The operating shell (Windows 98 or better).
- The effect of loading all the desired applications.
- How the hardware connectivity is accomplished.
- What redundancy strategies should be followed to ensure operation in a revenue and mission critical environment.

The PC Hardware

First, if your company follows the rule of buying the biggest and fastest machine available at the time, for a decent price, you probably have a number of different generations of machines, with varying speeds, disk capacities and software features. The minimum size required to run a screen based telephone appears to be a Pentium III or better, and a 333 MHz machine or faster. A minimum of 256K of RAM is necessary and a decent sized disk. The key here is that the additional process introduced by a PC-based telephone and integrated customer contact center desktop tasks must not increase call processing time or call length.

Second, most PC systems follow some recognized standards. After saying that, it seems there is no such animal as a standard PC, as each vendor adds particular features and design enhancements they believe to offer a competitive advantage over their peers often in the same model number without notice to users. While some of these added enhancements do deliver true advantages, they also conspire to defeat a true standards-based approach when installing telephony hardware. With Intel building more and more of the whole computer functionality into prepackaged motherboards, it seems PC may be moving closer to a predictable "standard" configuration, yet it's still somewhat illusory.

Any PC device selection involves the classic compromise of immediate utility and productivity gains, versus the device's inevitable obsolescence. This book does not purport to give advice on PC selection. However, in the writer's experience, buying a machine is like "never being too rich or too thin," i.e., your PC can never be too big or too fast — buy the biggest fastest machine you can justify.

The impact of non-standard machines on telephony interfaces means the designer of the hardware interface must make a decision where to place the telephone interface device. There are two strategies that providers of PC telephony interfaces pursue when they "load" the PC with the hardware connectivity or the soft phone: an internal bus-based telephone board deriving its power from the PC; or an external telephony integration device that derives its power from the telephone switch. The positioning of this hardware is significant for two reasons:

- The compatibility with your PC hardware (the external box typically means you are not tied to one PC vendor over the life of your customer contact center).
- The dependence on the uninterrupted PC operation for continued telephone instrument function. ACD telephone systems are typically built with a great deal more attention to operation in a power loss or other crisis. The vendors of PC and network systems typically did not anticipate these systems

being installed in mission critical revenue oriented environments.

The telephone device replacement hardware is attached to the telephone system as a normal extension. The hardware must then emulate the functions of the actual telephone instrument. Remember that there are no standards in purpose built ACD instruments either, so every device must accurately reflect the existing proprietary protocols required by the particular switch manufacturer.

There is also the issue of connection to the telephone system, the company local area network and the host or mainframe.

- Attachment to the telephone.
- LAN connection.
- Any legacy host based emulation.

We will deal with this when we deal with applications

The VoIP Connection

The PC device must be attached to a megabyte LAN network to even consider customer contact center quality voice. The buyer that cannot understand your agent clearly may not even remain a prospect.

The PC then must have some sort of soft phone application that will include code running in the background and some sort of hardware attachment for a hand or headset. This may be an internal card or external adapter.

☐ THE INTEGRATED DESKTOP

Both Microsoft (TAPI) PC telephones and VoIP interface standards have given substance to this market. These are early days but in both cases the suggested standards are gaining growing adoption and features enough to meet the requirements of a production oriented customer contact center.

The key to the successful adoption of integrated desktop work environments is twofold:

- demonstrable economic justification for adding PCs and dumping telephone instruments, and
- real applications that deliver measurable economic benefit.

VoIP integration accelerates this dramatically as this potentially eliminates added switches and layers of middleware necessary to make voice and data work together transparently.

WARNING: Establishment of a standard software PC configuration and "locking it down" is essential to the survival and sanity of your customer contact center management.

Agents are employed to handle the live incoming call component of a customer contact center. The other agent and staff performance goals are:

- to process as many calls as possible,
- as many calls in as short a time as possible, and
- as high a level of satisfaction to the caller as feasible.

From an agent's point of view, these goals can be reached more readily if:

- the terminal or instrument and system are simple to use,
- the terminal/instrument and system are unambiguous in their use,
- key strokes are kept to a minimum,
- nothing introducing any undue fatigue factor into the task, and
- nothing "steps on" the primary business transaction.

Many customer contact center CRM, help desk or telemarketing applications require the agent to follow complex scripts and enter data while conducting an intelligent conversation with the customer or prospect. It is important the terminal (ACD, PC or CRT) does not get in the way of the business at hand. Try it yourself — try to talk intelligently and type accurately at the same time. Next, add a dumb phone. Now try listening and responding persuasively!

Most leading ACD and computer vendors now provide ACD instrument functionality in a window on a PC. Many of the vendors discussing this approach to integration of telephony function have an objective that reflects their position in the marketplace as a telephony vendor. Either they want to maintain a "branded telephony presence," and thus implement a telephone model. (If a telephony vendor, they are telecentric in their focus.) Or, if a computer vendor, their position is datacentric and more applications oriented.

By promoting one or other of these two positions, the specific vendor hopes customers and prospects will buy more of their system offerings. Thus, they try to justify their offering on the narrow basis of telephone or computer hardware efficiencies.

The point here is that neither focus is correct, as it should be a combination of the improved user environment through easier operation, lower training requirements and overall productivity gains. The real leverage is the seamless integration of telephone and computer functions with the application. One keystroke invoking a macro command that effects changes in both the computer and phone state, but also follows a specific scenario required by the application or applications session automatically following the process, capturing call flow data and making the agent perform beyond their skill level.

Significant attention should also be paid in these applications to maintaining a voice and data audit trail collecting customer and agent behavior and performance much in the way "Internet cookies" work in a web environment. Often partial recordings are being made for compliance or quality monitoring purposes. There is so much market data and intelligence inherent in these transactions that is lost but for a little foresight that could change. We will talk more extensively about this in Chapter 15.

Unsurprisingly, the PC-based telephone instrument is not a particularly good replacement for a telephone instrument. Particularly with a device designed for high volume telephone call handling. The reasons are simple. The use of single telephone-based function keys is simpler and more direct than addressing these features via a combination of CONTROL + QWERTY keystrokes, a mouse or a keyboard-based function key. However, if this is how the comparison is made, the entire point of combining the voice and data functions of a call in one device is missed.

The integrated desktop PC call management scenario directly addresses workflow and application integration. As a result, it goes to the core of the customer contact center productivity problem by streamlining the process and resources necessary to serve a caller. The technology combination is designed to reduce the time, effort and skill level needed by the agent to "navigate" the technology, in the process of serving the customer. Focus is on the customer as the system "delivers" the required tools and information, or places it "just one key stroke away."

Application scripts can anticipate steps, deliver on-line tools and automatically gather call tracking and wrap up data. In spite of real advantages of a PC-based telephone instrument, many customer contact center ACD installations still use ACD telephone instruments.

☐ THE ACD AGENT INSTRUMENT

The early success of an ACD system was particularly dependent on the design of the individual ACD agent instrument. The number, speed and intensity of the transactions processed by the agent is a major element in the success of the application. The agent will spend many hours using the ACD phone. The ease of use designed into the agent phone has a significant bearing on your customer contact center success. Agent ACD telephones are not transferable between manufacturers. You get what you get with the ACD you select. The agent terminal should be a major source of investigation and comparison. Some models are limited, though some manufacturers have done a tremendous job in agent terminal design.

As ACD systems made their way into the market, agent terminals were nothing more than multi-button telephone sets similar to those used on electromechanical key systems such as the 1A2 (The system behind the classic big black phone with five lighted white line buttons and a red hold button).

It was not until the introduction of the first true digital ACD in 1973 that a manufacturer decided an ACD needed an application specific "industrial strength" terminal. Among other things, the use of a unique instrument increased the cost of the ACD on a per position basis. But, at the same time, it was clearly demonstrated the time to process a call was shortened, and customer satisfaction was heightened as a result of the ease of use of an instrument that allowed focus on the caller's needs. This efficiency was taken to new lengths with the introduction of integrated applications terminals combining telephony and data functions.

241

If a call takes less time to process, the time and cost of handling the call is reduced, therefore, the number of calls an agent can serve is increased. The application specific design of the ACD system and consequent increase in cost is more than offset in the favor of the user. As a result the market for ACD systems continues to grow.

The early applications of VoIP phones have been focused at distributed customer contact centers.

The Mechanical Packaging

In the high volume call processing world of customer contact centers and ACDs, poorly manufactured instruments wear out quickly. Weight and robustness of the instrument are important. Pushing the feature buttons on the phone so often causes the key contact mechanism to fail and the illumination source to burn out. The repeated insertion and removal of a head or handset plug into the jack can unseat and break the electrical connections if it's a poorly designed or manufactured instrument. If the connection is not cleanly severed, the deterioration of the connection will introduce noise, static or a loss of volume into the call transaction. The instrument will be used constantly, be moved around the desk intentionally and can be unintentionally pulled off the desk by an active agent. This occurs when an agent gets out of his or her chair to stretch, reach for a needed item or attract the attention of a supervisor. A properly weighted instrument is less likely to move and more likely to serve as an anchor for the retractable headset cord.

Discouraging movement in the agent work area, such as standing up and stretching while working the telephone, is less productive than allowing it. Many customer contact center managers believe that providing a longer cord than normally provided with the telephone reduces wear, allows greater motion and less restriction of movement. The underlying issue is one of allowing the agent a perception of some control of their work life. The more active the agents are the greater the wear and tear on the furniture and equipment. Given the big picture, however, hardware is less expensive than "live ware." This also applies to any VoIP agent instrument or attachment.

Analog or Digital Instruments

Today, typically the instruments provided with ACD and PBX/ACD systems are digital instruments. The audio signal arrives in a digital form and is decoded into an analog audio signal at the phone instrument so an agent can hear and talk to the caller. This means the instrument constantly needs power for the coder/decoder (codec) to operate and to generate all the command and control signals for simple call setup, breakdown and communicating to the system any expected system status messages. It also allows the instrument to have many features. It typically cannot operate if power is lost from the switch. To get around this limitation (namely a catastrophe in the case of a power outage), complete power back up can be provided to the switch and instruments. Or optionally, 2500 single line phones can be installed with duplicate wiring and termination devices, if the telephone company CO can provide analog circuit backup.

With an analog instrument (i.e. normal "black" phone), the signal arrives in an analog form and typically all call status signaling follows standard telephony voltage changes (as found in a simple 2500 series set). This allows the use of line power to accomplish all functions. This means in a bypass mode the instrument can draw line power directly from the CO, thus it can continue to answer calls as if a single line subscriber set. No secondary backup instruments or wiring are required as these instruments operate in bypass on CO provided line power, i.e. the telephone customer contact center can be in darkness, but the phones can continue to operate.

There was an alternative "hybrid" instrument, which used digital status signaling and analog audio. These allowed a feature rich agent instrument and the ability to bypass directly to the instrument in a power outage. These instruments provided the best of both the analog and the digital world, with default to bypass without a second instrument, yet providing fully featured operation. These instruments are seldom encountered, but they do provide a lesson in combining rich features and non-stop operation that we cannot forget.

There was a brief period when a number of instruments required power from an external power source. Here a separate transformer (like a desk calculator power module) plugged into the wall to provide local power to the instrument. Although a viable strategy to increase the functional power of the instrument, it also added to the complexity of the installation in increased power requirements, planning around failure, and expense. Occasionally this strategy resurfaces when the instrument power requirements exceed what can be supplied on standard telephone cabling. Occasionally these systems show up in the secondary market.

The "other shoe" here is that to stabilize and back-up the power to these commercial outlets, which supply power to instruments that have external power needs, you have to "bullet-proof" the wall plugs. If the switch is backed up and the instruments are not, who answers the calls? This can get expensive.

243

Cabling

Ideally cabling should be standard three pair telephone wiring found in a typical telephone installation. Using other than standard "skinny" wire introduces inflexibility and increased expense in planning and installation.

A new installation does not create a major problem, as new cable must be run. Where the system is replacing an existing system, the old cable probably cannot be reused, and when this system is replaced, new cable will need to be run. Cable installation is a relatively expensive, labor intensive and inconvenient exercise. The frequency of cable installation events should be kept to a minimum by intelligently over-cabling (i.e. putting more cable in and more phone jacks than you ever dream you will need) at every opportunity.

If a vendor promises to reuse existing cable in a new installation, have them test the cable, check all the runs to be reused and warrant, in writing, that the replacement system will operate as specified using the existing cable. REPEAT-GET IT IN WRITING. When nearing a critical deadline, it is most embarrassing to discover the installed cable will not support the new system. It is expensive and inconvenient to replace this while the center is in any state of operation. Repeat: it is better, safer and you will enjoy a more trouble free life if you run new cabling. Insist your contractor label and document every cable run, origin and termination point.

Ergonomics

As overused as the term "ergonomics" has become, it is still appropriately expressive in explaining the discipline of designing manufactured things to reflect the fact people are intended to use them easily. The relationship between a driver and an automobile includes using techniques to increase safety and reduce confusion and fatigue. Although the issue of safety is less dramatic in telephone instrument design or soft phone function, fatigue and confusion mitigation are not.

The instrument used in a high call volume application should not get in the way of the task at hand — responding to a caller's needs. The instrument should be all but transparent to the conduct of the transaction. If the instrument intrudes into the process, it should be minimal in impact.

Key Placement

It is important the keys on an instrument are logically laid out. Processing a telephone call can follow one of three scenarios. The designation of these events parallels the frequency of encountering the call type. The suggested frequency of encountering these call categories could be as follows;

Call Type	Frequency by percent of calls per day
Simple call	90% of the day,
Complex call	9%, and
Exception call	hopefully less than 1%

These suggested percentages have been arrived at unscientifically to suggest the frequency agent needs certain call processing functions and features. This relative percentage comment about transactions does not track to time involved. We will talk more about the time and cost of exceptions later in the book.

The features required by an agent fall into three categories: primary, secondary or exception functions. Again, classed by frequency of use. The key layout and use should follow this form.

A dilemma faced by a manufacturer as ACD instrument features evolve is how to add physical keys to an instrument. Because software is more flexible than hardware it is possible to "double up" functions of the various feature keys via a "shift" mode. This technique is typically encountered in PC software applications. As this occurs, confusion increases. So do training requirements. Enter the "soft key."

The soft key technique trend crossed out of the PC business into telephone instruments and ACD instruments. To dedicate a key on an instrument is an unambiguous strategy, but it also causes the telephone instrument key field to proliferate with every new feature. This is undesirable from a manufacturing and inventory point of view. The redesign of hardware takes time and costs a fortune in reengineering and inventory management.

Later entrants in the ACD market learned from the PC software market and made extensive use of dynamically assignable keys. These keys are located physically adjacent to a dynamic liquid crystal display (typically below the dynamic display line) that allows the current functions to be displayed and changed based on a predictable call processing sequence. A single key depression activates the necessary system steps to cause a feature to activate. This mirrors software use of application "function" keys that embed a series of keystrokes into a macro command. One key executes a complex function.

The placement of primary, secondary and exception function keys needs to be logical and pay heed to the amount of movement required by the user's hand from the natural rest position. The natural rest position on a telephone instrument designed for high volume use is center bottom. This position is typically described as the heel of the palm centered and at rest position on the work surface or on the frame of the device. On most telephone instruments this covers the 12-character dial pad.

In an inbound customer contact center this is one of the least used key fields, therefore an exception function. It also strongly suggests the proposed instrument was not specifically designed for a customer contact center and began life as something else. The design and tooling for a telephone instrument is among the most expensive and inflexible aspects of building any switch.

From the "home point" the movement to the primary function keys should be minimal. The movement to the secondary keys a little further and so on. There is an arbitrary physiological barrier or "fence" established at each extra stage of movement from the rest position. This causes a slight increase in effort and fatigue, yet reduces confusion and training, it's a functionally exceptional motion. Few ACD vendors have spent the time to research and implement the results with the goal of reducing confusion and fatigue. The underlying goal should be to allow the instrument use to become "autokinetic" or second nature. A better, yet superficially contradictory, goal is to make the instrument easy to use through unambiguous keys, labeling and layout. The more you customize the instrument to the application, the less generalized it becomes. With that goes flexibility — a dilemma for the manufacturers.

Again, with the introduction of PC-based integrated workstations, many of the training or confusion based objections are minimized as more and more calls are "scripted," based on the complexity and frequency of occurrence. But, the issues of obvious and autokenetic keyboard use are not made less important.

Key Labeling

In key labeling, acronyms are undesirable. Do you know what a key labeled "HD" does? Or worse yet, "Hld?" High technology has an image of arrogance, partly brought about by the insensitivity to ultimate end users of the products and services spawned by the technology. It is assumed that users have a responsibility for their own destiny. And in many ways they do — users take the course of least resistance...if it is easy to use, it will be...if its hard to use it won't be!

The majority of ACD vendors appear to almost ignore this reality in system interface design, particularly at the agent instrument level. Function key labeling should use complete, plain language labels. If a "shift" function is used so the key has two or more roles, use of different colors for each function label are desirable and helpful in reduction of agent confusion and to accelerate training.

There are three or four ways to imprint the label on the key. The issue here is making the label robust enough to withstand many hundreds of thousand activations and still be readable. In a word, wearability. The processes are in order of preference and expense: molding (the character/s into the plastic key cap), gas etching, printing or painting the label and finally use of a label decal. The higher the quality of the instrument the better and more expensive the process. The better the process, the crisper the key label is. They remain useable after five years.

Key Size

Key size is critical for three reasons. Agents move fast and are occupied with the primary process at hand — serving the customer. First, precision of motion should be a minimal requirement. They are usually interacting with a computer terminal and displayed customer and campaign data, so the ACD instrument is the least important device at hand. It also should be the least intrusive.

Second, female and male agents have different hand sizes and many women have exceptional fingernails. Large male fingers and precisely manicured nails do not take kindly to precise keystrokes on tiny keys with low-grade feedback.

Third, effective tactile range and "blind" autokinetic response does not work well with small precise keys. For this reason "membrane keys" (smooth surface keys like those found on inexpensive paper thin electronic calculators) are less popular on ACD instruments. We do not recommend the use of flat membrane keys despite their moisture proof advantages when a beverage is spilled. With the keys, The only major cost advantage is with the manufacturer NOT the user.

Tactile Feedback

There is a great deal of need on the part of a user for some confirmation that the action just taken, worked. That's why PC keyboards "click" even when they don't need to. It's why touch-tones feed back to a caller dialing a call and any one of a number of other minor events provide confirmation to the user that an action has been acknowledged. Without feedback, there is confusion and with confusion, a reduction in productivity. Take the example of dropping the ring in attempting to short cut caller delay; the caller and agent take the otherwise saved six seconds, to orient themselves due to the lack of predictable phone behavior.

Display Screens

There has been proliferation of the information displayed on an ACD and administrative instrument. Despite the advantage of information being available over no information, over the long term placing this data on a telephone instrument is the wrong solution. The first reality is users do not look at the telephone display screen if they are not obliged to. Doing so interrupts the current transaction. This transaction usually involves a computer terminal with a screen and keyboard. If there is a need to physically look at the instrument display, the agent must scan to the instrument display, refocus their eyes, read and comprehend the information, look back to the primary object (computer display), refocus and continue with the transaction. This is doubly complicated when you add softkeys that have dynamic functions as the whole concept of ergonomics and developing autokenetic behavior is diluted further.

When the primary object is a CRT screen, the light intensity and distance is different from the low power LCD of a typical ACD instrument. Eye fatigue increases.

The solution is to closely couple the application (CRT and ACD functions) so all the visual feedback/displays occur on the same screen. This has been elegantly implemented on PC-based integrated desktop devices with significant productivity improvements. Fatigue, particularly eye, is dramatically reduced.

Call Type Identification

Early ACDs supplied city of origin announcements or "whisper queues," a flash announcement heard only by the receiving agent or a visual identification display. This was the first call type identification and was a system-based feature, rather than individual instrument-based, even though it appears as an agent oriented instrument level implementation. This had important implications as a "repersonalization" tool in the customer contact center.

The goal is to equip the agent with initial data so as to provide the caller with specific information and thus avoid the inevitable and time consuming caller identification questions of where and why are you calling. Automatic number identification and screen population are a significant advance over this original ACD feature.

Now there's the growing popularity of dialed number identification (DNIS) and automatic (caller) number identification (ANI) services, voice response for account number collection and switch-to-host and automatic screen population ("screen pops") links. These features negate the need for the agent to ask who is calling, and reduces the necessity to ask why the call was placed. Also, since the call response is now targeted to an individual caller's particular needs, the calls are shorter. Nonetheless, the whisper queue had its place in the scheme of things, since it allowed the agent to receive some call personalization data without needing to look to the instrument for identification data.

Now with the integrated PC workstation, the former whisper queue data can be sent directly sent to the PC and the necessary work script invoked automatically, thus reducing the required analog to workflow "agent translation."

Head and Handset Jack Type and Location

The jack port on the instrument may range from a fragile modular plus (like an RJ-11, which is similar to jacks in your home) to an industrial strength dipole jack. The tradeoff is ruggedness and size against lower cost and fragility. It is noted, however, that manufacturers have increasingly adopted the less robust, smaller and cheaper RJ-11. The repeated insertion and removal of a head or handset plug into the jack can unseat and break the electrical connections in a fitting not built for this purpose. Although the RJ-11 modular jack is plentiful and inexpensive, it isn't designed for repeated connects and disconnects. When it is stressed or breaks, and if the connection is not cleanly severed, the deterioration of the connection will introduce noise, static or a loss of volume. The heavier the fitting the better, yet recognize the cost and size increases as you adopt increasingly robust fittings.

The location of the jack may be on the front, side or back of the instrument. In a few instances it may even be dismounted from the instrument and mounted in the knee well of the workstation. Some ACD specific instruments are equipped with two jack ports, one on either side. This is for left- or right-handed operation and/or training purposes. A trainer may work with the trainee at the workstation and plug into the instrument simultaneously. The most desirable location tends to be toward the front of the device.

A cradle or handset rest is desirable for supervisor positions. Headsets do not need this as they generally remain with the wearer during the shift.

A word on headsets. I am a 54 year old male who works with a telephone and computer much of my day. I could not work without a headset. I don't feel like a "telephone operator," it is a trivial "inconvenience" and I guess it messes up what is left of my hair. So what...My comfort and productivity are more important. It has worked well for me for the last 15 years. I will avoid exclusively using a handset again.

Instrument Cable Termination
The cable termination is also a weak point on the instrument; particularly if the instrument is going to be moved about the desk. Constant flexing stresses the connection. The big issue here is shoddy assembly as opposed to poorly designed connections. Make sure your maintenance contract covers the connection on the cable to the desk.

249

ACD Instrument Features
This portion of the discussion is organized to reflect the most used or primary agent features. This reflects also how frequently they are encountered in popular ACD machines.

Further ACD systems are becoming less feature rich as many features migrate to applications. Features that fall into the secondary and exception categories tend to be less and less available because these were originally developed as reactions to the inability of data systems to gather informal data easily.

During a traditional ACD acquisition phase some of these issues may seem relatively unimportant. They appear as features you can do without. A fact of life associated with the assimilation curve of any system, particularly an ACD system, is this:

As you become more familiar with machine and less intimidated
by the complexity of operation, the "little" features that seemed
unimportant during the acquisition process suddenly become
very meaningful as their business impact becomes appreciated.

Even though it is easy to befriend a particular vendor during the buying cycle, listen to all of the ACD vendors. Give them all a chance during the early stages of the ACD purchase process. The worst that can happen: time is spent to learn some clever new customer contact center management tricks. The reality is that many of these features were developed to satisfy a real user need. Take the time to understand the "whys" of the features.

With the advent of PC screen-based telephone functions, the feature sets have the potential to be enriched again. Here occurs an interesting phenomenon. Many of the features that have shown up on ACD systems, particularly the standalone systems, have been reactions to unresponsive data processing departments and systems. For example, wrap up.

Wrap up is an ability for the ACD system to allow the impromptu collection of ad hoc numeric data or codes. This data can be sorted after the fact to identify calls by type, origin, disposition, revenue, etc. The fact this data was captured on a telephone system is relevant only because it was part of a call record or transaction type hash count. A better repository of such data would have been the customer information or order entry system. In the past, however, this was not done easily. So the ACD vendors provided a fine interim solution. Now computer telephony both solves and complicates the whole picture again.

Primary ACD Instrument Features on a Traditional Instrument

The features listed here are arranged in alphabetical order rather than order of importance:

- *Add-on or Conference (in a third party)*: allows an agent to add another party, typically a supervisor, into the call for further help.
- *Asterisk and the Number (Octothorpe) keys (or the "Star" and "Pound" key)*: I recently heard these called the "snowflake" and "tic-tac-toe" keys! These take on a special role in an ACD. They are used to signify key fields (such as a pause) when any data is entered at the touch-tone pad. This may be wrap-up, call disposition or peg count data.

Note: There is a trend in the beginning of the generational shift, with the move from traditional analog to digital (TDM) switches and now to VoIP, that affects agent functions. When a new core technology surfaces to replace a mature core technology, typically the first thing addressed is core plumbing — the analog switch matrix to digital switch network and now digital switches to transporting VoIP. But, the engineers tend to address connectivity first and leave the "lower level" usability features to last. Thus, "Release 1.0" appear to "go backwards" in user friendliness. This is exactly what is happening with VoIP. Early iterations do not meet the feature maturity of older technologies. This will change.

- *Available*: an individually identified key or, on a lesser featured instrument, a "tap" or "flash" plus key sequence. Use of the "available" key by the agent identifies to the ACD system they are available to take the next call.

  There are three ways to implement the feature on an ACD instrument;

1. Manual identification to the ACD that the agent is available for the next call.

2. Automatic identification of the agent available status immediately following the release of the last call. This is called "forced available". Release is "hang-up". The system is smart enough to understand the use of the release key to terminate the call and return the agent to the available state.

3. Managed manual identification, which is used when there is recognition by the vendor that the agent may need time between calls for "after call work" or wrap-up. Using the release key, ends the call and places the agent in the wrap-up condition. Upon completing the after call work the agent activates the key to return to the available state. During this period the system has tracked the three agent states; the incoming call time, the wrap-up period and the return to available.

There is now another method made possible through the use of an integrated desktop workstation.

"Scripted work state management," which utilizes a Windows script. The individual call/transaction can be scripted to follow a certain workflow, i.e. follow predetermined steps, states and responses. At the end of the transaction, upon release of the caller, the agent is automatically placed in the wrap up state. However, given that a multi-tasking desktop operating system is in use, much of the need for post call work is eliminated as the system may be gathering wrap up data or performing wrap up tasks automatically.

Behind the simple activation of the available key by the agent, is a philosophy that needs to be understood from a management point of view. It would be simple to allow the system to totally manage the agents and control their change from work state to work state in the normal and predictable course of calls. This however removes the perception on the part of each agent that they have some control over their destiny. Remove it and "burn out" is accelerated. Manual control of the agent's state by the agent, coupled with real-time management control to avert abuse appears to be a humane balance.

- *Hold*: this works like any "hold" key to place the call in suspended state of silence, while some other activity occurs concurrently with the call being held. Redepressing the hold key re-establishes the connection so the conversation can continue.

- *Pound Sign or Octothorpe key*: see "asterisk key" mentioned previously.
- *Primary Extension or Call A*: this is the primary line appearance, extension or directory (DN) number of the instrument. This may be a hardware address (station number or directory number) in a PBX/ACD or a logical resource location awaiting use by an agent signing on with their agent identification number. Once an agent signs on to this instrument/extension location, this logical ID number supersedes the hardware address once entered into the instrument as the agent signs on. All agent properties and class of service conditions are attached to this logical ID or address. Calls are now routed to the logical ID currently resident at this hardware address or extension number. Once the agent signs off the position returns to an inactive hardware address until reactivated by the next agent signing on.
- *Secondary Extension or Call B*: this is the second line appearance at the instrument. This may be a second ACD extension number, administrative PBX extension number in a PBX/ACD system, or a DID circuit appearance.

  It's noted that most PBX-based ACDs require two different line appearances, one for the applicable agent group and one so anyone on the same PBX/ACD system may call this agent position and so the agent can make outgoing calls. These are physical port (extension) addresses. Most standalone ACD systems do not assign lines to physical addresses, but instead they separate call activity by combining transaction detail data by call type, assigning the service necessary and collating the resulting data accordingly.

  Be forewarned that this PBX-based feature means that outgoing calls or any other activity on the instrument extension are typically collected as separate events and not consolidated into the agent's ACD reports!
- *Ready*: this key is used to signify to the ACD that the agent is available to take the next call. It's used in the manual "available" mode described above.
- *Release*: this is the equivalent of the switch-hook or "flash" key. Depression of this key during a call disconnects the call.
- *Status Advisory Tones*: different manufacturers use different audible tones and signals on the line to advise the agent of certain conditions. These may be zip tones preceding an inbound call or reorder when an incorrect or illegal dialing sequence is used.
- *Tone Dial Pad*: self evident as the dialing mechanism for internal and exter-

nal call setup. The pad may also be used for data entry in conjunction with wrap up data entry.

- *Transfer*: depressing this key after an alternate station number is dialed, (e.g. a supervisor station), allows the transferring of the current call to the target station.

Secondary ACD Features on a Traditional Instrument

- *Enter*: use of this key alerts the system that the field of digits just entered via the touch-tone pad is complete. Either a dial stream, an ID number or data. This key can be used independently or in conjunction with the * or # keys to indicate pauses, backspace/erase or field delimiters. It also introduces a human factor element long missing in telephony: the ability to "edit" a dial stream before "execution" by the switch.

- *Flash or Tap*: the "Flash" or "Tap" key is used as an alternative to briefly depressing the switch hook to alert the ACD or PBX system to the fact the following key sequence is an operating instruction. The tap or flash key was introduced to resolve the dilemma created by using the switch hook to terminate a call and to alert the system that the next dialed numbers were a command code. The length of a momentary depression meant different things to different users and many calls were unintentionally terminated, rather than transferred, etc. Manufacturers solved the dilemma through use of the flash key. This key sent a fixed non-fatal switch hook flash message to the system that now placed the caller on hold and waited for the dialed command code.

- *Mute Key*: this provides an ability to lock out the "talk" part of the instrument voice circuit so that the agent can hear the caller, but not the reverse. Typically this is a feature of a headset, though this has migrated onto the more popular agent instruments.

- *Prerecorded Automatic Announcements*: this is a feature that places personal announcement capabilities with the agent. It may be inherent in the system or part of the instrument. It gives the agent the ability to pre-record their personal greeting: "Thank you for calling XYZ Company, This is Andrew. How may I help you?" The argument in favor of this feature is fatigue reduc-

Note: In the data entry mode is the tone generator deactivated? If it is not, the tones associated with the use of these keys will interrupt the conversation. The data entry mode therefore cannot be used until after release is activated and the caller disconnected. More modern ACD systems allow concurrent conversation and data entry as they automatically switch on and off the tone generator based on call/instrument status.

tion and consistency of greeting delivery throughout the shift. This is now being dramatically expanded through the use of sound card options in the integrated desktop PC or outboard instrument device.

- ***Repeat Call Type or City of Origin (whisper queue)***: because customer contact centers have the ability to serve different callers with the same staff, and do so with greater precision if the agent is alerted to the call type ahead of time, "whisper queue" announcements were introduced. This is a short audible announcement of the call type or city of origin. These are brief, often cryptic messages to alert the agent to a call type and subsequent handling and script considerations. They are often used in conjunction with or replaced by display messages. They have a strong advantage in the fact the agent does not have to look at the instrument to get the message. Again, this feature is progressively being rendered redundant with computer telephony links and scripted workflow at the integrated desktop workstation.

- ***Sign On, Sign Off***: this is an important function. No ACD system should be without this. This allows consolidation of management statistics by the identified individual and not the hardware position. Before the instrument can become active, the agents must identify themselves. In doing so class of service properties are extended to this position. Individual statistics are then collected and may follow the agent no matter which instruments, groups or splits they sign on (and off) and work during the course of an accounting period. This is a boon to management as there is no more need to separately track which agent was where for what shift, hour, etc.

- ***Unavailable***: activation of this key alerts the system to the fact this agent is temporarily unavailable for calls. The condition should be tracked and reported by the system. Returning to the available state is done through use of the available key.

- ***Work***: this is a key that formalizes the wrap up or work state. Activation of wrap up may be associated with use of release on some systems or upon release require the agent to physically enter this status by using the "work" key. It may also be used to track the agents' position when they are reassigned into an extended non-call related state, yet still being accounted for by the ACD.

Exception Features on the Traditional Instrument

These can be an individual key or keys that are dynamically assigned by the particular state the agent is in. These are called "softkeys" and while they are assigned that function, depressing this key activates the feature expected. This, however, means the agent needs to be able to follow the LCD screen and know which feature to activate while concurrently working at a data screen. This is an elegant solution to reducing instrument size while adding features, but it adds eye fatigue to the agents as they are constantly reading two devices with different display technologies.

- *Acknowledge*: when the ACD system has supervisor to agent messaging capability, some systems allow a single keystroke acknowledgment of message receipt by the agent.

- *Bad Line or Trouble*: as some line failures are not clear hardware failures (a "hard" failure), but rather a subjective or "soft" problem (e.g. line noise or low audio volume), agents must detect and identify the calls (thus lines) affected. Depressing the Bad Line or Trouble key does this. Administrative alarms and reports identify the components handling the call transmission that could be causing the problem.

- *Help*: use of this key is to provide the agent some instruction on the operation of the instrument. This is analogous to the F1 key in DOS or the Help button in Windows, but without the information depth possible on a PC.

- *Emergency*: use of this key typically activates a tape recorder, alerts a supervisor and initiates a time stamped hard copy error message. The airline industry introduced this to record threatening calls. Use expanded to other applications such as agent training and sale verification.

- *Manual Operation*: in those systems allowing bypass operation to the usual ACD instrument, this key acts like a switch hook. By depressing the by-pass key, upon notification of the arrival of an inbound call (ringing and/or flashing light), the agent is connected to the caller. Upon redepressing the key, the call is disconnected.

- *Mute*: depressing this key cuts off the "talk" circuit on the instrument so an "in person" agent to supervisor conversation can be heard without the caller hearing, yet allowing the agent to monitor the "listen" circuit for any caller comments.

- *Park*: like "hold", this allows the agent to put more than one caller in a state of suspense. This is done by activating the park feature and sending the caller into a particular park address, identified by a number. More than one call can therefore be simultaneously placed in this state by one agent position. To recover each caller, the process is reversed, again by using the "unpark" code and recalling each parked call by the assigned "parking address".

255

- **Ring Volume Control**: this may be a rotary knob or slide control that allows the volume of the ringer to be increased or decreased. It also may allow complete silence.
- **Supervisor**: depressing this key alerts the supervisor, typically via a console message only, that the agent needs attention. The supervisor may join the call as a third party, monitor the audio only or physically come to the agent position.
- **System Speed Number**: this allows the use of abbreviated or speed dialing by the agent. Frequently used numbers are entered as one, two or three digits. Again, with the advent of the PC-based integrated desktop workstation, alpha speed number lists can be presented in one of two ways: straight alphabetic speed number lists or context sensitive speed numbers lists based on "account type," "call or case state" and "help." The system only provides services and numbers germane to the account, the state of the account or the type of services needed based on a combination of the two.
- **Test**: depression of this key allows a local diagnostic sequence to test the agent instrument functions, keys and lights for problems.
- **Time**: depressing this key displays the current time.
- **Volume Control**: this may be manual or an automatic gain control circuit that can be used to keep the transmission volume at a comfortable level. This may be a rotary knob or slide control that allows the volume to be increased or decreased. Some agent instruments use a code then repeated pressing of the same key to incrementally raise or lower the call volume on the listen and/or talk circuits.

Data Gathering and Reporting

The leading ACD vendors quickly learned how to provide reams of quality quantitative data and the other channels, such as email and eCRM systems have vendors who are beginning to integrate similar tools. To make this data ultimately meaningful it needs to be used specifically to improve the performance of each individual customer contact, agent, group and campaign in a customer contact center, then manage the physical resources to serve this business. (The next chapter deals specifically with the qualitative side of this same data.)

From the beginning one of the major ACD values has been its information collecting, processing, reporting and presentation. With information from the present and the past, business progress can be tracked and understood — when, and how your business arrives as much as from whom; then, how well your business captures these opportunities by the way you serve your callers. Customer focused email requests are no different.

From a staffing and skill perspective, a business can tell its most productive customer service and sales agents, groups and supervisors. From this data predictions can be made about the future customer call and contact center behavior (short, medium and long term). When that's coupled with a good forecasting, staffing and scheduling program, you can plan and organize your trunks, staff and ACD facilities to optimally meet the business demand that callers and business campaigns place on your customer contact center.

The success of early standalone ACD systems can be almost solely attributed to the power of their ability to manage the gathering of information and then process, report and present the data in a meaningful manner. Standalone manufacturers, like Aspect and Rockwell, clearly owe their survival in a brutal cost sensitive marketplace to their understanding of not only the ACD application, but also the value of providing high quality MIS and ACD features. These manufacturers continue to successfully sell application specific ACD switches against generic digital systems (Avaya, Fujitsu, NEC, Nortel and Siemens) that cost much less. Mainly because buyers realize the cost of doing business with focused technology (managing the application and customer opportunities, staff and network resources), far outweigh the costs of technology versus more modern, but unspecific, hardware. Through their understanding of the uniqueness of the ACD application, the manufacturers and vendors can protect their customer base and profit margins from undue erosions in spite of fierce price and feature wars.

The first true ACD vendor managed to eclipse the Bell electro-mechanical technology of the mid-seventies, with not only digital technology but also their ACD MIS. Today, there are extensive advances being made both in the development of onboard ACD MIS systems and in the availability of external data gathering and support packages for popular PBX/ACD and CO-ACD (CEN-TREX ACD) systems. The continuing explosion in desktop computing and PC software is driving these advances. Color graphics and display techniques benefit from PC and computer game technologies, and these techniques have migrated into real-time customer contact center management tools.

These same PC advances have brought complementary advances in the forecasting, staffing and scheduling systems business. No customer contact center manager with more than 15 agents can afford to be without such a system, especially since many are priced less than $15,000. Implementing a good service level and staff scheduling package is worth a 15% savings in productivity through more appropriate reallocation of staff to meet caller demand.

New customer contact center management applications are being discovered almost daily. These range from in-depth media sourcing analysis, staff adherence, customer call and status tracking to common cross platform database administration, diagnostics and reporting. Essentially the customer contact center equivalent of network management tools like Netview and CA UniCenter.

The workflow and reporting lessons of the ACD were initially ignored by email system developers, as email was not predicted to become an ubiquitous and high volume messaging medium. The integration of business email response functions by a group dedicated to responding to these requests was almost completely ignored in favor of the consumer front end. Development for the email author was favored versus the business recipient until about five years ago when the email server was introduced. Yet, the majority of these were developed without understanding the disciplines and technologies that exist in a stationary customer contact center, albeit focused at telephone calls. That has changed. Some of the customer contact center models that you can now find in corporate "customer email service" servers are queuing, triage, skills based routing, and email specific features, such as request templating, automated response and FAQ generation.

☐ MANAGEMENT GOALS

There are four facets of management reporting on the service callers receive in a customer contact center:
- Data gathering.
- Calculations.
- Report generation
- Presentation techniques.

Management style and use of this data is most critical to the orderly management and long term operation of the customer contact center. Optimal data gathering and reporting is also vital to reaching the four basic management goals of any customer contact center:
- Effectively achieve the business objective(s) of the customer contact center.
- Provide a satisfactory response to the customer.
- Develop the long term business vitality of the customer contact center.
- Do all of the above, at as low a cost as possible.

The modern customer contact center needs to also include electronic channels reporting, such as email, text and voice chat, and web orders since only with multi-channel quantitative data gathering and reporting can management obtain the total customer contact workload. Today, this data is available in a disintegrated form, but we do expect this to quickly change with companies like Siebel Systems and Blue Martini, leading the way. For example, Blue Martini has combined its call center technology with Apropos management products and Siebel Systems offers call center solutions that include call management, sales and service automation, workflow and business process optimization.

Multi-channel Challenges

With email and web chat, the data and objectives are remarkably similar to that of a traditional call center. There is, however, one major difference; and there lies the crux of an unwritten customer service organization agreement: emails will be answered like traditional correspondence, i.e. not answered with the electronic immediacy with which they are delivered. Email users are currently accepting the notion of the batch nature of email — it must be composed, addressed, sent, and carried across distance, received, sorted, opened, read and understood and an answer composed. This notion of and acceptance that email is a hybrid of the traditional batch "snail mail" model WILL CHANGE. Users (your customers) will begin to recognize that email is more immediate than currently assumed and that their email arrives seconds after dispatched and can be responded to almost immediately, provided that the resources exist to respond. Web chat has already broken through this "batch barrier."

Fortunately, for the customer contact center manager, email hasn't joined the real-time communications free-for-all. Mainly because most email users:

- Assume that their communication takes time to arrive, so a 24-hour (or thereabouts) turnaround is acceptable.

- Realize that composing a text message is meant to take time and be more thoughtful.

Customer contact centers have not widely adopted call center real-time workflow models to serve email, but they are living on borrowed time. Many are reluctant to adopt this model, probably because they have learned that the better the service provided, the greater the service demanded. While they might be able to get away with this for a while, it won't last forever; the customer will eventually demand a near real-time response to their email communications.

Many customers and center managers find text chat is an acceptable alternative channel with some of the same informality of a telephone call, although it normally lacks the control of a scripted telephone call. Telephone calls are a person in queue, while an email is a question in queue that can sustain a delay. Text chat is somewhere in between. From a management point of view it's more like a problem than a customer solution because it is spontaneous and unpredictable. It is also hard to tell how many web chats will turn into money unless the customer contact center is able to "push" an order form to the visitor.

Measurement Criteria

To achieve the goal of measuring customer demand and customer contact center response, there are eight areas that must be measured — some objective, others clearly subjective. Mature customer call center management understands the importance of such data.

Objectively, the quantitative or functional elements are:

1. The revenue or result of the calls, and deviation from these targets.
2. Individual telephone agent performance.
3. The service level provided incoming callers.
4. Network performance.
5. System component performance (IVR/VRU, ACD, queue, route and host response).

On the subjective level, the customer contact center must be able to measure:

6. The quality of the experience (results) or the goodwill impact upon the caller.
7. The accuracy of the information exchanged or delivered to the caller.
8. The skill, attitude and morale of customer contact center employees.

These items are measured two ways: by gathering objective device or agent (unit) performance data and consistent observation of and research into caller and agent actions and behavior — quantitative and qualitative data.

□ACD SYSTEMS

True ACD systems, particularly the standalone systems have done an excellent job of objective data gathering, storage and presentation. The PBX-based systems have always poorly emulated the best-of-breed standalone ACDs. There are a number of data gathering quirks in the PBX-based systems that severely compromise the accuracy and presentation "spin" of agent data. For example: In the PBX-based system a call is counted from the time of answer, not from the time of appearance on the line. This means dynamic supervision and delaying

Qualitative Data

In the first edition of this book, *The Incoming Call Center*, we maintained that qualitative data could be gathered through market research and monitoring. The next edition, *Customers, Arriving with a History, Leaving with an Experience*, documented how one ACD vendor had begun to automate the collection of call recordings for the monitoring process. This was the scheduling and gathering of agent call samples using voice processing and recording technology so as to eliminate the need for a supervisor or auditor to be available to gather the samples. However, the subjective aspects of this function still lie with a live monitor.

The application of this technology meant a massive increase in convenience for management and a significant reduction in the chore of monitoring. The other major benefit was that the machine would keep trying to fulfill the required monitored call sample until the planned call count was met. We will discuss this more in the next chapter.

the call in this state is not only not measurable, but not possible. Because PBX-based ACD systems still insist on two line or extension appearances for an agent — one for ACD traffic and one for the extension (DID, outbound or intercom traffic) — thus, separate reports are kept for one agent and two classes of traffic. Though much is made of the ability to present combined reports, complexity is added to the process.

Check how your vendor does the following:

- When does a call start statistically?
- Are all customer contact center agent work calls collected in one "bucket" based on the role of the agent?
- Are email (web chat, assisted browsing) events collected separately?
- Can these events be integrated into a single report based on total time spent on customer transactions: calls, text, batch (email) and near real-time (web chat, assisted browsing and web call-back)?

Data Gathering

The ease and convenience of gathering and reviewing the customer contact center performance data allows time to be spent analyzing behavior data rather than collecting it to determine what happened. The business insight that contact center management can gain from this information allows strategic and tactical changes to be made to maximize business opportunities. The secondary sales benefits of management convenience and insight into strategic business opportunities are what ultimately justify the decision to invest in a state-of-the-art ACD system.

In 1979, Norris Tapp, a brilliant Datapoint ACD systems engineer first used the phrase "if you can't measure it, you can't manage it," in reference to ACD data measurement.

Today, it is universally acknowledged that good data is necessary to produce meaningful reports. However, the first question is what basic data do you need? Call records generated by a PBX, key system or CENTREX are a start, but they are generally inadequate. Why? The underlying philosophy behind call record collection in an administrative switch (PBX, CENTREX or Key system) is to provide accountability for call usage and provide usage data for network management. Here, call records are collected by station and trunk. The PBX business calls this process "station message detail recording." SMDR data is kept for every completed outgoing call. Sometimes counts are kept of inbound and internal calls. All this was designed to help with corporate accounting, distribute costs to the responsible department and improve telecommunications management. This data is fed through a process called "call detail reporting" to price the calls and produce reports. CDR software is typically found on an external processor. This does a fine job for the accountants and telecommunications engineers, but almost nothing for customer contact center management, particularly those with inbound applications.

A PBX-based system is generally less concerned about data since the main purpose for which it was designed was that of a switching facility. A true ACD system never seems to gather enough data, because the system purpose goes beyond switching to serving a primary business goal.

The majority of ACD systems, even those with origins as a PBX, now provide agent sign-on/sign-off features that allow the flexibility to assign and move staff to any position during the day without compromising the accuracy of the data gathering. This ensures an agent's statistics "follow" the agent. The alternative is to manually maintain a "who's on first" list and then sort the station data against this in an attempt to obtain personal report data from hardware statistics. The last generation of customer contact centers had many junior analysts, first to manually transpose this data on to accounting spreadsheets, then to enter them into a PC spreadsheet program. This is no longer necessary as most ACDs now gather data by individual agent identity and provide the availability of an onboard ACD (or closely integrated third party add-on) data processing capability.

As previously mentioned, many PBX-based ACDs have a separate extension for outgoing, internal or DID calls and that data is gathered separately. Thus, no consolidated reporting is produced to show the total session for the agent at that station. Add logical identification and agents moving from station to station or multiple shifts with a different agent manning the same physical position during different hours, and data normalization and accurate reporting are a chore. It adds real effort and confusion to subsequent daily center management. It is important all call transaction data be integrated into one uniform set of data in one database. Managing multiple formats on multiple databases creates needless management overhead, not to mention muddles the objectivity of critical people management data. This feature has improved as the PBX vendors began to understand the ACD system features they were losing orders to.

Peg Counts

Early systems, including some standalone ACDs, collected peg count data only and not the detail of every transaction. Any report is only as good as the data displayed. Garbage in, garbage out. Incomplete data in, incomplete and misleading reports out. Peg counts are essentially stroke counts of a particular type of event whether it's an inbound, outbound and internal call.

The time it took to complete these transactions is divided by the raw counts to create averages. From these counts and averages, reports are then produced. This is the type of data the early electro-mechanical ACDs collected. The Bell System developed FADS (Force Administrative Data System) and BOFADS (Bell Operator FADS) that used this data gathering technique. This is the type of report that gave rise to all the early controversy of customer contact center sweatshops. Average and inaccurate reports' being applied subjectively is no way to run a business.

Trace Agent

There is a clear preference among sophisticated customer contact center management for more data rather than less. It is almost impossible to predict the next piece of marketing and customer contact center performance data upper management will ask for; therefore, complete call detail captured and stored constantly is the ideal. To this day only standalone ACD vendors provide data with the flexibility required to manage a real-time business. One vendor calls this "trace agent," after the ability to reconstruct an agent's workday after the fact. In many cases, days after the fact. It may be a trunk, an agent, a call type, ANI number or any logically identifiable record type.

A successful PBX vendor has a feature they call "agent trace" that in no way resembles or is even in the spirit of the "trace agent" feature. This illustrates the confusion and apparent misrepresentation when a marketing department gets involved in explaining how things work. "Agent trace" is merely a data filter that provides the user the ability to identify a certain type of call AHEAD OF TIME and have the system track and record those transactions only! With this early strategy, every system required a crystal ball!

Transaction Detail Records

Transaction detail recording is essentially the collection of every event down to the keystroke level, so a complete date and time stamped audit trail is available for every circuit, system component, telephone rep working the system and every call processed by the system. A transaction detail record consists of events, which when assembled together, amount to a record, complete or incomplete.

Why would a manager wish to see incomplete records? Failed and incomplete call events identify all sorts of problems — system, trunk, people and training inadequacies. A simple peg count or "tally" of events does not adequately identify the "choke point" occurring in the call progress. A peg count is a raw number. Simply a quantity with little qualitative flavor. Little or no subsequent analysis can be done with these numbers, as they are conclusive or final numbers. In contrast, transaction detail recording cannot provide an answer but it can identify where calls fail. These records can pinpoint where the responsibility lies and provide the insight necessary to solve these productivity eroding and business blocking factors.

In a customer contact center, personnel and circuit productivity is measured by successfully completed transactions. If a call fails to reach the desired conclusion it still occupied a trunk, used the switch and took the time of a TSR. It is important to measure the total impact of all transactions, good or bad. For instance, you can determine if a specific customer contact center failure stems from:

- Inadequate network facilities.
- Inadequate ACD equipment.
- Poor queue management.
- Poor manpower planning.
- Lack of training.
- Poor morale.
- Poor scripting of the business transaction.

On the positive side, this detail can provide strategic insight into where staff assets lie so they may be used to the company's advantage.

From a personnel management point of view, transaction detail data is both individual and unequivocally objective. There is little room for management interpretation. There is no averaging at the employee level; thus, there is much less likelihood of the arbitrary and capricious conclusions that are possible with generalized or inaccurate peg count-based reports.

Any human resources department worth its salt should insist on complete transaction detail — as it is objective, equitable and legally defensible. In both a merit or discipline situation a complete audit trail is vital to document the good and the bad. The relevant events making up call records and then the individual agent audit trail are:

- Telephone rep sign on/sign off times.
- All event start and stop times.
- Call disposition for all complete and incomplete call events.
- Holds.

- Transfers.
- Three-way conference calls (supervisor assistance, etc.).
- The trunk used, in and out.
- The position used (calls to or from).
- The telephone agent identification.
- Any numbers dialed (out, internal or wrap-up) for call accounting, account and list management purposes.
- Any numbers received from the network such as DID (Direct Inward Dial) or DNIS (Dialed Number Identification Service) for media sourcing" or other purpose.

There may be other digits transferred to and from the telephone position to be recorded for management of data-directed transactions. The less information you collect, the greater the handicap you build into your future. Analysis can only be conducted on available data. The system cannot gather too much data. To effectively measure productivity of available resources, all relevant states must be tracked. Without complete data, true productivity management is as incomplete and ineffective as inaccurate reports.

Inaccurate or Missing Data

If you discover your reports contain inaccurate data, throw them away before your agents discover it. These reports will cause more damage to management credibility than missing data or absent reports.

Once an inaccurate report is distributed and potentially acted upon, the loss of subsequent management or supervisor credibility can be great. The energy and persuasion needed to restore the system to a semblance of its former credibility is extensive. Do not oversell the data gathering, processing or report production of the ACD. No amount of computing can "make garbage, gospel out".

"State" or "Record" Machines?

PBX systems tend to be "record" machines. That is, once a call is complete, a call record is assembled and shipped to the MIS gathering device. This is ideal for after-the-fact call accounting because the PBX only shipped completed call records to the call recording system, but this also means the MIS system receives dated data.

A standalone ACD tends to be "state" oriented. It communicates a change in a device state almost as soon as it happens. This is much more work for the ACD common control processor.

Traditionally a PBX system controller does not have the power to spare to manage call state data in absolute real-time so there is some latency between an agent's and the supervisor's screen state. The data delivered from a standalone ACD system is almost absolute real-time and in much greater detail. This is important from a supervisor's point of view.

If the data is not presented in absolute real-time, the screen view of their agent group does not reflect reality. If you can see an agent talking on a call, yet the screen says "available," there is an element of distrust toward the system. This happens more frequently in PBX-based ACD systems because the processor running the switch relegates data gathering and presentation as a lower task priority to that of switching calls. This latency, however, is encountered only in smaller (or older) versions of PBX-based ACD systems where the manufacturer stuffed all the tasks into one processor as a cost reduction ploy. One major PBX/ACD vendor had so underpowered its PBX/ACD that it actually stopped collecting data and generating reports during heavy traffic periods, when management needs reports most! Supervisor screens reflected what was happening 30 seconds to a minute ago.

Check with your vendor prior to signing the contract. Specify acceptable supervisor screen and reporting latency and make it a contractual term.

267

☐ THE HUMAN RESOURCE AND LEGAL IMPLICATIONS

All of this seems rather arcane until the realization is made that retaining customers, and the review, reward and discipline of employees is at stake. Traditionally this has been a big objection to every PBX-based ACD system as they typically have not captured every event in the machine and written it to an external device or stored it for later audit trail reconstruction. This is particularly significant when it comes to human resource reporting, merit, reward and discipline actions and challenging unfounded unlawful dismissal and unemployment claims. Today's PBX-based systems are more responsible, but it never hurts to check.

Millions of dollars a year is lost to customer contact center management in unjustifiable paid unemployment compensation or settlement of nuisance claims for unlawful dismissal. Historically, most companies have allowed this to become a cost of doing business since most ACD system documentation has been indefensible.

Many PBX-based machines report data in snapshot form from volatile memory. Once the report is created and printed the data is no longer available. And it was summary data to begin with, which any HR policy would define as less than objective. With the transaction reporting and after the fact audit trail detail from the standalone ACD vendors, and the new qualitative reporting systems that are becoming available, customer contact center management is beginning to be successful in their objection to and defensive of unjustified claims. This is money that goes straight to the bottom line, but this is only possible with complete, objective and accurate data gathering.

☐ REPORTING CAPABILITY

Achieve "good data" first. This is an overwhelming priority in the design of any tracking and reporting system. Report effectiveness is based on the depth of data gathered. Data gathering, normalization and storage of a usable database is everything.

There are two basic media to display data: real-time via a display (local supervisor CRT or access by an authorized remote browser) or by printing paper reports. These may provide tabular alphanumeric data or information in the form of graphs and charts.

Real-time displays present data "as it happens" or as nearly current as possible. There is a clear advantage to providing up-to-the-second data. Many systems, particularly the PBX-based devices, claim to have up-to-date CRT displays. On further investigation these turn out to update anywhere from 15 to 30 second intervals. The data on the screen is a rapidly aging "snapshot" of the reported period. In a large and busy ACD, by the time the data is painted on the screen it is already out of date. In some of the PBX-based ACDs, the "real time" displays and report become progressively dated or slowed down in a busy call traffic period as the switching of calls is made a priority over the administrative report functions. The standalone ACD systems build around this by using dedicated dual switching and reporting processors that do not share tasks. A load on one does not impact processing on the other.

True real-time displays are not repainted or refreshed, rather the individual fields making up the screen display change and update as the individual agent status, service level or count changes. Real-time means real time. What you see on the screen is indeed what is happening in the customer contact center.

A CRT display that is perceptibly behind reality lacks credibility and becomes unimportant to a supervisor. It reports agents as being busy, yet looking at the agents in the customer contact center, the supervisor can see they are now idle, and vice versa.

Each manufacturer has chosen one of two processor strategies:
- one central processor running the switch and the reporting, or
- a down line processor that runs the reporting system separate from the switching processor.

There are two issues arising from these strategies. First, how much horsepower does the processor have available for gathering, processing and displaying data. In the case of a second device, what is the processor overhead and response time penalty paid for the inter device communications and data processing and display. Modern technology and techniques have solved both of these problems; but some in the telephony business have been slow to adapt these to switching applications.

Nevertheless, technology prices have dropped and many of these price/performance advances are being used to update older systems to extend their life and provide more information than ever before.

Check with your chosen vendor which strategy they use (single or ganged processors) and which strategy they have chosen for the ACD machine you will be installing. Some vendors, to save cost (to remain competitive) load all the tasks on one central processor. For a buyer this is false economy because as soon as the installation is complete, you'll discover the drawback of a single monolithic processor, and it will be too late to use the original purchase leverage to reduce price. For obvious reasons, most contract center managers find an upgrade is less subject to discounting because the vendor knows that existing users probably cannot live without the upgrade.

Check your vendor's processor architecture and task assignment (at the size you are installing), then negotiate an extended discount on the original to cover any after installation "gotchas." An extended discount for twelve months to two years is not an unreasonable term to negotiate into a purchase contract.

Data Collection and Storage Processes

Even today, some ACD systems, particularly the PBX-based systems maintain a portion of the periodic call data in volatile system memory. Should the machine "hiccup" during the day, this critical data can be lost. No reports will be available for that period. If any of this data is in volatile memory the switch will have "lost" an hour or so of important data. Check how your potential vendor's ACDs collect, store and write data to disk memory. What are the periodic write intervals, time? Or number of call or event details? Ideally this should be by event, as it happens, with no batch write procedures.

Although we have seen a precipitous drop in processor and memory costs and explosion of memory technologies from mini-disks and flash memory, there still remains a cost to rewriting code to write this data to an external system. The telephone switch manufacturers have been reluctant to give up control of this data until the general adoption of predictable computer telephony standards. This meant they only had to present predictable data. Now any outside vendor or developer can provide reporting tools.

The answer you are looking for is that call detail is constantly and continuously written to disk so that data is available up to the instant of failure. The major standalone ACD vendors provide redundant MIS options, so that the switch, data flow and MIS processing is never lost.

Is storage of call records volatile (RAM) or more permanent, such as disk or non-volatile flash memory?

In most ACD systems the production of scheduled reports does not interrupt data gathering. Can the ACD manager create ad hoc reports from the stored data on the ACD or is this an off line process requiring another computer system? Most systems require a separate system.

☐ REPORT DEVELOPMENT AND PRESENTATION

The maxim, "if you can't measure it, you can't manage it" bears repeating. This is heavily reinforced by the fact "reports must be easy to use, or they won't be". Management style normally takes the course of least resistance because there are many other pressing tasks. The more readable, the less interpretation. The more self-evident a report is, the more useful it becomes as it is passed up to higher management. This can be a powerful factor in ensuring customer contact center management success. Computer generated reports have a certain, yet not totally justified, credibility.

If obvious and readable, without assistance of an expert, they carry more weight. "Mr. Controller, as you can see by the graph, as we are losing 25% of our calls, we are wasting enormous opportunity. We need budget approval for an increase in staff to capture the revenue potential offered by those callers. Our call to sale conversion rate is 45%, therefore those abandoned calls amount to $XX dollars!!" Seeing is believing.

If the reports are hard to read, or filled with jargon, they will be ignored. Transposing them to alternative forms, a memo, a spreadsheet or an Excel report, erode the power and credibility of the original ACD report. There is little reason not to use plain language in ACD reports. Even today, it seems the telecentric developers of these switches are doomed to jargon when the users remain ordinary business people. As a result this has created the opportunity for a boom in add-on report packages that repackage the data in readable formats.

Add-on Report Packages

These systems generally take the data already provided by the PBX/ACD system and present it in a more usable form. There is a great deal of activity in this market. There are two types of reporting packages:

1. Management Report Enhancement
2. Staff Management and Service Level Forecasting

The report enhancement systems all run on PC systems as client server applications and add real value to the voice switching system they are installed with. The machines they typically target are those most frequently encountered in the marketplace: PBX/ACDs and CENTREX-based UCDs. The most recent development in these packages has been to gather the quantitative data from the switch then compare and integrate it with quantitative data from the departmental data (CRM or customer service call disposition and results data) to arrive at customer contact center performance ratios and metrics. Computer telephony links have made this more easily accomplished and as a result there has been an explosion of available data. The more important question though remains the usefulness of this data?

The second group of software packages is the staff management and service level forecasting systems found at customer contact centers. These provide critical value and management timesaving opportunities. There is an increased blending of these functions. A number of PBX\ACD vendors in recognition of reporting and competitive advantages inherent to these systems are endorsing (or acquiring) software developers as partners in advanced customer contact center systems. CO-based ACD vendors have established relationships with external MIS system vendors, as an enhancement and competitive edge.

Report presentation offers another area for failure for an ACD developer, although this is less the case today as switch vendors have come to accept that the customer contact center application is as important as connectivity. Traditionally switch manufacturers thought that producing reports was less important than switching. The switch system was the focus of the effort and once complete, report data gathering and presentation was left up to the least qualified to develop, almost as an afterthought.

An understanding of the origin of most telephone systems helps to explain much of the philosophy behind switching applications. Technology drives the telephony business. Technology once drove the computer business also, at least until vendors found selling advanced technology only became a handicap. More time was spent explaining the technology, than the real business value the technology brought to a business user. Most telephone switching applications unconsciously consider end user management reports an afterthought and so the reports were designed and developed with little or no research. The dominant aspect of the ACD application at many companies was queuing, prioritizing and switching of customer calls. These telephony blinders created the opportunity for standalone or niche vendors who succeeded by focusing on customer contact center management issues as equally as the switching. Fortunately, we find a shift occurring in the telephony industry, although it came about more from competition as from a concern for the end user.

Report Style

Each vendor includes different data in the real-time displays. The single most important consideration is doing away with buzzwords, acronyms and codes — numbers and acronyms mean the machine needs experts, or highly trained personnel to use the device and read the reports. This retards acceptance, extends learning cycles and the use of the machine and reports by management. Demystifying the ACD reports is desirable. If it is not easy to use, it won't be! If it is not easy to understand, it will remain remote from the mainstream and certainly will not be exercised to its full extent. Alphanumeric data is more easily offered in real-time than graphical data. The latter requires far more computer power and time so the real-time slows down to anywhere from 2 to 10 seconds after the event. The tradeoff is increase in speed and capacity in communicating information by using color versus the slow down in real-time. Most of the leading ACD vendors offer down-line graphics reporting packages or interfaces to popular PC spreadsheet programs.

☐ THE CALL CENTER REPORTS

For simplicity, this discussion is basically restricted to a call center because at this stage we are using the ACD as a basic workflow model. The reports provided by the majority of ACD systems fall into three categories:
- Primary Reports
  - the service level provided to incoming callers: queue interaction

- the call applications: the trunks and the agents.
- Secondary Reports
  - system diagnostics,
  - component usage, and
  - alarm conditions.
- Exception Reports
  - call wrap-up or call disposition,
  - exception lists, and
  - ad hoc user reports.

Primary Reports

This is the first category of reports provided by the average ACD system installed in a customer call center environment.

Service Level Provided to Incoming Callers

Most well managed customer contact centers are staffed close to the optimal at most points in time. Service level reports are used to aid in the "tuning" of the balance of staff to demand. When unanticipated demand arises there is little that can be done to solve extraordinary problems. One call center manager wisely observes that current staffing and operations policy reflect the last crisis.

The basic reason for the use of an ACD is to ration your precious staff resources — agents and interactive voice response devices — across a larger universe of callers, which now can include Internet-sourced customer requests.

With the advent of the Internet and all it entails — email, web chat and web call back traffic, etc. — much of the original pure call center traffic forecasting and workload planning has been dumped. There was the assumption that the low priority email and web contact calls could be introduced as alternative tasks during low telephone traffic periods. Nothing has proven further from the truth — someone that takes the time to author and email or launch a web session and request a call back is equally or even more involved (read passionate) about a solution to their problem. The world just got faster. Fortunately, there are tools that can take care of some of these routine web and email requests, still, customer requests that are badly framed or out of the norm still require articulate customer representatives.

If callers believe you are not answering their inquiries in a responsive manner, they perceive you do not care. One of two things happens: The caller goes away forever or they call you more frequently till they get through.

Callers may even abandon their attempt to reach your customer contact center — they are now considered lost opportunities, lost revenue, lost goodwill. Yet, there were substantial costs incurred to encourage them to call. Instead they had to wait, or worse, not receive service. To understand how often and to what extent this occurs in your center, two report groups should be available.

Simple Queuing: As ACD systems become more sophisticated they have added queuing nuances that begin with simple group searches: Group 1, 2 and 3 or Group 1, or 2, or 3. Multistage or snap shot? The measuring of task assignment is vital, because typically Group 1 is the most desirable destination for an inbound caller, and cross utilization of Group 2 and 3, although desirable is not optimal.

Understand how "understaffing" Group 1, and so stimulating call volumes to Group 2 is vital. Often Group 2 is not the primary service group for this type of call — this type of call displaces their primary task.

Complex Queuing: Adding skills-based routing so that yet another layer of routing alternatives exist on top of the functional teams has significance from two potentially competing perspectives. The first is the root of establishing a customer contact center, and focusing volumes of homogeneous calls on a customer call center adequately equipped to serve most callers. The second is the ability to sort the inbound callers by specific demand and match these closely to the agent skill needed to answer their request.

The concern with such approach is analysis and anticipation of the majority demands. Narrowing focus of agent skill sets versus broadening agent training to offer larger pools begins to run into two potential problems. First, the added administrative effort (constant skill table maintenance and updates) and second, the breakdown of any economy of scale and reversion back to a less than efficient one-to-one service model.

Queue interaction reports are vital to understanding the service levels the individual caller (not the average) received. This is particularly important where skill-based routing is used. The good news in skill-based routing is the recognition of specific task frequency and volume, then the cost, value and/or risk this request represents to your business. The next step is understanding and matching the skill set necessary to respond effectively to the customer.

Now one of three things can happen:

- Recognize Frequently Asked Questions (FAQs) and drive a response and solution into a web self-service channel where power users can get significant support.
- Recognize a skill set that needs to be broadly adopted by expanded training.

274

- More targeted hiring of individuals with this skill.

Beyond skills-based routing and queue analysis, however, is the recognition of certain customer or prospect requests or call types that may be more adequately satisfied by other channel strategies. We will discuss this more extensively as we discuss call type analysis and displacement opportunities later in the book.

Incoming Call Waiting Information: The real-time CRT display should, among other data, display a count of calls currently being held in queue. This data, along with the status of individual agents (busy or not), allows a supervisor to make minor adjustments within the center to capture a few more calls. For instance, ensure as many agents as reasonable are available to serve callers, and/or adjust the group-to-group call intraflow so as to optimize capture opportunity.

The periodic reports provide the average delay and hopefully the longest delay encountered by a caller in the reported period. This data gives insight into how long the typical caller waits. A daily summary should give a spread of the caller delay encountered over the day and not simple daily averages.

275

Secondary Queue Reporting

Secondary queue reporting is a critical feature now that voice response systems are in widespread use in customer contact centers. Queuing structures can become a great deal more complex with self-selection via voice response.

Typically in a call screening or prompting scenario, the caller elects to first use the voice response system in lieu of a live agent. This may occur after the caller has already encountered a primary queue and been offered a hold message or self-selected alternatives. In a secondary queue reporting, the call scenario following the voice response device requires the caller to be sent to a live agent group. This agent group is currently busy thus queuing all inbound calls. This caller has now encountered a secondary queue. In most cases this caller, although reentering the ACD queue from a functionally separate process (the VRU), has already been queued and deserves some priority. From a management point of view the call needs to be tracked through the secondary queue state, but not as a new call. If the caller abandons and the call is not tracked as a caller encountering a secondary queue, the call processing scenario cannot be accurately measured. The primary queue statistics are inflated and the VRU data is fragmented as it is viewed separately. Few ACD vendors grasped this problem early on as the VRU systems were adjunct to the ACD. Even if they did, few had a coherent solution from a reporting perspective.

When talking to your vendor, be sure that the VRU is viewed by the switch in the same hierarchical role as an agent group, and the data is gathered, presented and integrated as if this was another agent group.

It is important to understand delay tolerance by caller type, by business type and by time of day. We are seeing increasing interest in analyzing call business type and delay tolerance based on simple geographical and chronological issues. The generalizations of a New Yorker being willing to accept less of a queue than a Midwesterner turns out to be fallacious. New Yorkers accept orderly and informed queuing better than a laid back Californian. This gives rise to another whole view of reporting, particularly when you consider the business type and value of a call. Add an interactive voice response device and there is another view needed.

Repeating an "old sore": with one foot in a bucket of boiling water and the other in freezing water, on the average things are quite comfortable. In fact the tyranny existing with averages is to hide the worst in a rather benign "averaged" number. The better ACD system designers realize this and give averages a broad guideline only and provide the "worst case" delay hold data also. The fact that transaction detail is available online allows exception reports to pull out even greater depth.

276

Incoming Call Abandoned Information: This data is rather meaningless when displayed on a CRT. The event is over and there is little now that can be done to correct for that call. The data should be provided as detailed "after the fact" reporting and provide the ability to determine what queue the typical caller will tolerate at different times of the day. Most desirable is a call aging profile showing at what point in the queuing process that callers abandon. From this data, agent groups can be expanded and contracted, announcements, script and timing can be changed, intraflow between groups within the ACD or interflow to other sites can be considered.

An increasingly encountered piece of information is the real-time service level calculation presented on the supervisor screen, which is typical of the more feature rich standalone systems. A service goal is established, such as "80% of the calls answered in 20 seconds." The system then calculates what percentage of total calls answered by agents met that goal. This adds to an already intense situation when little can be done to improve the immediate call capture rate. Histograms and graphic report representation are a great advantage in clearly showing the relativity of this data, both to the goals and to different periods.

The Call Applications

It is important to treat callers with a similar request with consistent service. For this reason call type reports are broken out by application, group or split. These are typically overview reports displaying the data by the originating trunk group — the physical ports or report categories.

If the system supports any trunk identification services, sorting calls by the inbound identification digits, such as the DID or DNIS lead number (i.e. the logical categories) is also advantageous.

The data provided should include the number of calls offered, handled, delayed and abandoned and should be expressed as relative counts and percentages. All trunks' busy data and average service levels are also handy.

The Trunks: Trunks are arranged in physical and logical groupings. Trunk group reports generally report like trunks. From an ACD management point of view, they should be reported as trunks carrying like call types, then arranged in like trunk types. This way the business responsibility is reflected first, and the engineering issues second. This is a subtlety, yet is critical for management convenience and insight. The line use data displayed on these reports, at minimum, should include call count, in and/or out, totals, abandoned, average length, trouble data reported by the agent or hardware problems, time out-of-service, time in use. It is desirable for the system to collect and display this data as individual trunk statistics and present it both as split or group reports and assemble it in logical and useful time increments such as clock and calendar.

277

If the system supports any trunk identification services, sorting calls by the inbound identification digits, such as the DID or DNIS lead number are also advantageous.

With the availability of Automatic Number Identification (ANI), there is now a need to analyze call source by geographic origin. The potential for advertising promotion, support and service level and market penetration analysis is significant, though the full extent of call source and marketing media effectiveness is yet to be realized.

An application for this collection and analysis of logical data is marketing effectiveness. By mapping the called numbers to the numbers calling, analysis is available for marketing applications like the advertising effectiveness. Now comparisons can be made of lead generation effectiveness of an advertising campaign running in one geographic area (call origin), versus another ad in another region. Significant coordination is required to do this "media sourcing," but data and value can be tremendous.

The Agents: Agent productivity is held out as the largest single gain to be had from these machines. This was probably true before it was rediscovered that most employees do the best they can while they are working. Give them better tools and they will do more. In fact the power of an ACD is in its ability to take an individual employee's average performance and make it better, without burning out the employee. Moving the quantitative bell curve forward. The real power of agent productivity reports is that the machine tracks all activity, does it in a constant and objective manner, then provides clear and understandable reports.

In a good ACD, no additional interpretation is needed. In this day of full disclosure, an agent is entitled to see these reports and many will act upon them. Most ACD systems collect data by individual. Management labors under an enormous handicap of uncertainty if they do not have this feature. To review individual productivity without being sure that the data reviewed is actually about that individual is fallacious.

Ideally agent report data should take two directions: as overall performance measurement and as individual activity. The objective in separating the approach is to enhance overall management oversight, yet lighten the management load of tracking individual detail unless there is such a requirement.

These machines have the ability to produce "blizzards" of report data. This is undesirable. Management has little time and even less interest in this detail, UNLESS THERE IS SOMETHING WRONG. Then the collection of transaction detail versus peg count data becomes incredibly important. A manager can only trust high level reporting strategies if they know, at the time a problem occurs that they can recover the "blow-by-blow" detail. Literally, "unpeel the onion."

A manager can scan an overall agent force productivity report and determine all is well, or note any exceptional trends needing attention. Individual activity reports are critical. These have value in merit and disciplinary reviews, as they provide a complete audit trail of all activity over the period in question. As previously mentioned, one standalone ACD vendor feature is TRACE AGENT, which allows the manager, once a problem has been discovered, to reconstruct an audit trail of all the events an agent was involved in. It allows this to occur after the fact.

Agent performance reporting should include:
- sign-in duration,
- count of all incoming,
- outgoing and internal calls,
- totals,
- average duration of all these events, and

- analyze the time the agent spent in the various work states while manning the system.

Agent activity reports are the individual call or transaction records for all events the agent was involved in over the studied period. Data should include:

- the event type,
- the disposition,
- start time,
- stop time or duration,
- call origin,
- call destination and
- all digits dialed by the agent, delivered by the network or entered by the agent as call wrap-up or cataloging information.

Again it is desirable for the system to collect and display this data as individual agent statistics and present it both as split or group reports and assemble it in logical and useful time increments such as clock, calendar and shift periods.

279

Secondary Reports

The second category of reports provided by the average ACD system installed in a customer contact center environment are concerned, in general, with the health of the system.

System Diagnostics and Component Usage

The system should watch itself and warn management of potential problem conditions. This is important from a trunk availability and quality point of view. Many conditions exist where a trunk is out of service, or the sound quality eroded to the point of effective failure. A hardware failure is clear to most machines and they will report it accordingly. A circuit quality issue, however, is more subjective and typically is only discovered when an agent receives or places a call. The agent should be able to report this via a "bad line" or "trouble" key on the instrument. A failed or poor quality trunk negatively impacts service levels. Other system components also should be tracked and reported to provide insight into the availability and adequacy of the system. As most systems provide a number of utility devices to support a given application (tone detection, recognition, generation, etc.) too few can lead to blockage. Tracking their operation and adequacy is important. The ability for a vendor to draw this data off the system via remote diagnostic capability assists in many ways to accelerate diagnostic and service action.

Alarm Conditions

Most systems provide a series of one-line alarm reports that are printed as they occur. Trunk out of service, failures to respond, "hung," all trunks busy, and line trouble conditions, to name a few. These ideally should be coupled with alarm notification at a supervisor CRT. This avoids constant checking of the printer that is often installed remotely from the supervisor consoles.

Exception Reports

This category of reports, useful in the analysis of different categories and sub-groups of calls, isn't universally available from ACD vendors like Avaya or Nortel.

Call Wrap-up or Disposition

The key to the success of this feature is twofold: First, the ability to enter data at the telephone instrument or soft phone that can be used at any time. No complex planning or programming is necessary to set it up. Merely notifying your staff that they are expected to add some data (obviously uniform numbers (keys) and fields) for each call via the touch pad. Second, that the results can be sorted and analyzed at will by customer contact center management without external data processing resources. This is an extremely desirable, flexible and powerful system feature. One standalone ACD vendor has gone as far as to "script" and validate this data entry so agents do not make mistakes or overlook this data entry.

NOTE: This is really the wrong place to collect this data as it is being captured on a closed database. But prior to the widespread adoption of Windows-based workstations and departmental systems, wrap-up was (and for some companies still is) an expedient and effective answer to ad hoc data gathering.

Exception Lists, and Ad Hoc User Reports

Most ACD systems provide standard report formats. Also, with the advent of PC systems, exporting call data has become increasingly popular. The advantage of being able to sort call records on some exception criteria allows greater understanding of the data. Few ACD systems include this feature as part of the ACD software — external systems bring this, provided the call transaction data was captured and is easily available to analyze. A number of recent innovations allow the call data to be exported in ODBC, EXCEL or other standardized format like ASCII comma delimited files.

Integrating Data from Other Customer Contact Channels

Now we have added web assisted browsing, web requested call backs or click-to-talk and email sessions to the call center. This requires adding new states and new complexities to data gathering and delivering reports. The universal switches led by Aspect, and more recently, Avaya deliver reports that use the ACD model to deliver a full view of an agent's day no matter what channel. The new challenge is measuring customer service levels as media switching goes mainstream.

The apparent solution does not lie in the contact controller. The question really lies with the customer reaching a point of their expectation being fulfilled. This now places a responsibility on some central customer experience mapping database. Now the reporting mechanisms need to develop data on the steps, sequence and outcome of each customer request.

Report Use and Management Style

Management is both a company and a personal style, therefore, the only comments will be about a few simple success strategies.

The most important issue surrounding report usage is how the personnel productivity reports are presented to the customer contact center staff. Are they a whip or are they a carrot? The most successful customer contact centers tend to take the latter approach using the reports to chronicle success rather than poor performance. Reports can be posted or copied and distributed to the agents. It is important that they are accurate, clear and self-explanatory and that they are distributed without initial management comment. The perception should be that the first people to see and study the reports are the people measured. If they are delivered to the subjects with management comment, the reports take on the overtone of "a report card." Self-motivation is generally acknowledged to be the best method of staff motivation. This is not a one-time event or a sprint, but an ongoing juggling marathon.

Using an agent's self image and peer acceptance in an enlightened manner is the least threatening method of having your staff meet your joint goals.

Report Presentation

Development in reporting strategy includes greater use of graphics and color as well as reaching outside the ACD reports to include "results" data from departmental sales and service systems.

Colored screens were an obvious strategy in the PC world, but were relatively slow to arrive in the ACD world. The lesson is interesting and applies as an illustration of the industry's reluctance to move away from core, and often narrow, disciplines. The cost of adding color is minor in the context of total system cost and minuscule compared to the impact. However, switch designers reluctantly added colored supervisor screens because of the additional overhead added to the management report process and the perceived increase in sales cost. The lesson repeats itself in system design: marketing, ergonomics, usability and management style are best kept from pure engineers.

Pictures allow more information to be communicated more quickly than columnar or narrative data. Hence, the popularity of business graphics and charts. Color adds a further dimension in speed of communication and depth of information understanding. Supervisors can judge customer contact center service conditions by looking at their screens from a distance. General appearance, screen hue and color can be scanned quickly and understood rather than reading and comprehending the individual screen detail. This can occur for a roving supervisor over a short distance. This is even truer in historic reporting when a graph can show the immediate relativity of otherwise dry numbers.

By using external PC systems to do the actual supervisor or administrative data presentation functions, much of the reporting load is shifted off the switch to an outboard processor. But with this the immediacy of a real-time picture is lost. This latency is because of two added processes that occur when the function is moved out of the switch processors: transmitting meaningful data to another computer and receipt and processing of that data into graphical reports.

Every processing step adds time. It is important to note the difference between "state" reports and historical reporting even though that "history" is as recent as the last minutes. A state report should not tolerate any delay, but create it now, displaying it, allowing for immediate revenue. A trend or problem occurs over a period as it builds or declines. This can tolerate some delay in reporting and trend analysis — again 30 to 60 seconds being about the most acceptable. Check with your vendor and understand the impact this will have on the ACD/MIS configuration you are being sold.

Artificial Intelligence (Genetic Algorithms)

There was talk of using "artificial intelligence" in ACD systems to allow the ACD to optimize itself based on prior successful treatment of similar conditions. Remember the old investment banking joke about the technology of "artificial intelligence." It was something you announced just before you took the company public!

The idea is still around, it's just morphed into what is now termed "genetic algorithms," as in machine-evolved code, that promotes success and "extincts" weakness, thus finding the best way to do things. This type of learning depends on accurate and "honest" reporting, thus further adding to the need for serious call centers to focus on standalone ACD systems. Applying predictive algorithms, as far as call handling is concerned, is still illusory as they work on statistical averages and not real caller experience. Further, they need high volumes of traffic to be theoretically effective.

There is a better way to do this and we will discuss this further when we broach the subject of customer experience mapping and management in the next chapter.

☐ IN SUMMARY

As customer contact center managers become more sophisticated and mature in their role they grow to trust the machine data and, thus, need fewer comprehensive reports. Everything they need to know for effective management is on one report — the ideal for an insightful manager. Trust in this type of "one-look" report is only possible if management is aware that they can go back to the ACD and recover past detail. Less is more, as long is more is still available if it is ever needed. Many PBX-based ACD systems DO NOT capture every event in the machine and store it for later audit trail reconstruction. This is particularly significant when it comes to human resource reporting, merit, reward and discipline actions and challenging unfounded unlawful dismissal and unemployment claims.

More attention is being paid to the quality of the customer experience at the "hands" of voice response technology and live agents. This makes for a new generation of reporting opportunities growing out of the traditional logging and quality monitoring features offered by another series of vendors.

Here is a dichotomy. As ACD systems matured, customer contact center managers wanted more and more reporting. This added a significant burden to the ACD system and vendors as they strive to answer all possible customer requirements. The obvious solution was to add outboard systems, relational databases and spreadsheets tools. Many customers remain uncomfortable handling their own data in this fashion and insist on packaged reports. Either way, access to your customer contact center data is desirable even though your system can produce 101 different report formats.

283

Customer Experience: Mapping and Management

The emerging market of tactical call center tools for compliance purposes and quality monitoring is being prematurely described as Customer Experience Management. The end game here, however, has more to do with the ability to richly index and map customer transactions for a host of other customer experience and revenue improvement purposes, than with literal voice recording.

The recording of customer calls arose out of the need to collect data for dispute resolution, such as a trading error in the stock brokerage industry or public safety emergency. This essentially grew into call logging, which captured the call center industry's attention as the best way to accurately receive and handle high risk or high value calls, with the implication being that a mistake was costly in terms of a financial error or an endangerment to human life. By the late eighties there was a growing frustration within the call center industry due to the pure telephony reporting focus of ACD systems, which only gathered data on productivity, i.e. the numbers — the quantity of calls handled by an individual or a group in as short as possible time.

The major ACD systems generate masses of quality quantitative data. As explained in the previous chapter, the data was first provided as summary data in the manner of "peg counts," then evolved into more detailed call records, rolled up into summaries of events processed by the system. These showed calls arriving on trunks and trunk groups, by call class, through the call or workflow routing preferences, and to the individual agents trained to service the customer's general or specific requests.

This data enabled call center management to determine that although certain agents took longer to process a call, they achieved more sales, while the agents that handled more calls actually closed fewer sales. It also brought to light a dilemma: there were specific agents who processed not only more calls, but achieved more sales, so what did they do better than the other agents?

Call Logging: Keeping a sequential recording of EVERY call made to or from a specific telephone that charged with accurately processing a high value financial transaction or dispatching emergency services. The first generation of call logging devices were large industrial strength analog tape recorders that sequentially laid down a magnetic record of each and every call on or to a device (trunk or telephone terminal). Banks, brokerage houses and public service agencies were typical users.

These archiving recorders were progressively made more efficient by using voice-activated recording, thus eliminating silence or "dead air." The tapes were physically labeled and stored with little thought of future review unless there was a problem. This worked fine for the exceptional dispute or question. A challenge or question required management to sit down and literally playback and listen to the entire tape to find the offending call, and thus the relevant analog voice content.

ACD Emergency Recorders: Most ACD systems calls were not considered of high value or risk so did not need archival recorders until the airlines realized it was important to be able to record a threatening or abusive caller. The most graphic illustration of the emergency recorder use is recording the "bomb threat" call. These limited recorders were later pressed into service to record everyday calls for the purposes of training and other sampling purposes.

The Quality Monitoring System: The next phase of management insight occurred when ACD users and vendors recognized that ACDs limited their data gathering to quantitative data. What was missing was some way of capturing the call content and agent delivery. Thus arose the need to focus on recording some of the agent calls for the purpose of agent performance review. These systems used techniques similar to voice response and voice mail systems and collected approximately 1% of the customer contact center telephone calls.

Mirror Data Screen Observation: Concurrently a small group of companies had decided to separately address real-time supervisor observation of the data screens used by these agents in the course of serving a customer call. A "mirror" data screen was developed to allow a supervisor to view the active agent screen as a passive observer. It was inevitable that these screens were recognized as being more meaningful when observed simultaneously with the call's audio.

In pursuit of an explanation, call centers began to request that vendors provide a way to directly address call content - the actual conversation. This was one of the reasons that technology that allowed manual "service observing" (a supervisor or team leader listening in on the call), which later became known as "silent" or "call monitoring," was introduced into the call center environment.

While, the initial goal of call logging was employee review, reward and discipline, it gradually morphed into what we see today, which runs the gamut from customer contact center recording and quality management to customer experience analysis.

The customer call center management goals changed from policing individual behavior to assuring a good customer experience, this took on the notion of "call quality assurance," with particular attention to valuable calls and delivering a good experience for a customer or prospect.

This application was originally all about agent call processing skills, although initially the efficacy was greatly over-debated because of the simplistic strategies previously employed to motivate large groups of call center employees.

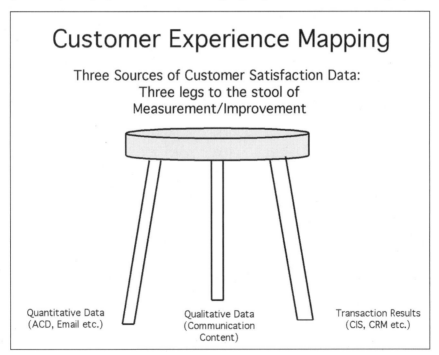

Figure 15.1.

Unfortunately, some of the fears of employee abuse and misuse of report information were not unfounded. Just take a gander back into the history of call centers. The old phone companies were known for running operator centers with working conditions just two or three degrees removed from "slave ships."

In the 21st century, widespread abuse and misuse are generally unfounded; but as always, a few extreme incidents trigger prospective legislative and union reaction. A bad manager will probably always be a bad manager and may depend on subjective data and data gathering to prove their biased point.

It was this collection of philosophies, platforms and background that in the early 90s gave rise to recording and monitoring technologies. This leads us to the subject of supervisor monitoring. The current discussion is whether a sampling or total recording system is preferable.

☐ SECOND GENERATION LOGGING AND RECORDING TECHNOLOGIES

Today, if you take the recording vendors' marketing positions seriously, call recording has evolved so as to be completely integrated with customer relationship management systems. The industry is still far from this "holy grail." Currently, it's only offered as a "strategic marketing distraction" for a collection of limited second generation recording technologies offered by many logging vendors as they strive to compete with "tactical quality monitoring" vendors.

The current second generation of offerings from the major logging or quality monitoring vendors are technologies based on early architectures, some of which are ten years old. For instance, first generation call logging systems were based on industrial strength tape recorders with a basic sequential indexing system. The second generation took this model and moved the recordings onto proprietary recording servers with almost the same basic indexes. These disk-based systems closely resemble traditional tape recorder behavior with a somewhat richer indexing system.

Eventually, they incorporated screen capture and CTI data that requires multiple servers for collection of computer telephony data, data sessions and the index collection and normalization across multiple files or devices. This adds an additional layer of indexing routines, which enables various levels of data export and sorting.

The key objections to second generation logging remains the rigid separate sequential indexed, closed proprietary databases and a process that cumbersome-

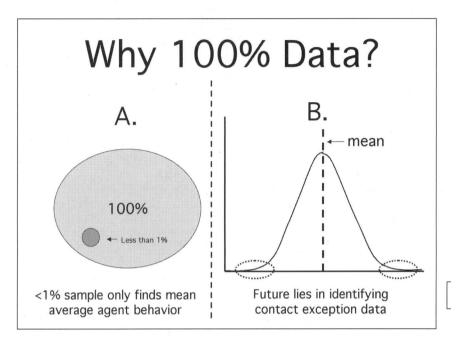

Figure 15.2.

ly maps and synchronizes voice to other records such as the index and screen sequences. The frequency of lost or scrambled review data because the clock or index is unsynchronized is unnecessarily high and inconvenient for the user.

The nightmare for the vendor is unmapped call components (mismatched audio, screen or index record) because of a clock error rendering the index by time inoperable. That record (or component) is lost until the offending field is identified and corrected on each separate index AND server! The usual solution is to discard the call as a sample, yet it may contain that vital nugget of marketing data.

Nevertheless, today's loggers have a major advantage over quality monitoring devices because they record all calls thus getting away from monitoring samples.

VoIP

The introduction of Voice over IP simplifies connectivity dramatically for any call recording vendor. The connectivity is accomplished by an interface (presumably standards based — though as of this writing CISCO differs from 3Com) that allows the recording server to "sniff" the LAN for relevant voice and data packets, collect, organize, index and write these to a single server.

If it can be done cost effectively, total recording has a huge advantage from a statistical purity perspective and for gathering historical intelligence and after the fact review of customer behavior.

Once the logging companies break out of their rigid second generation systems, they can allow dynamic call mapping to recognize "out-of-range" conditions and identify broken processes, badly trained agents, "at risk" transactions and customers. Still, with the state of most of the recording (sampling or logging) architecture today, the transition will take time.

Today, every second generation logging vendor has an engineering program to move to a third generation standards-based open platform technology that can cost effectively record every call, email record and customer web event in an open format with an SQL compatible index.

It is noted that recording and monitoring applications have generally evolved in relation to the needs and capabilities of the customer contact center industry. Today, that means tracking and recording one customer's contacts across multiple touch points.

As you now know, most ACD system reporting is limited to quantitative, not qualitative reporting. But the market segment that is most dependent upon ACDs, the customer contact center, is battling with a market differentiator that the ACD fails to address: the quality of its management tools, particularly when focused at content quality management. Hence, the thrust into recording, analyzing and reporting the quality management process.

☐ IT'S ALL IN THE NAME

Happily, tools do exist that can gather objective data, allowing a contact center manager to position its monitoring as an "objective sustained customer focused quality assurance program" that's vital to the success of any and all quality campaigns. Yet there needs to be a careful selection of the terms used for this process. In some companies, monitoring is the preferred term since it eliminates any ambiguity in what one of the main goal of the process is all about — monitoring call handling by agents — but this can give rise to the possibility of a "them versus us" management problem.

A better term might be "quality audits" since the process also can be used to ensure that customers get quality service and accurate information in a manner consistent with the policy of the company and the mission of the customer contact center. Call auditing or quality reviews are in many ways better "labels" that won't attract the attention of your legal department, unlike the term "monitoring."

Consider this, if Detroit can inspect their work product, why can't a service company — your customer contact center? Maintaining quality is essential to profits and the survival of your business. Recruiting your agents to this cause is the best method of deflecting any criticism. "Why would we want our customer contact center to do a bad job?" Back this up with objective data gathering and enlightened use and there are fewer problems.

☐ RECORDING AND MONITORING GOALS

It is critical to the monitoring process to have clear objectives. It is also important that the goals and the use the data will be put to is clearly communicated to the agents that are to be monitored.

This may be done at the time of employment; i.e. having the agent agree to accepting monitoring or service observation as a condition of employment. Informed consent is a sound basis upon which to begin this process.

The goals of recording calls are:

- Initially the role of logging is keeping a record of high value or high-risk calls.

- Ensuring the quality and consistency of the contact the customer has with the company and the customer contact center. Making sure the information exchanged is adequate, accurate and delivered in the manner intended. Training and customer contact center script guidelines are important starting points for any monitoring strategy.

- Ensuring any orders taken are genuine. Many customer contact centers pay incentives to employees upon a customer order or commitment, long before the customer pays for the goods or services. Order "verification" is essential to prevent "soft" orders or outright fraud. Monitoring is a deterrent to this type of activity.

- The maintenance of agent productivity in a customer contact center is an ongoing task. To monitor intermittently and give the agent force the impression this is a halfhearted effort is to invite an erosion of quality and productivity. There are few secrets in a customer contact center and as soon as monitoring is stopped or is restarted, the entire staff will find out and react accordingly.

- Individual performance is the basis of most merit raises and compensation plans. The attitude the agent projects to customers on the telephone is a component of this review and cannot be measured without observing the agent on the phone.

☐ REGULATORY BACKGROUND

There remains a minor degree of uncertainty to silent monitoring or recording. Each state has differing legal requirements that can be obtained from The Direct Marketing Association at www.the-dma.org.

State and Federal legislation falls into three categories. Laws based on privacy protection, against illegal wiretapping or interfering with confidential transactions. Your legal department will probably do what most legal departments do — go and make a project of researching the subject to justify their existence. They will comeback with a resounding "maybe" when they find there is no clear prohibition of monitoring, only a suggestion that recording of review sessions might be illegal under laws whose spirit and intent was to prevent illegal behavior like unlawful wire tapping, not ensuring quality customer treatment.

When the issue of customer contact center quality assurance was made known as the real objective of monitoring, not interference with operator privacy, the FCC and the states enacted legislation with varying levels of restriction aimed at preventing illegal behavior. Nonetheless, state utility regulations and maybe even the state penal codes may also say something about illegally monitoring a confidential conversation. Again the spirit and intent of this type of law was generally directed at *prima facie* illegal behavior, not quality assurance and consumer protection. Your legal department may trot this out as a peripheral objection, but there are hundreds of companies in your state who have already set the recording and quality assurance precedent in similar industries and customer contact strategies.

Despite many legislative attempts to circumscribe this business right nowhere in the United States is it banned outright. Where there are regulations, varying from requiring one-party or two-party consent to the monitoring, there appears to be no absolute prohibition. Check with the DMA website if you are uncertain.

If you are required to seek consent, it's generally of the person to be judged by the monitoring session NOT the calling customer. That is your agent. This is easily done by making consent a condition of employment. Gaining consent from the caller when two-party consent laws apply is more troublesome. The simple approach is to add language such as "to assure the information you receive from Acme Computer Company is accurate and given in accordance with company policy, we may monitor this call" to the announcement at the point of the greeting announcement.

In the past this pleased the legal department, but it had a potentially negative effect because of two issues: the additional time added to a call to give such notice when long distance was significantly more expensive, and the "kill" effect it may have on your business. Callers can make a decision to hang up, so you lose a prospective customer or alternatively, stay on the line then involve the agent in a discussion of the fact that your customer contact center monitors calls.

Alternatively, you may add a monitoring disclosure term to a contract the customer may sign "that for quality assurance purposes, any calls you place to our customer contact center may be monitored from time to time." Adding a disclosure to invoice language or other customer communication is also an option. The problem the lawyers have with this approach is that it is passive and does not give the customer the opportunity to act on the information, short of discontinuing their business with you.

Either way, announcement or printed disclosure, if people don't like it they are going to go elsewhere with their business. The most logical strategy is to disclose in written form that the objective of monitoring is in the interest of giving the customer quality service. It has a lesser "kill" effect on calls and it does not lengthen calls. The good news is that this has become such a common practice, customer objection is the very rare exception rather than the rule.

In this writer's humble and impatient opinion, the point of monitoring is to ensure quality customer treatment and consumer protection. Not monitoring, based on legal confusion and uncertainty, is more of a business risk than a legal risk. The question becomes can you afford not to care and not do anything about quality assuring your customer contact center's work product!

☐ CALL RECORDING

There are three points of origin of call recording — two began with audio and one with data session recording.

Call logging and quality monitoring began with specific purposes in mind.

- Call logging to keep a record of high risk or high value calls in the case of future questions.
- Call monitoring began as a manual and real time process, i.e. a supervisor listening over the shoulder of an agent. Technology later allowed this to be done randomly and anonymously when voice response technologies became affordable and flexible enough to sample agent calls without human intervention — well, other than to establish the initial monitoring/sampling

schedule. The quality-monitoring machine went out and dutifully collected the sample.

With manual audio monitoring it was possible in some older computer environments to create a mirror data session so that the supervisor could simultaneously watch the same screen the agent was working from. Initially, only the data process was visible until it was obvious that only viewing the data, like the audio only, was only half the story. However, recording a data session for later playback with a voice recording requires two different technologies and yet a third matching process. This conundrum has basically been solved. Call center recording/monitoring systems can match the audio and the data session. This voice/data gathering and session matching has created a significant industry niche with about ten different vendors with significantly different technologies, generations and approaches to addressing customer contact center recording applications.

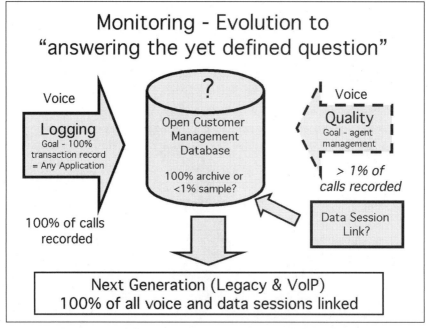

Figure 15.3. This figure shows the origins of the two major tactical applications, logging and quality sampling. Both try to play in each other's market, but this will not succeed until both applications adopt a common standards-based open platform. Then none of the application boundaries, which apply to the current 2nd generation, will apply.

☐ THE PROCESS: SERVICE OBSERVATION BECOMES QUALITY MONITORING

The whole objective of service observation or call quality monitoring is tactical and agent centric. What types of calls could be handled better and what agents need improved process and training to achieve this? Inwardly focused quality monitoring is important. There are three points of a call where observation and measurement needs to take place:

- the call data as gathered by the ACD,
- the actual call audio, and
- the call data or screen session as used by the agent to serve the caller.

These should reflect the quantity of calls, the quality (sampled), and results of customer call transactions.

Service Observation

In a manual environment, a supervisor or service observer plugs into a supervisor instrument and waits for an appropriate call to occur. This connection can be observed on a supervisor screen and a silent conference set up by the observer.

As the transaction happens, numeric data is collected and reported in real-time or with after-the-fact statistics. This comes from the ACD. The service observation or monitoring refers to the recording or listening to the audio portion of the transaction between the agent and the caller. The final element is the use of a mirror of the data screen session the agent is working from to serve the caller. Ideally the monitor or reviewer needs to see the state of the agent being monitored (the supervisor screen), and the data session they are working from, while listening to the call.

In most call centers the last element, i.e. sales results or a service request outcome, or however your customer contact center defines a "successful" call conclusion, is not included in this measurement or transaction. Consequently, there is an incomplete "performance picture."

If monitoring is done manually, in real-time, concurrently with the agent receiving the call, a supervisor or reviewer can select the particular agent's data screen in a mirror form. They can also concurrently watch a supervisor screen so collection of status and time counts can be written down from this screen. The monitoring supervisor should then keep a detailed record of whom, and for how long, each agent was observed. In a manual environment, be extremely careful to give each party fairly equal attention. If there is a problem and one agent is identified as needing further attention, document this before resorting to extended observation.

All this implies that the call was monitored in real-time. That is, a supervisor or reviewer must be available at the same time the call is being received by the agent. The problems with this real-time approach are five fold:

- Monitoring is no supervisor's favorite task.
- Scheduling monitoring sessions is not simple.
- The full call is almost impossible to catch.
- There is always a more pressing task that interrupts this chore.
- Extensive manual documentation needs to be maintained to make the review worth anything, particularly if the review has merit, compensation or discipline implications.

There is a bank in California that has a policy of monitoring 12 calls per month per TSR. There was no way they could do this for 1500 agents per month easily, so policy was seldom met. The agents knew this, and worked accordingly. The call center management realized they were doing a less than stellar job in this area. They also wanted to be able to add an incentive portion to their compensation plan. They instinctively knew there were opportunities they were missing in training. They were acutely aware of the haphazard state of their documentation procedures. More in a moment.

Scheduled Call Quality Sampling

Alternatively, the call can be recorded. But this means the real-time view of the agent state is lost. So is the real-time mirror of the data session, unless this can be recorded and played back concurrently with the voice recording.

The next generation of call center quality assurance products were designed by Dr. Gene Swystun in 1987 and introduced in 1990 by a customer contact center vendor. These products were so successful at filling this niche that there are a number of manufacturers now selling these quality-sampling products.

There are three products to the typical customer call quality assurance suite and these are:

- The call sample scheduling, gathering and playback system (data gathering). This can be either a sampling system or, more recently, a total call recording system or logger.
- A companion, call template builder, scoring analysis and documentation subsystem (review scoring and analysis system).

- A data session sampling and recording system that matches a call to the data session (voice and data session synchronization).

This approach assures a scheduled call quality sample and that all agents scheduled will receive similar sample gathering treatment. Predictability and certainty are two key elements to quality assurance. The system releases the supervisors to the immediate role of supervision so that the review role can take place at some later and less pressing time.

The call scheduling and sample gathering system is a cross between an intelligent "tape recorder" (even though it is a digital recording on disk) and a voice mail system. This computer-based system allows management to schedule the agent session samples based on their work assignments. A few sample calls are then recorded and played back at some later time by the reviewer. A supervisor review of these samples is then conducted with a form or computer template that allows some sort of objective scoring process for the purpose of individual review and peer comparison.

During this review, the reviewer can "cut and paste" certain call segments into a voice mail-like message to be played back to the agent as necessary. This frees the reviewer from the necessity of being available when the agent is available. It obviates the need to wait, listen to silence, and deal with partial calls or the wrong type of calls in order to conduct a meaningful review.

The advantages of intelligent recording are many:

- It frees the reviewers from the tyranny of having to be present to listen to a customer or agent call, then "catch" an eligible call, meanwhile listening to all the other things that transpire in an agents day. The net effect of recording a session and then playing back everything, skipping or cutting out incomplete or inappropriate calls, with the silence eliminated, reduces the task to 50% of the time it formerly took. No exaggeration.

- Any measuring and scoring system only works when it is perceived to be certain, objective and repeatable over the universe of subjects to be measured. In many customer contact centers, monitoring has been a policy that is intended to be carried out, but inevitably takes the lowest priority in a manager's day due to more pressing tasks. Monitoring defaults to an expendable function when any customer contact center crisis occurs, no matter how small. Suddenly the necessary notions of certainty, objectivity and repeatability are lost and the value of monitoring degraded. Just as with ACD system reports, any error or inconsistency debases the principal of accurate reports.

297

So, too, with quality measurement.

- The use of such a system assures that monitoring and sample plans are met. The system will blindly gather the required session sample until the quota is filled. Human resource policy is met. This goes a long way to dissipating the flaw in most measuring or monitoring practices — incomplete or substandard data samples.

- The old gardening adage applies: "just walking around the garden with a pair of hedge clippers makes everything grow better!" So it is with systemizing the gathering of monitoring sessions, which can assure the agents of predictable and certain reviews.

- Finally, schedules, samples and reviews are rigorously and automatically documented. This is a key issue for any human resources review. If you have a policy it must be adhered to and applied equally across the applicable employee universe. Then it must be fully documented.

A recent case illustrates this most dramatically. A large California bank using this technology successfully challenged a former employee's unemployment compensation claim. Historically this bank has never challenged these claims as its monitoring process almost never met the policy plan. As a matter of course the bank could not show that this cumbersome process was equally applied to all employees. Now with a system that collected these samples with certainty, and documented time and date gathered, time and date reviewed, and the time and date the subject employee heard these annotated review sessions, the bank had evidentiary quality documentation. They were able to stand before the unemployment compensation review board and absolutely refute the dismissed employee.

In the past an employee, based on manual practices, could claim they were subjectively and casually reviewed, seldom counseled and never disciplined. The bank could now show the time and date of every sample, management and employee review. They could also replay the sessions that showed the employee violating policy. For the first time in the history of the bank's Teleservices department, they won an unemployment hearing and didn't have to pay the former employee, as had been their habit prior to employing this technology.

Attachment to the ACD

A quality sampling system requires an active attachment to your ACD and this may be less than desirable as it passively and actively displaces switch capacity. Most switches, particularly PBX-based architecture have a finite number of "service observation" or supervisor ports. The quality sample-recording device

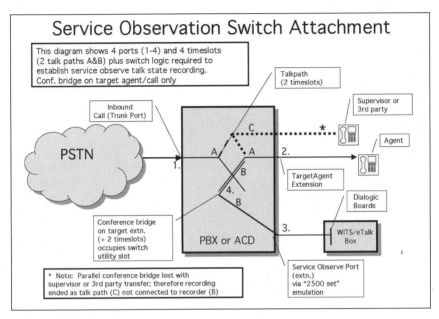

Figure 15.4.

attaches to these ports and emulates a supervisor telephone instrument.

The agent sample gathering scheduler commands a port on the device to set up a service observe port conference on the switch selecting an agent call to record. The quality monitoring system seizes a service observe port and establishes a conference call with the intended agent, without the agent being aware the call is being monitored.

There are four significant disadvantages to the attachment approach that become particularly critical if a customer contact center is large or expands the notion of analyzing recorded call and data session data beyond the tactical focus of addressing agent quality. These are:

1. The attachment and conference call model occupies two added ports and a conference bridge during every sample recording session. If you have a small customer contact center and ample switch capacity this may not be a problem.
2. The scheduling system cannot ensure the call recorded is the type of call you care about monitoring so more calls are recorded than required to gather the number of samples necessary to find specific calls with quality improvement opportunities. No call is ever recorded from the absolute beginning.
3. The only state that is recorded is the actual connection to the agent and if the

call is transferred away from that agent the call is lost to the recorder. However, any hold state or a conference with a supervisor or other third parties are included.

4. Because samples are the source of the data, luck becomes an important component in gathering broad enough samples to capture the extremes in a typical performance bell curve. The future of a customer contact center lies in these anomalies.

Logging Systems as a Source of Data for Quality Management

As quality monitoring became a significant application in the mid-nineties, the logging vendors awoke to find quality monitoring vendors installed beside them, gathering duplicate samples of the 100% call data they already had recorded.

They followed by offering the logging user the ability to select specific samples from the calls they had recorded which amounted to 100% of the data, beginning at the beginning of each call. This worked for a company that had a compliance requirement and had already installed the logging system.

At first proposing a logger as a quality data gathering system made little economic sense as the "plumbing" was more expensive than a quality monitoring system. The logger required connection to every trunk or agent to be monitored, a codec and a recording path to the storage media.

However, there is a distinct advantage to being able to get to any accurate sample for monitoring and review from 100% of the call data and to follow a call from any perspective: customer, process or resource (VRU or Agent) or through any call state. Revisit Figure 15.2 on the advantage of total record versus a less than 1% sample

Logging systems require a simpler passive attachment as they can be attached in parallel with the trunks or the agent instruments. There are some tricks to this due to the types of trunks and/or proprietary instruments (and signaling) offered by the popular ACD trading turret and PBX vendors. Most of major logging vendors have answers to these quirks.

The original objections to using logging hardware to accomplish quality monitoring are fast departing. Most of the second generation recorders based on closed databases and proprietary servers of the early loggers from Comverse, Eyretel and NICE are being replaced with the third generation of digital recorders. But, many of these systems are developed in overseas environments and are behind in the latest technical developments out of Silicon Valley and ignore American business culture. These cultural gaps are NOT bridged by email and the phone and show up in slow development and time to market.

The later US-based vendors' products are based on the latest open standards (NT and UNIX) and servers available from almost any source, with open SQL indexing in the native data and inexpensive storage. This also negates any economic value pure quality sampling systems may have had. This means that third generation open platform vendors like Mercom and Voice Print spend development time on the applications, not on maintaining basic operation of the coordinating indexes and recording record and plumbing.

Selection of the actual database engine is also critical to the number of records that can be managed by simultaneous users and whether or not this can evolve to real-time "at risk" customer analysis. The key is a massively robust SQL like Interbase or Oracle. Avoid MS Access-based applications, as they cannot grow to meet the demands of a busy recording application.

Data Screen Capture

This may or may not be required by your customer contact center but takes on greater significance as you expand your after-the-fact call analysis from simple quality monitoring to workflow analysis.

In a quality monitoring application it is generally acknowledge that listening to the voice component and looking simultaneously at a view of the agent's current data screen is a valuable strategy. Gathering sequential data is not a simple process. There is software on the "client" workstation gathering the data and sending it to a server-based recording routine. There are two methods of doing this.

First, these systems capture the actual bit mapped image of the initial screen. This is done in its entirety. It is then updated in one of two ways:

- "Snapshots" of the image are made as changes or data is entered. Early systems captured the entire screen but quickly realized production processor power and LAN capacity suffered with this back office function. The interim step was to capture a fraction of the screen (1/4, 1/8 or even 1/16) as required or in 5 or 10 second increments. The problem that arose from this method was that we were leaving the "actual" and offering a "facsimile" of the session, i.e., a "slide show" when a "movie" was desirable.
- A more recent strategy is to take the original snapshot and update the character changes by collecting the changes through the command graphic interface routine in a Windows session. This means that significantly less data is being streamed from the client workstation to the data screen-recording server.

The next step is to take this data stream captured at the workstation, and compress it and throttle the client to server messaging and data transfer so it could be sent in bursts in lower LAN traffic states. This should occur in the background so it does not displace valuable bandwidth from customer facing applications.

When you look at your data session capture provider you need to understand the data capture method, and how and when the data crosses your network to the storage server. Is it compressed and throttled to be network friendly?

Further, exclusively bit mapped screen capture requires an additional layer of bit to character conversion which means indexing by screen content is additionally problematic.

Review Scoring and Analysis System

The critical component of a suite after a voice sample and/or call screen data is selected is the review system. This requires session playback.

In the case of the early quality monitoring systems, the reviewer may listen to the call session samples from any telephone anywhere, using a voice mail metaphor. This is adequate until you realize that a call is being juggled on a touch-tone phone and a review is being conducted on a separate screen, with possibly a separate session for the data. The move is to consolidate the session control to one screen for voice playback (out of the PC speakers or a headset adapter), the screen session monitoring and the review template.

Additionally, the early voice mail-based review systems that required the service observe ports were initially telephone centric so that playback was only over a telephone network. This meant that any remote playback required some sort of telephone network and incurred any attendant long distance cost. By moving these recordings to data files, the data can be compressed and shipped more flexibly and cheaply.

Quality Review Template

The quality monitoring application allows the construction of a call review template for each call type or class of agent. This template is basically a list of all the call or script elements your company deems important or necessary when talking with a customer or prospect to execute a particular type of transaction. These issues may be the relatively objective presence or absence of a script component or a more subjective, value based judgment. Binary values are assigned where a "yes/no" measurement is applicable such as "did the rep use the standard greeting?" or "did the rep identify themselves?" Other issues such as agent attitude and script delivery may require applying a more subjective value. Here degrees of subjectivity are associated with the issue. Different templates can be set up for different call types, customer applications, agent types or skill levels.

As this proceeds, the call type template is applied and each relevant call milestone or issue is noted in the review for presence or absence or, where more subjective, assigned a value in a range of values, typically a one to ten type of score. The simple addition of these values creates an overall score for the observed quality of this transaction or session.

As you well know, a busy agent may indeed meet productivity quotas, but at the expense of company quality policy or sales objectives. The opposite may also be true. The charm of this type of product is its ability to rank the quality performance of the subject agent, draw productivity data from the ACD MIS and balance the productivity (number and average length of calls handled), against the quality scores and even sales results. This results in a scatter chart to compare performance across agents, groups and even the monitors or reviewers.

By taking calls by agent or group and placing the data on an X/Y/Z chart it is possible to determine where training is most effective or, alternatively, where shortfalls occur. The goal of the reviewer is to find the best practices in the call center from a script, objection handling, close or customer mollification perspective, and "clone" these best practices across the customer contact center.

This is where the real value of this system comes to the fore. It allows a manager to analyze the strengths and weakness in customer request handling and determine the most profitable, effective and customer friendly way of reaching a conclusion that benefits the customer caller and the company.

Without a disciplined and automated way of gathering these samples and integrating them with ACD MIS call handling data, the function simply does not occur in a manner that can be effectively scrutinized and used to optimize agent behavior and productivity.

In summary, when quality review was first introduced in 1990 it was truly breakthrough technology. The shift from quantitative data collection to providing tools for qualitative evaluation was significant. Certain vendors recognized that software-based value engineering in customer contact centers was important. The limit was they almost exclusively focused at staff evaluation and reducing management effort in gathering, selecting and reviewing this data.

Total Recording versus Quality Sampling

A low-grade debate continues between the logging vendors and the quality monitoring vendors. In the past the quality monitoring sample argument scored the most points because of storage economics. Why collect and store 100% of the calls if you are only going to review less that 1%. The exceptions are public safety, compliance or dispute resolution. If you want to look beyond agent quality to workflow improvement, then total recording wins the debate since it's impossible to predict where the items of interest might lie.

However, with the cost of disk storage plummeting to the point where a terabyte (in excess of 200,000 hours of recording) can now be bought for less than $25,000, the debate is now moot. Add the traffic impact on a traditional ACD or PBX of an active service observe switch connection, and passive total recording is a foregone conclusion. The fact that all of the quality-sampling vendors now offer discreet logging tools is simply delaying the inevitable demise of these standalone devices. Sampling and averaging also has the potential to mask the critical data, such as best and worst practices. Again, review Figure 15.2.

The cost of total recording platforms is dropping to the point that they are now less than 20% apart in price from a quality sampling system. If the recording platform you consider today cannot expand to record and archive 100% of your data, it will probably be obsolete before its accounting life ends.

☐ LIMITATIONS WITHIN THE SECOND GENERATION ARCHITECTURES

Pandora's box was opened with the statement that customer relationship management systems are a natural extension of call recording. Various recording and CRM vendors have publicly announced partnerships. This does not automatically mean these systems can elegantly integrate to produce meaningful data about a specific customer. Despite the current "plumbing" limitations and lack of tools the long-term value of this insight is dramatic.

The immediate problem is the current second generation architecture sold by the major call logging and quality monitoring vendors. With its current proprietary file formats and recording servers, it cannot expand to the requirement of open standards. The same is also true for the quality sampling systems. Add the attachment limits and the traffic intensity from a data gathering perspective and there is no upgrade path. Every installed system (Witness, eTalk and other systems) using service observe connectivity (typically Dialogic or other popular switch terminal emulation card) cannot be expanded to gather larger samples without the replacement of basic architecture.

The major logging vendors persist in using proprietary "token" based voice recording structures that can then be mapped to an external SQL index. Separate data (indexes and often servers) is gathered from screen and CTI links. The indexing, sorting and synchronizing of these "closed" data sources adds significant administration that has proven unnecessary in open data and system standards.

Expect the major logging vendors to replace these systems in 12 to 18 months of this publication.

Voice logging philosophy is also not particularly friendly to their new target market, the customer contact center. That is because high value, high-risk call logging assumes there is one application maintaining data record integrity and playback. Customer contact centers need to be able to get into the databases and add and delete records as well as append and strip data for various process analysis applications associated with sales verification, collections and post call documentation. The closed sacrosanct world of logged voice records does not lend itself easily to the rough and tumble of customer contact center process analysis.

With the price of storage dropping radically, while costs of poor process and liability are skyrocketing due to badly handled calls and badly handled employee reviews and dismissals, the argument for total call recording is over.

☐ TACTICAL APPLICATIONS FOR CALL RECORDING

Customer behavior, the revenue they represent and the value of customer retention can no longer be ignored — the evolution of data warehousing and data mining concepts now touch the call center. The question becomes: if there is a record of every call (data in the customer relationship management or sales force automation systems) and audio in a call logging system, why can't a company look for high revenue, high risk or marginal customers through key words or carefully designed informal scripts to illicit important individual customer data, such as satisfaction, media influence and response source? We are reminded of the early chart showing the expense of the actual call (a factor of 1), versus the cost of generating the call (10:1) versus the revenue potential (typically 100:1 and above).

Arising from this is an instant realization: A quality monitoring system that samples and records less than 1% of the total call and data traffic is a tease at best. But there are other immediate tactical applications for call recording. Let's take a look.

Sales Verification

This is an application where a third party, either individual or regulatory, requires independent third party confirms of the sale. These are typically outbound consumer calls made with a predictive dialer.

With these open third generation logging and quality platforms there are a number of features that allow users real process savings through applying flexible recording to pure telephone verification applications. Significant labor cost reductions are immediately available and present an accelerated investment return in verification environments.

More importantly, no interruption occurs in the telemarketing script where a forced verification transfer breaks presentation and often generates an arbitrary loss in continuity. This frequently causes the caller to hang up and a sale to be lost.

With total recording of all customer contacts (voice and text streams), verification recording is an extension of the existing recording functionality. This offers the immediate advantage of reassigning expensive staff in either local or third party verification groups to primary sales or production tasks. Unlike the recording of verification events driven by business rules-based "recording on demand" (manually or automatically) as on some second generation platforms, this process requires no system decision, added process time or transfer of the connection by the solicitation staff to launch recording. No interruption or transfer of the call to the verification group for confirmation of a valid sale is required (unless regulated). This is now one seamless conversation.

Additional differentiation occurs in the back office use of the open verification data. This now exists in an open database for replay by the verification staff, third party client reporting or customer accessible documentation.

An example of the repurposing of recording data is to send an email to the customer thanking them for their business. Attached to the email would be a wav. file confirming the sale; or within the body of the email message would be an 800 number and perishable PIN the buyer could use to replay their commitment (as a personally accessible documentation).

In the case of the third party client, the positive and verified responses can be sorted from the null records and transferred to them as a final completed daily sales file. In this manner only valid sales are maintained as valid response records.

The record indexing is as rich as a client wishes to make it with completely open SQL compatible meta-tag fields. This enables the contact record to be identified and recovered based on any appended meta-tag collected in real-time from the telephone network, switch (TDM or VoIP), predictive dialing engine or solicitation database and order entry system.

The real value is to move most verifiers into primary sales roles in all but legally required applications. These individuals earn in the range of $12 to $15 per hour so the payback is immediate

Collection Commitments

You can also use the same recording technology with an outbound predictive center. Collection departments regularly speak to late and delinquent debtors and in most cases a reminder is all that is necessary to collect payment. Many collectors use a logging system to track that these collection calls are made in compliance with state and federal laws. Now this logger, provided it can perform the function described in sales verification above, can be repurposed to record a debtor's commitment to pay as agreed or under a new schedule. Again, the recording can be sent as a personal email or access with a personal perishable PIN for reviewing audio documentation of the agreement.

The immediate value in this application is the "lift effect" on collection success and payment velocity.

307

Miscellaneous Voice Documentation

There are many other transactions where after the fact audio documentation of the conversation could be informational or provide formal terms to an agreement. The notion of automatically launching emails with .wav attachments or information on client access to these voice files as after the fact documentation has real charm as an application for reducing paperwork. The value here is the potential to eliminate extra paperwork and process.

☐ CUSTOMER EXPERIENCE MANAGEMENT: THE NEXT STAGE IN THE CONTACT RECORDING EVOLUTION?

All seems relatively settled between the two applications of monitoring for archival or quality assurance purposes. You either chose a call logging system that gathers 100% of the call records or a call sampling system for random quality assurance calls. Both rather dull and tactical applications that four public companies NICE, Comverse, Eyretel or Witness need to make more exciting. As a response to this, they have singularly announced, then collectively claimed that they have moved into the customer experience management business. Now these companies claim this to be a strategic application. The good news is in a couple of years they may be right, but today, given the state of their architecture and record mapping and recovery, it's only a gleam in the eyes of financially questionable companies.

Add to this the fact that the data is being gathered in the wrong place and it is pretty clear to see these proprietary systems will be supplanted by open architectures within truly dynamically applications assignable customer data from central database systems.

Mapping an Individual Customer Experience

The logic behind being able to map an individual customer experience is elegant and has very real application in being able to capture the "context" of a customer contact. The example is "the angry phone call following the three unanswered emails."

How these recording companies are going to do this is essentially theoretical (at least currently) as they are just beginning to build the links needed between their recording systems and switches, email, web applications and CRM systems and other vital data. The problem they face today is the cumbersome architecture and index restrictions. Simple file export is not adequate, as this must be a near real-time application.

What is possible now? The ability to identify certain values related to the call type, agent, disposition code or other key field gathered from one of the systems involved in the call process. But, this has yet to be extended beyond phone calls; although with the correct database connectivity, there should be little objection to mapping all contacts related to a specific customer or case in chronological sequence.

The best that is available today is an ability to present this in a visual form so that patterns and anomalies can be spotted in massive amounts of data. The leader in this space then allows replay of the call by clicking on the graphical "index" identifier.

At this time, the whole discussion about customer experience management or data mining in the context of call recording is fatuous, since all that these systems are currently capable of doing is gather data. This is even more limited because of the sampling strategy of quality monitoring systems or the closed indexing of the logging systems.

A couple of the second-generation logging vendors offer a table translation technique for down-line mapping and recovery of calls. The few companies now offering third generation digital recorders deliver native SQL indexing on standards based platforms. Once this transition is complete, true customer experience mapping can occur, and with it the opportunity to analyze and act on the data.

Improving the Customer Experience

The question becomes "what to do with the data to improve the customer experience?" This gives rise to a whole next generation of call processing and workflow improvement tools. As of now, the only next step that is occurring is the application of training or elearning tools to agents after a skills gap or deficiency is identified manually. This may be aided by some sort of data calculation that indicates agent A is less adept at call type B. That is all, but it offers significant potential for the future.

Matching Recorded Voice to Recorded Data Screen Sessions

The reviewing of matched call recordings to recorded data screen sessions offers greater insight into call process and workflow. One recording vendor has gone as far as mapping and matching the energy wave of conversation and the keyboard activity so screen and field changes can be observed. The opportunity here is to iconize the voice recording component with screen and field "milestones" that indicate where in the workflow the call was as indicated from a script or data entry point of view. This has not yet been achieved as it means combining mixed media (analog and digital expressions). The impact of achieving this from a workflow analysis perspective is powerful.

Contact Recording Application Advances

The next major advance in contact recording will be incremental as it relates to applying available recording techniques and technologies to real world applications. This means simple things that have a high return on investment like improving call flow, labor displacement or revenue improvement.

The business applications of telephone sales verification, collections or complex customer support can benefit now. The immediate benefits are a significant cost reduction by displaced labor, lifting collection rates, eliminating post-call agreement documentation and presenting an accelerated investment return. An open architecture digital voice recording system can execute a number of additional routines beyond index data gathering and simple voice recording. For example, launch an email with a .wav file attached, that's a transaction confirmation indicating the voice documentation can be reviewed by dialing a number and entering a temporarily assigned PIN to review the content of the call. This can apply to verification of the collection of a debt (with debtor access to the recording) or confirming information delivery (service and support) or terms and conditions of an agreement. The voice documentation can be accessed remotely or launched as an email attachment to the customer.

Behavioral Trip Wires

This assumes these contact collection and index systems are capable of near real-time recognition and identification of the critical data that identifies a customer as going "at risk." The goal here is to be able to set "behavioral trip wires." This is so we can detect "out of range" behavior of customer contacts, product sales or performance (service issues), agent (individual or group) performance or processes are detected early.

Call Recording and Voice Recognition

As the major logging and quality monitoring vendors searched for differentiation, they first embraced data mining and customer experience management with little understanding of how to deliver real value beyond a marketing smoke screen. Now they are actively discussing the potential of voice recognition as it applies to recordings. The problem is that theory has gotten ahead of delivery. The major logging vendors have yet to replace proprietary platforms and indexes with open systems with native index systems such as SQL, but they still want buyers to believe voice recognition is available in the near future.

Voice recognition is a rapidly maturing technology that requires sophisticated software and powerful computers dedicated to the analysis of ONE conversation at a time to recognize key phrases that indicate critical data. It's a well-researched and stable technology.

The problem in logging data is that in most systems it is captured and compressed at the trunk or extension and carried in a compressed form to the disk for storage. There it is written concurrently with thousands of other calls in a (often proprietary) compressed state. To analyze this data, it must be decoded (compressed data to uncompressed), then speech to text, then search on either key word or phrase to get to meaningful data. You need massive computing power to do this down line. And for what purpose?

Nowhere in the current logging architectures is there a easy place to intercept a call in its native form and apply real-time speech (or phrase) recognition. This is further complicated as the conversations in traditional telephony are monophonic. This means there is no distinction between the voice sources (both parties as speaker or listener). VoIP makes the initial speaker separation simpler as the voice packets are stereophonic, meaning there is a separate packet for speaker and "listener." Though there are other problems with analyzing IP voice.

During the sales cycle, do not allow a vendor to distract attention from dated architecture with "gee whiz" voice recognition features when they have not made it out of a closed second-generation architecture.

Shedding Call Center Workload

This approach requires rich call process tagging so that call volume can be granulated a number of ways. The immediate application for call recording is identifying the call by: type, values, costs, risk, and total and relative volume. This then allows root cause analysis of the workflow.

Victor Nichols, CIO of Wells Fargo Services Company, made an astute observation, "no one calls our retail banking call centers for fun. They call because something broke." (Twenty-two million times in 2000!) "The bank branch failed in some manner, an erroneous statement data or ambiguous marketing message or confusing website creates the need for our customer to call." His request: to analyze the cause and content of these calls and help short stop voice calls and divert them to self-service or other less costly communication strategies.

☐ IN SUMMARY

One interesting fact that is coming out of call center operations is that unless it is a pure direct sales response, most calls happen for less than positive reasons.

There are serious benefits from the analysis of why calls arrive at your call center and identifying the "something broke." By applying the lessons learned, this could dramatically improve customer service. Bad marketing messages,

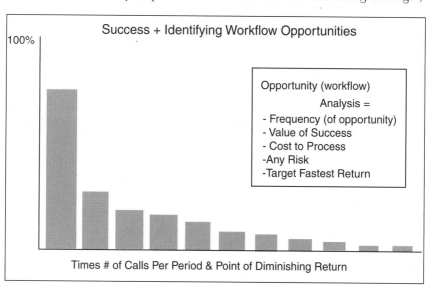

Fgure 15.5.

poor retail service, ambiguous instructions and documentation, product quality, and so on. Beyond incremental changes in recording applications, this is the future of customer contact recording, which means being able to tag and analyze root cause, outcome and other relatively binary data elements so that simple root cause and workflow analysis can be applied to existing contact loads. Load shedding customer contacts through improving upstream process is a major opportunity.

The next opportunity is understanding customer information requests (the what and why) and delivering the best and most acceptable solution. Understanding these issues probably means a significant improvement in self-service strategies by improving customer self-help websites, content, FAQs and integrating channels like fax back and voice to accommodate those who are less web articulate. With the right call data, quantitative, qualitative and content, much can be done once these call recording vendors deliver third generation open architectures and rich database indexing. Expect significant developments in this space.

CRM Within the Customer Contact Center Environment

A return to the personal service that's offered by the neighborhood merchant is an economic impossibility. The edge for the local business is its limited customer base and the associated knowledge base that allows delivery of personalized service. This is normally not possible for a national company; yet it's widely recognized that personalized service can grow sales.

There is also the growing need to effectively attack the rising cost of sales. As a counterattack, many companies have turned to "inside" sales channels. They are also linking their corporate databases to customer call centers, turning these centers into powerful sales and service support systems with the help of sales force automation (SFA) tools.

To accomplish these types of transitions is significantly challenging. Surprisingly, the challenge occurs more within the cultural make-up of the organization as a whole, rather than within the implementation of customer relationship management (CRM) technology.

But where does the call center's leading technical feature, the ACD, fit within the CRM realm? An ACD has great value in handling high volumes of like calls, but its original focus of homogeneous treatment of volumes of callers is increasingly perceived to be a disadvantage in this day and age, where sensitivity to individual customers is considered a competitive edge.

Database technology (the underpinning of all CRM technology) allows personalized treatment to be delivered to a much larger base of customers on a grander scale — the inherent promise in any CRM proposal.

The trade-off when integrating CRM solutions within a contact center environment, however, is that progressively sorting or granulating the call load down to a specialist begins to defeat the early notion of contact center economy of scale. Although the introduction of self-service channels can solve much of this dilemma, these self-service solutions still must be backed with default to a call center populated with generalists who know their way around a company, its products and procedures from the customer's point of view. As a best of breed example, take the USAA survivor relationship team we discussed in the Preface. USAA sensitively concluded the bereaved shouldn't need to analyze and understand the company's corporate structure and silos, so they intervene with a specific relationship management team to help them in their time of need.

☐ CALL SOURCING

Much of the necessary personalization data can be provided by telephone network identification data. In the case of inbound calls, call type identity is being increasingly enhanced by the use of expanded DNIS (Dialed Number Identification Service), calling party number or ANI (Automatic Number Identification) and voice response technology to request more specific identification (account or social security numbers) from frequent callers.

For example, DNIS service can be used to identify a call type by publishing this particular number as a source of information for a particular need. Alternatively, the DNIS digits can be translated to another value based on, say, the location of the caller. This brings us to regional caller identification and sales territory management. Hewlett-Packard (HP), a computer manufacturer, uses DNIS to solve a cultural problem and to boost productivity through inter-regional competition within its customer contact centers.

HP advertises one number nationally for its supplies department — the use of one number or "one number presence" is important to HP and many other companies. But it isn't a panacea. The customer supply center needs to know what region the caller is calling from to assign the call to the sales team responsible for that particular region. HP could have advertised multiple telephone numbers across the country, but this would create confusion and contradict the marketing image HP wanted to project, i.e. "it's simple to do business with HP." DNIS couldn't be used since all callers dialed the same four digits. AT&T came up with the solution. It provided a digit translation service that could identify the region from which the caller was calling. This allowed HP to take advantage of its marketing campaign, the ACD to route the call, and the customer supply center to efficiently assign regional sales responsibility to the appropriate sales team.

The technology allowed the transparent assignment of the call by eliminating a step in the call (connecting to an agent for the "where" inquiry), thus shortening the call length. The real win was gained in promoting sales competition between teams while maintaining accountability. A single number is published by HP so marketing confusion is kept to a minimum and no economy of scale is lost through partitioning trunk groups by offering separate customer call center numbers for each sales region. This is a most intelligent application.

Where further identification of a caller is required, interactive voice response (IVR) technology can provide absolute identification. IVR technology can greet the incoming caller and ask for the touch-tone or spoken input of some personal identification number. This may be an account number, telephone number or other identifier. The IVR device can check this against an associated database, validate a "paid up" customer, determine the product and model in use and route the caller to the correct person or voice response script to satisfy the request.

315

ANI was originally anticipated to be a breakthrough in caller identification but due to the lack of database matches or enrichment did not work out as planned. ANI has proven to offer more of a physical security layer (an authorized phone caller) rather than a precise logical identification as there is no assurance the account holder is the caller on a physical telephone number, or vice versa.

Originally the only way to identify a particular inbound call type was to terminate trunks of a specific type on a fixed set of ports (a broader application of older DID technology). DNIS changed this. Now an organization could promote a specific number, coupled to a specific promotion, product or service, allowing callers to be identified by a specific promotion or need, i.e. all calls arriving on these ports are associated with a specific call or business type. Although in the beginning this was just one number related to a physical port. At this point in time, DNIS can be used to provide a certain predetermined type of service to the caller: such as to treat them as priority callers or provide other personalized treatment or collect statistics about a particular type(s) of call.

AT&T then introduced expanded 800 services, which included DNIS, allowing the sending of the last four digits dialed by the caller to the customer call center. In doing this AT&T effectively repeated all the advantages DID had provided, but on an 800 inbound WATS trunk group. This meant the last four digits were received at any port, so a logical identifier accompanied the call and enabled more media source numbers than trunks.

Now a trunk, physically tied to a specific port (with a physical hardware address), no longer needed to be dedicated to handling one type of call dictated by the port/hardware address and associated call treatment table. With trunk-based table definition, calls are deemed to be identified by this address. Currently DID, DNIS and other identification services allow logical call-by-call changes to the identification data and thus call treatment or routing. The system expects some network-derived data (logical dial number versus physical dialed port (number/address) to identify the call type. Upon receiving this data, it reads it and does a caller treatment table (logical) look-up. This is described as software defined as opposed to hardware (dedicated port) defined. If this identification is absent the system defaults to the physical address table.

The downside is that the advertiser must manage the publication of unique numbers in each medium and relate that to a sourcing table. These must be coordinated with the customer contact center to ensure measuring the individual effectiveness of each medium (print, radio, TV) by the number of "hits" or responses to the dialed number.

Media Sourcing

Call sourcing can become media sourcing, where the call or web response is tracked back to the advertising medium generating the response. While this is relatively straightforward with web and email sourcing, telephone calls are less sim-

---

Caveat: Unlike DID, DNIS service requires the user to have more circuits than assigned numbers. If an 800 subscriber wanted to use up to 50 different numbers in different markets, a minimum of 51 800 circuits must be terminated on the customer switch. A second feature included in the service is non-rotary hunting. The inbound caller would dial 800-NNX-0030 (the thirtieth # in the trunk group sequence) and arrive with the DNIS designation 0030 on the thirtieth or subsequent trunk. If 21 callers had arrived earlier, and all trunks from 30 to 51 are in use, this caller will not hunt from the 51st trunk back to 1. Instead a busy signal is returned to the caller, even though trunks 1 through 29 are idle! Most carriers have changed this anomaly, but check directly with your 800 vendors.

One work-around to reduce the impact of this problem is to assign the heaviest usage to numbers early in the trunk group. This requires sophisticated understanding of your anticipated loads, clever grading and constant monitoring to tell whether or not these decisions remain correct. This is a value and insight a sophisticated ACD system provides the owner.

ple. Mainly because identifying individual media channels is more difficult when they are more numerous than the technology can discern automatically. A request for such information from a call center agent to the caller is often forgotten or deemed less valuable than getting to the next call, so it is skipped. This is still an open issue. To give this type of sourcing a chance to work, there must be consistent scripts and proper execution by the individual agent. Even then, often the caller cannot recall or just doesn't care where the advertisement was viewed. Unfortunately, the data is vital to intelligent advertising media management.

The ACD

What all of the above means is that the flexibility in all call center systems must be increased. Also, with the new inroads into call sourcing, ACD systems instantly needed to become smarter. Many manufacturers have provided the necessary upgrades or are working diligently on these issues.

☐ INTERACTIVE VOICE RESPONSE

An expanding trend in leading customer contact center installations is the meaningful computer integration of voice and data systems. Linking computers and telephone switches makes sense. Linking computers to interactive voice response to identify callers and deal with simple inquiries is even more powerful.

Call length is cut and callers are treated more personably and provided more options. Other issues driving this are the necessity to do more with less staff (often less skilled staff), increased time pressures on customers and customer call centers alike and the need to be competitively different by delivering better customer service and support.

There is still tremendous undeveloped potential for the integration of voice response technologies, especially as self-help tools and as superior and less costly response intervention.

The real value, however, is the seamlessness of the integration upstream of the live call center. Merely attaching a IVR device to an ACD as an alternative to a live agent has triage value, but integrating it into the call processing sequence so as to enhance the convenience, speed and quality of the callers experience brings real value to the call center.

Overflow

Among the many applications IVR devices are able to fulfill is taking a message for later review by the intended party. IVR can perform this function in an ACD, relieving the queue of some of the waiting callers. Once the inbound call volume has dropped off, theory has it that the agents will call the IVR device to collect the messages and, if necessary, call the callers back. It doesn't work as well in practice as the vendors would have you believe. The first thing an agent does not want to do after working a busy period is to immediately start making calls to a voice response device to get more work. There is little incentive to do this, particularly if there is manual phone number transposition required. There is no short term reward: The harder they work the more work they have to do!

This, however, is how many ACD vendors integrate IVR messaging. When a decision has to be made by an agent to make an outbound call, there is a philosophical obstacle placed within the process. Also, pure call back messaging doesn't work seamlessly, or even that well, because of two facts:

1. Most customer call centers are staffed marginally to begin with. The busy hour load engineering regularly accepts abandonment and permanently lost calls. To now add message retrieval and call back tasks to the inbound staff would place the center in permanently understaffed condition.

2. Call backs must be performed promptly as many callers are calling with impromptu requirements such as impulse purchases, some immediate help through a computer program, whatever. If the promised callback does not come quickly, what started out as a service improvement strategy comes back to "bite" the customer call center. You are perceived to be providing worse service than not answering the original phone call at all.

A leader in the ACD market recognized this and integrated the IVR in a manner that addressed the decision points to make them as transparent as possible. This vendor, like all the other vendors, recommends IVR messaging as an alternative to queuing, but this vendor also solved many of the problems related to call back. The solution: erode the difference between inbound calls and outbound callbacks, i.e. the vendor made the call backs appear to be almost the same as an inbound call, at least as far as the agent was concerned.

For example, upon encountering the queue, an incoming caller is given the option of remaining in queue to reach an agent or leaving a call back message. If the option to leave a message is taken, the caller is prompted to leave their name and any information about their request. This is done with conventional scripting and recording. When they are asked to leave their telephone number, they are prompted to dial their callback telephone number on their telephone tone pad. The IVR device decodes the tones and records the digits. The caller ends the call expecting a call back. As the customer call center inbound service level begins to improve and fewer incoming calls are being presented to the system, the IVR feature begins to insert callbacks into the incoming call process. They are identified as callbacks to the agent. Simultaneously, the callback telephone numbers are displayed on the instrument display. The agent can listen to the audio message, then originate the out call through a single keystroke, (which causes the displayed digits to be dialed) or dial the call and listen to the audio message during the few seconds of network call setup. This is intelligent integration.

These dynamics also apply to "click-to-talk" requests, as these must be responded to near real-time or the business and your credibility is potentially lost.

Today, approximately 65% of web hits do not result in a sale, however this does not compare unfavorably with "the just looking thank you" phenomenon in a retail store. The reality — not everyone buys! The numbers relating to the propensity to buy or conversion value of click-to-talk contacts on the web inquiry, click-to-call, or order by web or phone are still fluid. However, allowing any process objection to the natural flow of the buying process almost ensures a lost order.

The basic question still remains "how do I make money at this?" The answer is not surprising. Rather it's a matter of looking at the data and tracking by steps the inquiries from start to close and then analyzing the attrition rate (i.e. customer experience mapping). If you track web hits to conclusion, the fall out rate can indicate where the "script" or process "breaks."

☐ DATABASES AND SYSTEMS INTEGRATION

The majority of calls placed to or from a customer call center by your staff are an event in a chain of events. To be an effective communications channel all the staff in the center must have access to customer history and status database. Hence the birth of the customer relationship management (CRM) category.

We know that customers will call anyone and ask nearly anything to get what they want. The requirement is that every person who may possibly encounter a customer (or a co-worker helping a customer) must have access to the tools necessary to serve the customer in the best interests of the customer and the company. This could be seen as an argument for CRM. Actually, it's really just an argument for a decent customer database that allows departmental systems to inquire and update their particular interest in the data so all customers can be dealt with in context.

Look at the example of the USAA Survivor Relationship Team related in the Preface. They did not need a CRM system. Rather the need was for customer database access and continuity across each corporate discipline and skill set, consolidated transparently to the customer. The databases must be accessible to all of the potentially customer facing applications, which differs from one application holding the database as was envisioned by the original CRM sales tool architecture.

Increasingly, this CRM system communicates with the phone system, so certain events occur automatically. Customer screens appear with the call, customer screens follow a customer call if they are transferred to another department and the center can even allow the customer treatment choices with interactive voice response (IVR) without your staff being involved. This makes for more satisfied callers; provided that these systems are logically connected and integrated into your inside sales and service center.

The first examples of these connections gave rise to some odd and inequitable results. For instance, most voice response devices are designed as standalone devices. The builders assumed they would pretty much process calls without heed to any integration with live operators on an ACD — a call arrives at the ACD and then is automatically routed to the IVR system. The caller conducts their transaction. At some point, either intentionally or as an IVR script default the caller may be switched back to the ACD and a live operator.

When this happens, two conditions must be recognized:
1. Is this a new call (to the ACD)? The call passed through the ACD once and then was routed to the IVR. Now it has been returned to the ACD.
2. Should it be queued differently as the caller has been possibly been queued once on initial arrival and is now being queued again? Double jeopardy.

Few implementations has overall knowledge of the contact status or control or relationships with open cases, such as an outstanding email inquiry. This makes for disjointed treatment at an event (micro) and a relationship (macro) level. The dynamics of web, email, ACD, IVR and human operator interaction with callers is lost in the pattern of transfers between different machines.

To build better management and service solutions the logical approach has been to couple the customer database and computer system managing this to the call center switching system. A major benefit initially proposed for linking a switch to a computer, and automatically delivering the account record to the agent was a reduction in call length. This can run a minimum of 6 seconds, though it's regularly reported to be as high as 20 seconds. The real value goes well beyond shortening calls and delivering screens automatically into automated data gathering and workflow improvement.

The goals of the major vendors of these types of links seem to be one or all of the following:

Better, more logical and faster call flow. This benefits both the calling customer and the customer call center owner. Also, calls are shorter.

Giving callers more options. Many of the simple information request calls can be diverted to self-service channels like a website or voice response machine. But, in addition, there's the opportunity for more personal or focused treatment of the caller. This should also means shorter calls, by giving one machine the overview, call status control and data gathering responsibility. Vendors approach this from their particular bias.

These two objectives were initially addressed in a variety of ways. One solution pitched by major manufacturers as the ideal solution was to give the host computer the job of overseeing the entire operation. This ignored history, i.e. what about cost, response times and software development? Also, distributed processing had eliminated many burdensome mainframe tasks that could be relegated to minicomputers, which also discredited that approach.

The next solution many vendors presented was to shift processor intensive decisions and look up steps out of the customer data system into the gateway or the switch. All the switch or gateway system needs to do is to issue a command to send customer "record A" to terminal "address 333." Many found that approach appealing. Rather than provide the data system with raw information it must digest, refer to a table and issue a record routing command to the data terminal matching the extension number found in a host based table.

In many early implementations of switch-to-computer links, this was the strategy adopted by the ACD suppliers. Computer response times became less critical and no extra code was developed as the commands issued to the host over the switch-to-computer were by way of terminal emulation.

Simple terminal emulation may be frowned on by many MIS and engineering professionals, but it achieves major results quickly. Measurable business results are often needed quickly in the integration efforts as they justify investing in the next step. Get results then go for elegance in subsequent peer-to-peer integration.

Linking the incoming digits with a departmental system. This is accomplished through call and caller identification. There are three sources of call identification.

- Dialed Number Identification or DNIS identifies a call type by capturing the specific digits dialed by an 800 caller.
- If this is a local call, Direct Inward Dialed (DID) digits can be used in much the same way. A common trunk group also saves on circuits as logical identification (by dialed digits) means you no longer have to set up a different trunk group for each call type.
- Another technique for call type identification is IVR or automated attendant. An IVR device is put in front of the customer call center so callers can select the service they want. "For sales press 1, service 2 and software support 3." There are two sources of caller identification.
- Automatic Number Identification (ANI), which may be provided as a subset of ISDN in an out-of-band format or as an in-band signal. This offers great promise; but users have discovered this is not as precise as envisioned. Anyone in a family or company could be calling your 800 number.
- A better, more precise and less controversial way of doing this is to use IVR and ask the caller to enter some meaningful number such as personal ID, social security or account number.

Once the DNIS, ANI or other identifying data is captured by the ACD a number of things can be done with it.

- Triage and routing,
- transaction specific screen pop, and
- measurement.

The most simple is the routing of the call to the intended agent group. This group can be apprised of the call type through whisper queue announcement or a numerical display. The more intelligent devices will expand this digital data into plain language alphanumeric descriptions of the business type or pop a transaction specific screen. Most systems supporting DNIS also collect data about each logical call type for subsequent administration.

☐ INTELLIGENT NETWORKING

With the emergence of voice over IP and real-time online chat, the Internet has taken call centers to a new level. Managers are now scrambling to transition into IP-based computer telephony integration (CTI) and advanced IP-based customer contact centers so they can serve their customers better. Two challenges remain in the transition of agent culture to new methods of dialog: the unnatural aspects of real-time chat and the sudden need to read, comprehend and write a response in a previously almost exclusively aural environment.

In distributed call centers, agents from different locations are all connected by a network, which can be a WAN (wide area network), Internet, extranet or VPN (virtual private network). Data and processes are shared across multiple sites for multiple tasks. IP networks offer a preferred environment for integration. All types of messages, whether they're received via phone, fax, e-mail or snail mail, can funnel into a centralized contact center. Agents now need to respond via any method the customer requests, while simultaneously viewing complete on-screen information on the customer. Channel types, relative volumes, priority, value, cost and risk are vital issues to understand in deciding how to emphasize response methods and priorities.

In the past, networks were analog, meaning they were slow and somewhat error prone. As digital data transmission entered the mainstream, there was an explosive growth in private T-1s and even higher capacities that could host private, high-speed digital service — usually a VPN. But with this higher digital speed comes an increased need to pay attention to reliability in every layer of the system.

323

General Motors, Citibank and many others have built their own corporate networks, and quick corporate growth has come to depend on managing these networks to match business demand. How to control and use efficiently such expensive digital services? This question led to the Intelligent Network (IN) concept. No longer do you have to own or lease the physical circuits to get all the benefits of an intelligent corporate network. You can participate in the benefits through software-based control and dynamic capacity allocation. IN is built around Signaling System 7 (SS7), which is an integral part of ISDN. SS7 controls the telephone network from "out-of-band" — on a separate channel.

Today's signaling is in-band, with the information transmitted (silently) on the same circuits as those that carry the voice conversations. There are three basic problems with this in-band signaling: First, some 35% of all calls aren't completed, yet expensive circuit time is used to setup and manage these incomplete calls. Second, it's relatively rigid and offers no dynamic bandwidth and third, it is vulnerable to fraud.

The IN alters the way the telephone network delivers services to end-users. Today, the call-processing and decision-making functions for the network reside in the switching system. The IN distributes these functions to a different set of elements.

There are four key elements to the IN:

A *Service Switching Point (SSP)*: The switch, which receives, sends and processes signals, and controls the interconnections of transmission facilities.

A *Service Control Point (SCP)*: The switch manager that, along with the Service Switching Point, controls other network elements. Traffic data and other network support information are collected in SCP.

A *Service Manager (SM)*: The database and customer database access device that controls the databases associated with the network. This can be the entry point for end-users to control their network and the information about their services.

An *Intelligent Peripheral (IP)*: Non-switch resources, such as announcements, music and other network utility devices. The IP interconnects with the SSP and provides specific information about services.

These newer strategies as espoused by vendors, such as CISCO ICM allow end users direct access and control of network functions through an ISDN interface. The benefits of the IN are simple.

Before the IN, changes in a company's service were stored at the switch level. That is, every switch point in the network. Many switches, many databases, each of which had to be coordinated. With the IN, these service requests can be controlled and stored at one single point, the SCP, which is accessible to the ISDN end-user.

These intelligent routing efforts most likely will morph into the control intelligence of multi-channel packet-switched customer satisfaction network control of the future, which may be beyond the skills of corporate IT and network managers, not to mention customer contact center management.

Integrating the Internet Into a Traditional Call Center

When the Internet first arrived, promoters predicted the speedy demise of the call center. They also predicted the death of print media, as we know it. However, print and paper, which survived the telegraph, telephone, radio and television before the Internet, will also survive the Internet. The spontaneous aspect of human nature and the overwhelming desire for convenience ensures traditional media's survival, i.e., bed, bathroom and beach don't fit well with limited battery life, bandwidth and screen technologies.

The pundits were also wrong about the demise of the customer contact center industry. Customers are like water and gravity, in that they will find any path to the resources that they believe they need from your company.

Convenience, not displacement of call volume, is the reason an organization should add the Internet as a communications and customer service medium. The Internet channels offer extraordinary convenience and economy for both to the customer and to the owner of the customer contact center alike.

People do not call a call center for fun. Typically it is because something "broke."

Think about it. Have you picked up the phone and called your retail bank or your own call center for the warm and rewarding experience of speaking to a live agent. No, normally this type of event is driven by the failure of some other process that requires the intervention of some other human who may, or more time than not, solve the problem.

The reality is the fact that today most calls made to call centers should probably not happen. When the root cause of the call is understood there are a number of things that can be done better to ensure the customer has a better experience, negating the need to place a call for help or additional information. Placing a call to a call center, encountering a voice response machine, being queued to wait for a live agent to be given aural information is a partially satisfying transaction at best.

With the Internet entering the picture, the goal in many customer contact centers is to get callers off the phone to less costly and more satisfying channels. Also, many centers are mandated to analyze the reason for the call, and if it is a process failure, resolve the failure to reduce the possibility of a reoccurrence. Here is where web-based channels excel.

☐ CALL ANALYSIS

Before you commit to these new communication channels, you need to understand why people call your call center. You must dig into the core of your calls to find out:

- The reason or root cause for the calls.
- What these calls are about.
- The call volume.
- Relative frequency of the calls.
- The value of the calls.
- Cost of the calls.
- Risk.
- The expectations the calls represent.

  Then chart the most frequently encountered calls by type, then prioritize by volume, cost, value and risk.

  After that, ask:
- Could these inquiries, or orders, be better served by integrated web processes?
- How should these channels be backed up with traditional call center services?

  With this analysis it is relatively logical to see where the focus of workflow repair and self-service solutions efforts should apply.

☐ RELATIVE TRANSACTION COSTS

You have the results of your call analysis in hand, but you still aren't ready to make the move to a multi-media contact center. You·now need to analyze the various cost structures. The relative response costs to service various customer contacts are very clear:

- A web "hit" that is responded to by information on a website costs approximately a $1.00 per hit since the total cost of the website are spread over a universe of visitors.

- The cost of an unexceptional email request or an order entered on a standard screen template that limits customer creativity by forcing the request into specific subjects and categories and responded to by an intelligent email or order system is about $3.00 per transaction.
- An email request that needs a customized or personalized answer balloons to around $10.00 per transaction.
- An IVR handled call also can be processed for about $10.00 per call.
- Click-to-Request a call back costs in the range of $20 plus the hit cost
- The cost to handle a conventional telephone call (click-to-talk) begins at $21.00 per transaction, and rises from there.

The interesting corollary is the speed of the required response. The web hit expects information now, but in the context of an automated system. Assuming you are following and enriching the FAQs, essentially this real-time request is responded to with an automated real-time answer. The form email request or order is expected to be responded to within a reasonable time (12 to 24 hours). The customized answer has the expectation of a similar response time, while a call center "has operators standing by" to answer the customer in absolute unforgiving and expensive real-time.

The following table sorts the medium by channel cost and customer response expectation.

Generally speaking as you move the transaction from the top of the table in Figure 17.1 into richer and more complex media channels three things escalate, though not in lock step:

1. The cost of the transaction

Communication type	Cost	Response Expectation
Web hit	$1.00	Now
Templamatic email	$3.00	Today
Custom email	$10.00	24 hours
IVR handled request	$10.00	Now
Call-back-request	$20.00 + web cost	30 minutes
Telephone call or click-to-talk	>$20.00	Now

Figure 17.1.

2. The expectation for a speedy response, and

3. The client's emotional attachment to the content.

Also consider this: someone stopping to take the time to compose and mail a letter is significantly more attached to a thoughtful response from your company than an form email request. The volume of traditional correspondence may be dropping, but it is no less important in the scheme of things.

☐ SELF-SERVICE

A viable substitute for delivering expensive real-time service with live people is self-service strategies.

The first experience customer contact centers had with self-service was interactive voice response and this technology proved to reduce labor in a number of whole or partial transactions. Delivering the checking account balance, or allowing a caller to support the agent by entering their account number, are perfect examples. The results were generally positive, provided there was well thought out, logical and simple scripting.

An IVR or VRU has two significant values: first, either can be deployed on any voice service as no special technology, short of a touch-tone phone, is required by the customer. Second, simple aural information can be delivered to the caller. However, the information can't become so complex that it requires note taking by the caller. In other words, the IVR and VRU have restrictions to the amount of information that they can deliver or extract from the caller. Fax-back and email following the VRU request can be a partial answer, but still don't allow real-time dialog.

We have all visited those websites that are positioned as helpful, but really are merely marketing exercises. These types of misconnects create more calls than they displace.

FAQS

The trick in any call center workflow strategy is to analyze the total completed call volume by call length and look at the content of the short calls. Typically you will find many result from simple, and often repetitive, questions that can be put into self-service channels like website FAQs and VRUs.

There are two ways to find out: ask your agents what they believe to be frequent questions. Or better still, if a call recording tool is available, analyze the data by allowing the agents (through ACD wrap-up codes, CRM call type or disposition fields, or even paper tally sheets), to tag calls by FAQ. This will allow you to get to the count data and then content that needs to be "pushed up hill" to the website self-service pages.

The Web

The traditional customer call center is no longer the ideal. Rather it must be a cross-channel, cross-discipline, cross-application customer contact center. The web offers an infinitely better channel for callers who are web enabled at a computer or web phone. Now data can be delivered as an exchange of complex data that can be held in device memory or printed out. Dialog is possible.

Much has been written on the success of websites as self-service tools and the disciplines involved. However, as with any self-service strategy, there must be a commitment to delivering a meaningful substitute with richer website content and truly helpful self-service tools. These tools should be positioned to allow a customer to choose the depth of service that best suits them — under every possible circumstance: website with FAQs, customer downloads and other tools. Coupling these with real-time text-based chat, "click-to-talk" (VoIP or traditional telephone callback) and agent assisted co-browsing is sometimes essential. At minimum, publish the 800 number on the site.

The following are just a hint at the numerous titillating statistics available to support a move to a multi-media customer contact center:

- The cost of a web inquiry or response is anywhere from 1/10 to 1/3 the cost of a conventional inbound customer telephone contact. This makes the web a significantly more attractive channel for most businesses.
- At least 65% of buyers visiting a website to purchase goods or services never complete the transaction. However many prospective buyers abandon the web session and call the 800 number to complete their transaction. The problem with a disjointed multi-channel transaction is that the opportunity for an interruption and an incomplete sales cycle is high.
- The overall consumer-to-business call volumes have risen 30% since the advent of the Internet.
- When a call center is coupled with a website the two work better in tandem than they do as discrete channels.

Web Integration

Customer browsing of a website is commonplace, as is the ability to place a web-based order. A minor inconvenience can arise, however, if a phone call is necessary since in most cases the customer must use a conventional phone to make the call. That is a process objection. There is also the opportunity to lose control of the sale. The desire to couple everything into one seamless transaction so the order is not interrupted is absolute.

Most consumers have a phone collocated with the web-enabled computer and many home installations still use one line for both voice service and Internet access. To place the order call, the customer must disconnect from the Internet and reconnect the phone. Millions of dollars of impulse buys have been defeated by this one simple fact. Even today, many of your customers' computers are agnostic to the type of transaction — text, voice or video. More computer devices are moving in this direction, therefore, the customer contact center also must move to accommodate this plethora of channel options. The "plumbing" necessary to couple a web position and a call center do not differ if this is a traditional circuit-switched call center or one that has migrated to VoIP.

With an active website, most of the issues of integration relate first to marketing. Publishing an 800 number on the site or alternatively a "click-to-talk." The original implementation of click-to-talk sent an email to the call center to request a callback with callback data automatically launching or requesting an agent callback. Now, the click-to-talk function allows the customer VoIP web-enabled workstation to call the 800 number just like a traditional telephone call. Coupling the IP address to the VoIP call means that co-browsing, pushing pages and chat all can occur as a bundled transaction.

☐ UNIVERSAL SERVICE

One of the "holy grails" of the customer contact center has been the universal service center and the universal agent. Ideally this means that any agent can handle any type of customer generated transaction, web inquiry resulting in a chat session, co-browsing session or click-to-talk — fielding and answering personal-

---

VoIP

VoIP is currently one of the hottest topics for telecommunications professionals because it allows transmission of business voice over managed data networks, local and wide area VPNs. The benefits are significant as VoIP can reduce telecommunications spending, increase the flexibility of workforce resources to match employee requirements of flexible hours and decentralized virtual offices as well as round the clock customer demand by providing a ubiquitous platform for new telephony applications. VoIP is the most direct path toward voice and data convergence.

ized emails as well as customer calls. Unfortunately, as Bill Durr (*Navigating the Customer Contact Center of the 21st Century*) so aptly puts it, the universal queue requires a universal agent. Durr descriptively defines many of these transactions as "deferrable work." Integrating these tasks has real charm in staffing a busy call center optimally. The problem, as Durr so elegantly states it, is not the technology but the skill set and culture of agents. Call center management can solve the interleaving of transactions easily with the current technology. The problem arises when a center randomly offers different task types requiring differing skills to agents, who, to begin with, are hard to find and hire. The solution takes us back to root cause analysis, load shedding and self-service.

Leading vendors in the customer contact center business have been handicapped with their ability to deliver a universal service platform while telephone calls, faxes, emails and web visits essentially followed different formats and used different carriage methods. VoIP has made many of these distinctions moot. Customers in the 21st century have needs that can be better served by automated service applications than by live interaction.

331

These platforms are a first generation solution to consolidating the contacts on one device, while aggregating many customer support resources. This is an elegant first step; but predicts the next generation of corporate applications that will give customers managed access to an array of services, when and where needed, without focusing on a single server or access point. Both can give customers the convenience, constant access, and discretion of self-service with the constant option of default to live agents when self-service won't do or doesn't match the personal preferences of the customer.

The first generation of these convergence tools are the front door to customer databased solutions that allow customers to use the touch-tone telephone and speech recognition and text-to-speech for telephone inquiries. Other options like e-mail, fax, and the web are also immediately available to perform self-service requests. Within the limits of the customer device supporting the dialog, a customer can switch from one medium to another during a single transaction.

These flexible open systems deliver many of the systems-delivered advantages of reduced staffing costs, less employee turnover, shorter call queues and talk times, while expanding customer flexibility and choice.

Media Switching
Media switching can be best illustrated with an example: A prospect hits the website looking for a specific product and decides to make a purchase, yet can-

not find the answer to one essential question. How easy is it, without leaving the website to ask the question aurally and place the order? There is a real need to remove the prospect from arbitrary media choices such as leaving the website and going to a phone to place the call. This gets really ugly when a consumer has one analog phone line for voice and Internet connectivity, which for much of the world, is the norm. Ideally switching channels should be transparent.

☐ VIDEO

There are two applications for video technology in the customer contact center. The first is as an information distribution channel in place of an information broadcast board or call status display. This technology has migrated to less expensive, strategically placed TV tubes that can carry all manner of other information and inspirational programs as well as inbound customer request load updates.

The second and less understood application for video is as a customer communication channel. Two-way video is close to the most over hyped channel that has ever existed. Look at the minuscule stock values of any of the video companies. The only real survivor in this space is Polycom who was first in the teleconferencing business with a $1000 hi-fidelity speakerphone, followed by a similar integrated videoconferencing device. Their success is based on low entry-level price and most importantly, not attempting to make or change culture.

There have been a number of failed initiatives in video collaboration in customer applications and even delivering video calls to the customer contact center by way of kiosks. These have almost universally failed unless targeted at a high value application such as a worldwide trading floor or a high value nationwide commercial loan origination network.

As a customer channel to a call center, video has universally failed due first to the fact it is not a naturally understood tool like talking on the phone. How do I look on a video camera is a frequently asked question versus the seldom asked question of how do I sound on the phone? It is a plainly unfamiliar tool. Add a costly problematic technology, a voracious bandwidth appetite, and the whole video thing is something the average company is unwilling to test in other than well-proven financially justifiable applications.

The early experiences in retail banking and retail catalog kiosks are cautionary. A retail bank tested an extended hours video branch that was visited and tried out by all manner of customers with transactions that inevitably could have been transacted quite well without the benefit of vision. Those transactions that could be enhanced by eye-to-eye contact were counter-intuitive to a semi-public location as they involved personal financial details.

In the retail catalog test, a video kiosk was installed at a major airport where most of the bored airport visitors who tried the "TV thing" turned out to be too young to own a credit card necessary to complete a purchase!

This is not to say that video will not make it into the mainstream. But to date culture, technology and costs have defeated widespread adoption in all but high value business applications. Customer contact centers are mostly about reducing the cost of doing business, not adding costs in anticipation of pioneering the next great channel. There are special customers. Private and commercial banks have installed video conferencing technology for some of the best customers because the relationship was worth it. These clients will probably never call your call center using video.

Ubiquitous low cost video is a promise the Internet brings and when this becomes a widespread and natural channel, an integrated standards-based customer contact center will easily expand to embrace IP video. Then the next challenge will become potentially extended transaction lengths while sound and picture quality are adjusted on a call by call basis to conduct the transaction.

333

In Summary: This all supports the argument for earlier consideration of the next generation of integrated customer contact center. Here the environment is completely agnostic to the communication structure or medium, cross-channel, cross-discipline and cross-application compliant.

The Technology
Acquisition Process

Throughout this book, we have discussed the applications
and issues involved in using a customer contact center system, particularly an
ACD as the center of a customer contact center workflow system. We have spent
little time focusing on the technology acquisition process. The politics of buying any system, particularly for a customer service center are complex and
unique in each situation.

335

A hallmark of this book has been to show that saving money by buying the
least expensive systems is illusory. Spending a little more money wisely at the
outset can save and win vast amounts of revenue and goodwill for your company. This is revenue that far outweighs the relative few dollars invested in the system most appropriate for your application.

☐ STRATEGIC BUYING

The process of buying is more difficult than selling. There are many potential antagonists in this process; particularly if you are buying on the behalf
of a modern corporation. There are also political issues to contend with in
the company.

First, there is recognizing the need and gaining the blessing of upper management for the project. This in itself may be a hurdle. Information is provided
throughout the book to help you calculate the costs of not only upgrading your
contact center, but also the costs of poor service levels; then leveraging those figures against the cost of the resultant lost business opportunities.

Second, there are the antagonists that belong to the other interested operational entities within a company. These may be the telecommunications, data processing, administration and financial departments, all who may get involved in any acquisition process. This book provides information to assist the end user in navigating this process.

The customer center management has to live with these system decisions everyday. The results produced by the center are directly affected by these ACD and contact system choices. The reports, which flow from these systems, document that success or failure. From their point of view, there are two, maybe three, prospective allies or "obstacles" to obtaining what the end user needs to get the job done right. The client department head (sales, service and marketing) is the greatest ally. The technical departments (telecommunications and/or MIS) will be involved because these systems involve telephone line and computers. However, most sales or service managers typically have few opportunities in their career to acquire large capital items for their corporation. Conversely, few technical buyers have the opportunity to buy contact and call center equipment throughout their careers.

Finally, the financial and/or administrative departments who oversee capital expenditures and administer the buying process must be included in the process.

NOTE: Expect to see any human resource group in your corporation to begin to take an increased interest over the next five years. Quality, "quick response" and "high reliability organizations" all depend on well-trained and responsive people. The issues of training, quality control, responsiveness, morale, incentives and compensation depend on management and their ability to gather and analyze quantitative (objective) and qualitative (subjective) data. A sales or service representative is becoming a highly studied employee. This raises employee relations issues which human relations departments care about. As this is written, the sales and customer service employees of a major telecommunications provider have just settled a labor dispute with a a new contract guaranteeing 5% per annum raise. This has a "knock on" effect throughout the industry.

☐ THE POLITICS OF PURCHASE

There are three distinct groups involved in the purchase process involving any large capital expenditure. Traditionally the buyers have been the *end user buyer*, the *technical buyer* and the *economic buyer*. Each group has their own set of objectives. Each contributes a different skill to the buying process. Each is goaled and judged differently.

The User Buyer

The user buyer or user department often initiates the idea to buy an ACD. Often they are the recommender and the source of the funding for the purchase. If the budget is under the control of the user the acquisition is much simpler. However, because this is incorrectly viewed as just a "phone system," the budget can lie with the technical buyer. The end user is the actual user and ends up living with the results of the purchase on a day-to-day basis. The big question for the end user is how will this ACD work for me. How much more business can we win and keep. How will it make my life more convenient and how much operating expense can it save.

The Technical Buyer

The technical buyer is generally a telecommunications or more recently a information systems or information technology manager. These individuals are under enormous pressure to keep costs under control and keep systems compatible. The time and effort spent keeping abreast of their respective fields leaves little time to keep up with each end users individual tasks or problems. Technical buyers understandably have little time to really get into the customer call center application needed to buy the best solution for your day-to-day needs. The technical buyer is often the recommender, the technical "referee". Seldom does the technical buyer say yes to the purchase but may deliver an absolute veto due to technical incompatibility with the company's technical direction.

The questions asked by the technical buyer are:
1. Will the system do the job?
2. Does it meet the quantifiable specifications?
3. What are they?
4. Are these compatible with our corporate technical direction?
   a. Vendor(s),
   b. Computer platforms,
   c. Operating systems,
   d. Applications languages,
   e. Communications protocols, and
   f. Local area network standard.
5. Do we know the vendor?
6. Does the vendor meet any corporate "selected vendor" purchase and support policy?

These policies and specifications are heavily influenced by what the technical buyers know about particular applications.

If you have a "one vendor" telecommunications policy but that vendor provides a mediocre ACD, why blindly adhere to policy and compromise your business? Here is a major educational challenge. The end user buyer must intimately understand their business and business goals. There is an obligation on the customer call center management to learn what customer call center systems do. The better vendors are excellent at educating customer call center management. Build alliances with members of the technical departments. Educate these individuals. This can have a great effect when it comes time to select any end user technology.

The last decade has seen sales and marketing departments buying more technology than ever before, particularly with the advent of the CRM industry. Become an applications resource to the technical buyer and provide intelligence and insight into the application so features and functions take on real meaning.

Many end users have been thwarted by failing to recruit the technical buyer early in the process, winning their confidence and assistance by understanding what the minimum technical standards are, then investigating within those guidelines. If your selection of vendors meet the company guidelines, your job is easier. The recommendation comes with no fundamental surprises and the technical buyer cannot claim to be blind-sided. In the selection of call center technology, look beyond corporate guidelines as typically these favor big telecommunications manufacturers. None of the major telecommunications manufacturers offering "comprehensive solutions" have technical leadership in customer call center systems.

On the other hand, the technical buyer may claim this area is their exclusive domain. Influencing their applications understanding is then the default tack.

The Financial Buyer

The financial or economic buyer is often the ultimate decision maker, at minimum, the casting vote. This individual is aware of the larger issues in a corporation and applies special knowledge to this and other decisions being made in the corporation. What impact will this purchase have on the system generally and will it provide a return on investment within current corporate guidelines?

There may be a fourth element in a corporation assuming the user buyer is not already part of this power bloc. The sales and marketing element of a corporation has a tremendous interest in winning and keeping customers in the most effecting way. Recruiting these groups as allies can greatly enhance the process of acquiring the most effective system.

The dynamics of this buying committee is the balancing the end user's absolute need with the technical buyer blessing the technical acceptability and the economic buyer getting the "best deal for the money". There are three votes and they are loaded in favor of the "best technically qualified system for the money," not the best for the job.

And after a decade of systems being sold that have often fallen short of productivity expectations, more scrutiny than ever is being placed on these requirements and purchases.

As an end user buyer you have the most to lose and the most to win by managing this process well. Good Luck.

☐ THERE ARE TWO SIDES TO EVERY STORY

Can you play a round of golf with only one club? The answer is yes, but you won't do a great job and you won't have much fun.

As always there are two sides to every story. A vendor, who tells you their solution has advantages over another, while makers of that other solution are enjoying robust sales, should immediately raise suspicion. If PBX-based ACDs had such clear advantages, why are sales of standalone ACDs so strong? VoIP offers the same confusion though leaders in the ACD specific applications have been quick to embrace the VoIP packet-switching models in their systems.

PBX vendors forget the lessons of data processing. Mainframes did not provide all the solutions users needed at a realistic price. So began the minicomputer business and then the PC business. These offered limited solutions at lower prices. PBX vendors followed the same pattern, but began with unrealistically low prices. This meant they could not afford to build or support applications specific solutions like ACDs and predictive dialers.

Here are the arguments you'll get from a vendor explaining why his solution is better. We are using the job specific ACD versus the generalized PBX as the example. Remember the vendor is selling a telephone system. A smart ACD vendor is selling a business system and a better way of doing business.

1. "A PBX/ACD is an integrated system."

In all but a small application (20 agents or fewer) this may be an advantage. What is given up for the "integrated solution" is an ACD specific machine. If a PBX was designed as a PBX to begin with, you give up architectural, common control and database flexibility needed in an ACD.

2. *"The dialing plan is universal."*

This is true for administrative stations. An agent must have two extensions, one for production (ACD) calls, one for administrative calls. This requires two ports on the switch for both the ACD directory number and the extension. A different number must be dialed for an agent group. This can even be an outside line. Real messy.

3. *"There is no need for tie lines between the ACD and the PBX portion in a PBX/ACD."*

Tie lines are a commodity that consumes a port per end. Tie line implies ongoing monthly telco costs when we are only talking about a pair of wires in the switchroom. The trade off on a PBX/ACD is buying increased station capacity. A minimum of two per agent. This is not true of standalone ACD systems that allow logical addressing, and therefore only needs one physical extension per agent.

4. *"Calls flow easily from the ACD portion to the PBX portion."*

First, why is this happening. If this happens to most calls there is something wrong with your business flow. Five to ten percent is acceptable as there may be some non production (call) reason. Beyond this your business is not flowing smoothly. A good ACD will show you this.

5. *"Administrative extensions can backup ACD groups."*

This is true and a valid reason to use an administrative group, but be careful this is not a crutch for bad planning.

6. *"Reporting is integrated."*

This is not true on most PBX/ACD systems. All ACD traffic is reported in the ACD group and station reports. Any outgoing calls or internal extension to extension traffic is reported on the SMDR (station message detail reporting and call detail reports) in the telecommunications department that has nothing to do with the customer call center.

7. *"PBX/ACDs can grow easily."*

True, if the system has lots of capacity. An ACD call is five to seven time more complex to process and because it has revenue or goodwill associated with it is infinitely more sensitive (to blockage) than an administrative call. PBX systems are designed to ration trunk resources across a large universe of administrative stations. Ten or more stations per trunk. ACDs work the opposite way, not to mention common control (computer running the switch) capacity. When you take 20 or 30 ACD positions off a PBX you get the equivalent of 100 to 150 PBX stations freed up. Conversely putting 20 or 30 more ACD positions on your PBX dramatically changes traffic demands on the switch. Watch what happens if you are near a shelf or cabinet limitation!

8. *"All the PBX features are automatically included."*

You don't want these at production agent desks. Most ACD stations are there to support production tasks. Fewer worthwhile features are necessary. Few keystroke and no codes to distract from the business at hand. Remember most PBX only know how to use five features easily and maybe speed dialing.

9. *"Below 50 agents you don't need a sophisticated standalone ACD."*

Maybe your business is not worth protecting. If American Airlines loses a customer call center they estimate they loose $16,000 a minute. They use a true bulletproof ACD. What's your business worth.

10. *There are also the cost containment aspects.* Queuing theory is poorly implemented by most PBX/ACD vendors. If they do not use "multistage queuing with simultaneous lookback" you are hiring extra staff to compensate for this. Three extra bodies for every 20 positions. How do they sense and answer a call? If they do not dynamic answer, retarded on a call by call basis if no agents are available, they are costing you 10 to 20 seconds of billed 800 time per call.

341

You thought you got a great deal when they included ACD software with your PBX at no charge? What you don't know can hurt you badly.

These same sorts of issues apply when generalized approaches are taken to VoIP or email server purchases when the customer contact center is not considered as a primary application, rather just another user group. As the customer contact center leader, it is your specific challenge to end this cavalier approach.

☐ WINNING CONCEPTS

A clear business mission can focus to a buying process. Focus on a common business goal can keep the political dynamics at a level that avoids feature requirements and necessity battles. The buying committee is kept to a mutual goal. Examples of these goals may be solving poor customer service, meeting a new challenge represented by increased call volume or it may be doing more with existing resources, providing the same level of service with less resources, or reducing costs of operation.

☐ JUSTIFYING A BUSINESS SPECIFIC SOLUTION

In reality there are three justification issues; each occurs at a different level. Each is harder than the next to prove in the acquisition phase; yet becomes clearly the justification for making the purchase in retrospect.

- The system can save money.
- The system will allow increased ease in getting information about business performance and allow convenient insight into the business.
- Insight into the business will allow strategic and tactical decisions to be made to allow the customer call center to become an even more effective tool for winning and keeping customers.

The system may cause you to spend more money. Once the new system is installed it may graphically demonstrate how poor service really was and how much money and opportunity was being lost.

All the major ACD vendors have morphed into integrated customer contact center vendors and they have extensive experience in building detailed financial justification models for you. Accept at least three and scrutinize them extensively. There will be bits in all of them that can provide help.

Good Luck and may you always "make your numbers."

The Trends

The major trends within the customer contact center industry center around explosive technology growth, which has led to a growing library of open architectures with industry standard interfaces, more sophisticated tools residing in smaller centers and more component applications involved in each transaction. There remains a need to work with legacy applications as the traditional call center evolves into a modern customer contact center to answer the customer's demand for more integrated data (ACD, IVR, web-based, etc.). Nevertheless, rapid migration to a state-of-the-art customer contact center is critical to many businesses that want to reach or maintain profitability.

☐ EVOLUTION OF THE CUSTOMER SERVICE INDUSTRY

The technology of serving customers from a centralized point has been the typical role of the traditional call center. This is evolving in many directions.

1. The facilities that once just served to respond to the customer telephone call has now become the focal point for serving any contact channel a customer may wish to use. The customer call center has now been redefined as the customer contact center.

2. This contact is not simply a transaction, but a milestone in a customer relationship AND a customer's experience. It is not a finite event but part of a delicate ongoing process.

3. This customer or prospect "arrives with a history and leaves with an experience." This fact cannot be ignored. The one exception may be when the cus-

tomer contact is viewed in a very limited sense — as a one-time transaction and not as a key way point in a business partnership. Few companies, however, have the luxury of taking a customer contact out of context and focusing on content at the expense of continuity. Selling tickets to a rock concert or other over-subscribed event is maybe one of the few exceptions.

4. That all contacts with the customers within an organization are not only by telephone at a centralized and stationary customer contact center, but occur across the enterprise and therefore all the other "touch" events need to be woven into the contact database and process.

5. The wide acceptance and deployment of self-service technologies like the web and voice response technology.

6. The surprisingly fast adoption of the Internet and all that it implies in terms of rationalizing networks (such as VoIP) and customer-driven accessibility and contacts that need to be intelligently managed and responded to.

7. The recognition that a common customer database is no longer a nice idea but an imperative in our increasingly mobile global market.

8. The recognition that any part of the company that touches a customer needs to be part of the customer database continuity. Whether it is accessing customer history or recording the action taken toward serving that relationship.

Everyone basically agrees that consolidated customer records available to all employees that are likely to interact with customers are a good idea. Building a company philosophy and supporting a vertical and horizontal communications infrastructure is a challenge and a necessity for the information systems professionals (and the customer contact center management). Understanding the business requirements, however, has never been a strong suit for technologists, although it's now more important than ever. The customer contact center management must help to educate the IS department in this area.

□CUSTOMER-FACING

There are a number of significant issues driving the whole customer-facing aspect of a company and it has greater implications than simply the customer contact center. There are fundamental, strategic and tactical changes that need to be embraced by any corporation wanting to grow to the next level of meeting and exceeding customer expectations. Some are philosophical, other organizational and a few technical.

1. We simply cannot afford to "man the phones" as we once did.

Evolution from the customer contact center to the cross channel customer satisfaction network where customers based on their profile will find self service anywhere in your enterprise. Live operators will become the default condition, in all but culturally intense new sales and escalated service exceptions.

2. The drive to offer customers satisfying, self-service alternatives to traditional channels is both a cultural and economic imperative. The reality is many people prefer web and email dialog and these channels are significantly more cost effective than live operators.

3. The need to understand the relationship between your customer contact channels cannot ignore default channels for people that still prefer traditional means. Internet use confounded conventional wisdom and has stimulated more traditional contacts than ever before.

Voice still accounts for more than 85% of all customer contacts despite the prediction by early web promoters that email and other Internet-based processes would replace the call center. Call volumes have exploded while web only companies imploded due to their refusal to acknowledge the gregariousness of humans and the traditional infrastructure needed to build total customer sales and support.

"Bricks and mortar" are as strong as ever. UPS and FEDEX have had banner years shipping product from dotcom warehouses. The major hardgoods and food retailers like Walmart, Kmart, Home Depot and Safeway are winning the channel war against pure dotcoms as they learn to provide an integrated experience and delivery process. Print and broadcast media have benefited as web advertisers realized that web banner ads were a very limited avenue to markets they already reached. Success is clearly in orchestrating all these channels.

4. "Back to the Future."

To succeed, commerce channels and products will circle back around and enrich the understanding and implementation of traditional media and customer channels to the benefit of customer contact center owners. There are a number of significant lessons that have been learned in media performance focused at websites that will be expanded to traditional media and customer contact channels.

But the backside of the benefits that the web and ecommerce have brought business is the failure of dotcoms who lacked respect for the past and processes that arise out of culture. Death of newspapers and shopping malls didn't happen because of the way human beings prefer to work, shop and spend their leisure time.

5. Customer Experience Management will come of age.

Screen pops to "threaded experience," "trip wires" and "at risk customer intervention" across any medium. Self-service evolution will push us from the seldom-returned mail-back warrant card, to "wired" customer experience.

As much as I dislike omnibus terms like CRM and Customer Experience Management (CEM) they are broadly descriptive of a collection of tools that address significant industries.

The business that is call logging is going to recreate itself as providers of information for customer experience management.

The problem with the logging companies addressing this market is that they are primarily hardware companies that know how to move iron, no matter how many times they tell you they sell "enterprise solutions." Their representatives know little of the customer contact center culture and even less of the backend workflow disciplines a customer contact center uses to serve customers.

These are the real marketing, sales and customer satisfaction strategies, process and human capital aspects of a customer contact center. This is about to change as these systems deliver the "god's eye view" across all of the telephone (circuit or packet switched), VRU, computer telephony and/or CRM systems to capture multiple data fields for vector and out of range analysis. From this, customer defined "trip wires" can be set within the customer contact center that can detect positive micro events on a contact-by-contact basis. With manual identification of an undesirable trend, a filter can be established using common programming tools to automatically alarm when a contact or campaign develops a positive or negative trend.

This is the breakthrough content sensitive analysis that call centers have been talking about for years.

6. Development of universal workflow tool kits.

Arising from the analysis proposed above are obvious recommendations and process improvement that can be implemented either as training programs, or more practically specific "skill or knowledge support" tools in popular CRM or service applications. Elearning is a partial solution; but it has limits based on an employee "audience-of-one." There is little leverage to be had on a one-to-one basis because of irregular and subjective implementation by the employee. Many customer contact center employees are transitory so the equity in human capital is lost as they move to their next job. Driving more of the process into the application as skill support delivers a better multiple. This, however, is not to say that elearning is unnecessary or unwarranted as a complementary tool.

7. Open standards-based systems will support every system.

Love him or not, Bill Gates has marketed Microsoft standards as *de facto* standards that make the systems world a more predictable environment.

Because all of these customer contact center systems and subsystems are primarily business systems, the underlying technologies must readily adapt to the business environment they serve. This means open standard-based systems to the point that even closed operating systems environments (such as switching software) are going mainstream. Aspect has dumped its UNIX operating system for NT and redefined their voice only ACD switch as a cross-channel portal. Management information systems have delivered us significant productivity improvements over the last decade and this is expected to continue, provided these systems can handily integrate at every level, control and data exchange and collection.

As is true with any systems integration improvement — first the attachment occurs and this was primarily CTI integration, where systems never intended to connect were tricked into believing they were dealing with native terminal devices. This was accomplished through common translation layers. In a customer contact center the business value of CTI was clear in the form of "screen pops," coordinated call/data screen transfer and other coarse workflow improvements. These connections were accomplished with considerable pain and performance impact. Now these companies are circling back around to offer the next generation of systems on open standards-based platforms that reduce or remove the requirements of extensive translation layers, rather than simply adding another system partner or data destination.

8. VoIP

Telephone systems as we know them are going away, and at the same time, VoIP is improving the way stuff works at a dramatically lower cost.

VoIP will replace the discrete premise telephone switching matrix/network. This has been anticipated in the world of telephony by assuming the switching device would become dumb and dumber as more intelligence migrated to external servers that manage the tables and rules that make the basic connections within the switching server. This also assumed a discrete telephone system will survive.

This is not expected to happen as was predicted, as circuit switching has become packet switching, and that can be accomplished relatively elegantly in the context of a universal buss such as envisioned by IP on a LAN/WAN. The voice gateways are now servers that take and normalize VoIP voice and place it on your network. There are a number of outstanding issues getting this to work at the volumes required in a call center, but all of the major vendors are spending billions in research and development to replace the world's telephone infrastructure with packet-switched voice. The applications that justify this migration are almost exclusively customer focused.

If the next telephone "system" in your future is not VoIP, then it should be only an inexpensive interim or bridge strategy using used equipment.
9. The Chief Customer Officer.

Someone somewhere in ever organization must wake up and understand most companies are not organized for the benefit of the customer. Despite the apparent juxtapositions of stockholders and employees interest to that of the customer, if we lose customers we are done employing people to serve customers for profit and stock value! This is a significant issue for any company.

□ MARKET DRIVERS

There are significant market growth trends and market drivers prompting the growth within the customer contact center industry.

1. The customer contact zone is being much more broadly interpreted to include all possible business to customer contacts.
2. The various channels a customer can choose to contact a business are broadening the pool of staff that can possibly serve a customer. These formal contacts: paper mail, FAX, email (specific personal electronic address) and web requests (non-specific email contacts) need to be managed with the same tools as a real-time customer telephone call.
3. Corporate personnel (who reside outside the customer contact center environment) who may interact with the customer need to be equipped with basically the same database access as the customer contact center agents. Not only to maintain an accurate customer database, but their roles and associated workflow require the same sort of productivity management tools that have long been used in formal call centers.
4. In sales, a direct sales force or distribution channel is capable of significant productivity improvements when coupled with inside sales and marketing support. Deployment of any sales force automation (SFA) or CRM tool will force review and growth of the sales and support model leading to automation of more inside sales and support functions. While most vendors have rec-

ognized this, there are only a few that provide a good solution to gathering and synchronizing communications channel management and SFA with CRM or other databases and reporting systems.

5. A continued tight labor market requires inside sales/service skill support and productivity tools, since many times businesses are forced to hire less skilled employees to serve their customers.

6. Ecommerce changes the distribution model by often eliminating wholesale layers. This means adding electronic (web, email and what was formerly defined as EDI) and real-time customer contact centers to serve prospects and customers. This is highly influenced by the need for improved response and cost of sale models.

7. Cost of sales, tight labor markets and the need for fast market feedback has changed the launch model for new products and new companies. These new entities and products are now typically launched with direct marketing and response tactics that normally were only employed by large, mature business-to-consumer companies. These include mail, telephone, fax, email and web tools, all of which require robust list, campaign wave triage, response queuing and workflow management to be effective.

The customer contact center enabled suite needs to offer a manageable solution for small to mid-sized projects, especially where effective lead processing and measurement is vital to success.

349

☐ THE CHALLENGES

The integrated customer contact center is a challenge at any level. Positioning the center and its people to evolve with your customers' changing needs is a huge cultural hurdle. None of the major cross-discipline opportunities are in the skill set of the average telecentric or datacentric company because these issues cross strategic, process and people issues. However, there is a growing body of knowledge and skills on how to go about this.

The specific challenges are:

• Using real performance metrics. Why is a company in business? Take time to identify what you define as success from both your business and customer perspective. Reduce these criteria to manageable and measurable metrics. Gather, measure and track real metrics that define your performance — loyalty, satisfaction and productivity — rather than quantitative data like hold times and abandon rates.

- Synchronize customer communication channels. Identify every possible customer contact point or channel. Synchronize these communication alternatives to provide consistent, reliable and timely information. Know who your customers are, what they are buying and the channels available to them.
- Minimize the need to escalate an issue to a telephone call. Only after someone becomes incensed do they write a letter of complaint. This is macro data. Making a complaint call is easier for the customer, but significantly more costly for your business in the short term. These thousands of calls and other contacts hold the micro data and keys to the reasons for the contact, the reason for the escalation and offer the ability to intercept at an earlier stage in the customer relationship or process. By performing this analysis, it is possible to anticipate requirements or intercept those with lower cost response processes earlier in the cycle, and achieve the same or better levels of customer satisfaction.

The big win is to take masses of micro data and present it in a manner usable for improvement in strategic, process and staff performance.

- Shedding or improving unproductive contacts can dramatically affect result ratios and cost reductions. Capture and use customer experience data to provide proactive upstream solutions rather than reactive service.
- Be proactive. Identify at-risk customers and business. Intervene with save, "win-back," and margin protection processes to improve the contact center results ratio. Invest your resources by segmenting customers based on their profitability, propensity to buy and risk factors.
- Think long-term. Develop sustainable processes that result in long-term gains. A one-time fix will not drive growth. Leverage technologies that enable intelligent routing to facilitate better service and allow for flexible work schedules. Develop logical self-help opportunities for your customer to reduce incoming calls and related costs.

The source of this knowledge and skills are the many great magazine and books, as well as some of the more reputable consulting companies that focus on your customer satisfaction first, rather than offer a foregone technology answer.

To repeat, a customer contact center crosses *strategic, process, technology* and *people* issues and each one of these four component must be considered equally to achieve a sustainable business solution.

Epilogue

Rationalizing customer, employee and stock- holder interests is a constant dilemma. Great service is more expensive than rationing service so that a few cents can be eked out for a better earnings per share (EPS) ratio for the current quarter. But what of next accounting period? Maintaining customers is vital to the long haul, and retaining that customer is always significantly cheaper than finding a new one. The feedback possible from a "customer facing" customer contact center, is a nearly instantaneous "early warning" to deteriorating customer metrics. Skimping there can have enormous and long-term effect.

There is one constant that must be emphasized: customers have more choices than they have ever had before. They will choose to contact your business any where, any time and any way they please. Will you be easily available to win and keep their business?

☐ "MOVE A REGISTER, FORGET HISTORY"

The author has been involved in delivering real business applications solutions through a number of generations of technology. Through these cycles I discovered a really interesting phenomenon: acquiring any new business with a young development staff, inevitably leads to a huge rift between the older team and their technology generational alliances and the new group. By virtue of their later arrival in the profession these newer developers benefited from newer technologies. "What did the old guys know?"

Well the answer is that they often knew where the skeletons were buried. How did the particular industry or application evolve and why? Winston Churchill stated that he who ignored history was destined to repeat it. He might have been directly addressing the application of technology.

New technology generations often introduce completely alternative connectivity and applications approaches while ignoring all of the vital auxiliary management tools that made the prior generation usable. One great example is the introduction of voice response technologies as competition to call distribution systems. An extra five or so years was spent bringing these to broad market adoption as they were almost completely void of the robust reporting suite standard in call distribution devices. Similar experiences have occurred integrating net tools like email, web chat and "click to call" features. Particularly when you consider that early "web Nazis" positioned the Internet as the death knell for call centers, and print and broadcast media. Look at them today, nothing could have been further from fact. In reality, as a result of web confusion, more traditional calls are generated daily by ordinary people looking for an answer from a live person.

Now we are on the verge of the next cycle of throwing out all of the traditional telephone systems and replacing them with Internet phones or VoIP, which should take five to ten years to complete. Then many managers will look at the traditional call center workflow and realize that a multi-media contact center will require a reengineering of its workflow processes. This type of reengineering is fundamental, disruptive and downright scary, but to limit customer contact systems integration to the realm of "plumbing" is missing the entire point of a customer contact center versus a call center.

☐ THE CUSTOMER IS KING

As call centers grew in popularity and use, the traditional methods of winning and serving a customer base (field sales forces) proved increasingly inadequate and expensive. The same cycle is now occurring with traditional voice calls and Internet-based self-service with a preferred cost ratio of 1:10 for self-service.

Nevertheless, the less than stellar "customer friendly performance" of web implementation have almost single-handedly increased traditional call volumes by 30%. This opens a whole new discipline of customer experience mapping and analysis as the basis of workflow improvement.

Marketing and advertising disciplines are also seeing dramatic changes as results data flow in from the sales and support activities. Philosophies such as "sales navigation" dramatically change the way advertising and direct response work. As some anonymous advertising wag once commented "50% of my advertising budget is wasted! I just don't know which 50%." No longer can companies afford to squander advertising budgets nor do they have the luxuries of the 80's when any strategy seemed to work. Now every lead is vital to an immediate sale, a future sale or a marketing database entry. Customer experience mapping and analysis is automatically generating vital media source and message effectiveness data.

☐ COSTS OF SALES CONTINUE TO RISE

Every year the McGraw-Hill sales cost study reports another escalation in the cost of acquiring a customer by a direct sales visit. In 1987, when the first version of this book was written, the average cost of acquiring a new customer rose 9.5% over the previous year to $251.63 and business to business calls averaged $291.10. Many sales take a minimum of five calls to close the business. In 2000, this number had risen almost 10 times and is rising faster because technology makes markets finite faster.

353

Unsurprisingly, it costs five times as much to win a new customer than it does to keep an existing one. In the 80's Walker Research studied Citibank retail checking account sales and service in Manhattan. They found that the bank was winning 2000 new retail checking accounts a week, while the back office was losing 2500 during the same period. The key to stopping this net loss of 500 customers a week and the attendant cost was linking all customer contact events into a relationship that could be managed by teleservices. Get the customers out of the branches and onto the phone. Use the telephone as a servicing channel to augment the already successful use of automatic teller machines. The winning idea was linking the "front" and "back offices" in everything from mission, systems and staff compensation plans.

Integrating this thinking to accommodate customer preferences, economic realities and deliver the best possible result cost effectively over the longest term possible is your challenge. Technology is here to help, but only if it is in service to your business goal and never hijacks that objective just because it is technically possible.

Thank you of your time and good luck achieving your mission. We hope this book plays a small part in helping you reach your goal.

Appendix I

Request for Information: Computer Telephony

1.0 Introduction

1.1 Purpose of request for information

First National Bank of Axxum (FNBA) is soliciting information about potential teleservicing vendor solutions to enhance its retail Customer Service Department (CSD) systems with the intent of:

- Increasing the speed of transaction access and execution.
- Reducing transaction complexity.
- Reducing the amount of training required for FNBA Customer Service Representatives.

The response to this Request for Information (RFI) must describe features, equipment, software, vendor services, and expected FNAX accountabilities to complete the project.

The intent of this Request for Information is to describe the various attributes required of the new or enhanced Customer Service system. Responses will be used to select finalist vendors. Finalist vendors may be requested at a subsequent time to formalize their responses and bids for final selection.

1.2 Type of proposal solicited

FNBA would prefer not to undertake a large scale development effort using internal resources. Ideally, few changes to FNAX Mainframe and Voice Response Unit (VRU) applications will be required. FNAX is looking for a prime contractor for all software integration and customization outside of current FNAX systems. However, it is imperative that potential vendors clearly identify how their product and services allow FNAX to be more efficient at implementing or maintaining a Customer Service teleservicing solution than using internal tool sets. Thus, responses should clearly indicate both whether a vendor can support a specific feature or requirement and how that function would be customized and implemented.

Vendors must make a clear distinction between requirements and customization which will be satisfied by their standard package or tool set and those which require modifications and/or development. In each instance where modifications and/or development will be needed to satisfy a mandatory requirement, vendors must indicate a completion time frame to provide the new function. In addition, vendors must clearly identify where co-development on FNAX systems (e.g. Host or VRU) will be recommended to satisfy a specific requirement.

It is suggested that vendors review Section 5 to gain an understanding of the different scenarios and options to be estimated in vendors responses. Note that in addition to the scenarios involving a vendors participating as prime contractor, a response is also requested from vendors regarding how they would be willing to work in partnership with FNAX and other vendors should we choose a 'best of class' approach for the different components of a CSD automation solution (especially the Computer Telephony Integration portion).

1.3   RFI Format

Section 1 - Contains background information and sets forth instructions for participation in this RFI process. This section also includes background technical information on FNAX's current environment.

Section 2 - Defines the various high level functional requirements of the Customer Service System.

Section 3 - Sets forth preliminary minimum FNBA contractual expectations.

Section 4 - Is a questionnaire which solicits vendor information regarding a variety of topics.

Section 5 - Identifies the financial and resource information required for financial analysis and sizing of vendor responses.

Section 6 - Are the appendices which contains specific information, including:

- Transaction volume data
- A description of the top 15 transactions to be re-designed
- A document describing 'key concepts' which will be the basis for developing detailed business requirements

- A detailed set of preliminary requirements for a representative transaction
- A sample Terms and Conditions document which is typical of FNAX technology acquisition contracts.
- A glossary

1.4   Format of response

Vendor proposals are to be submitted with materials arranged in the same order and sequence as this RFI. Identical numbering schemes should be used as in this RFI. Instructions for each section precede the material and should be read carefully to insure that the type of response requested is fully understood. Failure to follow instructions may result in disqualification of the vendor.

The vendor should provide FNAX six (6) copies of their response materials.

Proposals are to be submitted on standard 8-1/2 x 11 inch paper (8-1/2 x 14 inch is permissible for charts, spread sheets, etc.)

Proposals construed as proprietary should be so marked.

FNBA will not be responsible for any costs attributable to proposal preparation or submission.

357

Responses are to be received by 5:00 p.m., 01/02/02 in the office of:

Phillip Smith
Senior Project Analyst and AVP
First National Bank of Axxum
12 Main Street
Axxum, ZZ 12345

1.5   Current operational and system environment

1.5.1   Current processing environment

FNAX primarily operates two mainframe environments for on-line and batch processing. The IBM environment handles the majority of traditional applications. The Tandem environment is the processor that drives our ATM / POS system as well as the Wire Transfer application.

> **IBM Environment:**
> 288 MIPS configuration provided by;
> IBM 3090 600S
> IBM 3090 400S
> Amdahl 5990 - 1400
> (Plans are to upgrade to a 350 MIPS configuration)
> DASD of 1 Terabyte primarily on HDS and IBM devices

MVS/ESA release 4.3
IMS release 3.3
CICS release 3.3
DB2 release 3.1
VTAM release 3.4
TCP/IP release 3.1

Tandem Environment:
Six Cyclone processors for ATM / POS processing
Eight processors for Wire Transfer.
ACI Base24 release 4.0

Front-end Network Processors
Four IBM 3725's (NCP release 4.2)
Two IBM 3745's (NCP release 7.1)
Enterprise Processing Volumes

1,500,000 IMS transactions per day supporting Branch On-line, ATM/POS, CIS and Hogan applications.

1,200,000 CICS transactions per day supporting eMail, Credit Card and Retail Loans.

3,000,000 paper items processed per day of which 60 % are on-us transactions.

1.5.2   Customer Service Department environment

The Customer Service Department is physically divided into two sites; one in the Xzzum and one in Axxum. The Axxum site is open 24 hours a day, 7 days a week. The Xzzum site is open from 7:00 am to 9:00 PM and is closed on Sunday's. Of all calls received 43% are handled by the Axxum VRUs , 32% are handled by the Xzzum VRUs, 16% are handled by the Axxum CSRs and 9% are handled by the CSRs in the Xzzum . 800 numbers are primarily routed to the Xzzum site. The Axxum site handles all local western Washington calls. In both sites the VRUs are the initial point of customer contact. In December the CSD VRUs handled 75% of all calls. Workstations for CSR, training and managers are divided as follows; 120 in the Xzzum and 173 in Axxum.

Multi-queue call handling

Agents are organized into three groups:

1) The primary queue answers and responds to 90% of general information and account requests.
2) The special queue handles all customer requests requiring research or additional customer assistance and takes calls transferred from the other representatives.

Special queue also handles queue overflow when there is heavy volume.

3) The business queue responds to calls from business clients for such things as check verification.

Note: There is also a Reference Center that provides support for all of CSD. The Reference Center performs Teller functions such as not on-us credit card advances, G/L functions such as fee reversals, letter writing, research, microfilm requests and a number of other support activities.

1.5.3   Telephony

Two Aspect ACDs, one located in the Xzzum area and one located in Axxum.

The Aspect ACDs are currently running on software release 8.0 (proposal in process to move to release 9.0).

The ACDs are linked together by T-1 lines so that they can balance calls between the two CSD sites. (a T-1 equals 24 lines)

Since the ACDs in both locations are linked together they can support each other during scheduled or unscheduled downtime.

Each ACD has independently programmable "routes" that control how calls are handled. An example of "route" flow would be:

1. Call comes in to the Axxum ACD
2. ACD checks for available VRU in Axxum
3. After a specified time (perhaps 3 seconds) the ACD will look to see if a primary queue representative is available in Axxum.
4. The ACD continues to check for an available VRU or primary queue Customer Service Representative (CSR) until another specified time period passes.
5. Next, the switch will start looking for an available special queue representative (all the time continuing to check for a VRU or primary queue CSR)
6. A search for a business queue representative would start if VRU, primary CSRs and special queue CSRs are not available.
7. At this point the route might tell the Axxum ACD to contact the Xzzum ACD and make a reservation in its queue. (This starts a "route" on the Xzzum ACD that will also have a programmed series of attempts to find a VRU, primary queue, special or business queue representative)
8. Finally, (and this may be as little as 12 seconds later) the ACD finds an available special queue representative in the Xzzum area.
9. The call is routed over the T-1 lines from Axxum to the ACD in the Xzzum and then to the available special queue representative.

Note: The "route" also includes information about the type and timing of announcements or music to be played while the client is on hold.

There are currently 190 programmed "routes" in use by the two ACDs, each "route" can have up to 256 steps.

All representatives are defined to the ACD by userid. These id's are attached to defined groups. Each group has a class of service, it is the class of service that determines what function (e.g. primary or special queue) is performed by the group.

A representative can sign-on at any station. The sign-on tells the ACD which group the CSR belongs to and therefore which class of service to allow the CSR to perform at that station.

Each ACD provides detailed reporting about the calls handled. The information reported each day includes but is not limited to: calls offered; calls handled; calls abandoned; calls transferred; average delay time; agent minimum, maximum and average call time; and wrap up information. These reports are produced in both half hour increments and daily summaries. Samples of one of these reports are in Appendix A.

1.5.4   Voice Response Units (VRUs)

The Aspect communicates with 9 VRUs dedicated to CSD teleservicing functions (7 are CSD and 2 are Pay-by-phone which is a CSD function) of which 55% reside in Axxum and 45% in Xzzum.

The VRUs operate from three servers, all Pentium III machines

The servers operate under Micorsoft NT software.

The servers and the VRUs run Intervoice IFORM software.

There are approximately 60,000 calls per day processed by the CSD VRUs.

1.5.5   Host network

CSD connects to the IBM Host through an IBM Server on a Token Ring Local Area Network.

The Token Ring Controller connects to an IBM 3745 Data Communications Controller which interfaces with VTAM on the IBM mainframe.

There are a total of 300 IBM 3270 type terminals in CSD.

Each terminal has a four session capability.

1.5.6   Applications

Access

The Access system is supported by the Stairs/VS Release 4 Level 3 application.

Access contains information on:

1. Branch location, hours, phone numbers and other miscellaneous information;
2. ATM locations;
3. Automated Clearing House (ACH) transaction activity.

There are more than 50,000 Access application transactions performed each month.

Bill Payer (PBP)

A FNAX developed application which allows customer to direct the payment of bills through a VRU session.

Bill Payer is an IMS DB/DC application developed with a Telon code generator.

There are approximately 1000 Bill Payer sessions each day resulting in about 6000 payment transactions.

Branch On-line (BOL)

BOL is the IBM host portion of our Branch On-line system. It supports both teller and admin (platform) transactions. BOL is the repository for most current day memo-posting activity. Eleven BOL strip files, which are files created daily based on information extracted from the various systems of record, are used by Teller, Admin and CSD areas as the primary source of customer account information. CSD sessions are part of the BOL on-line application. Each processing night BOL sends paperless memo-posted transactions from CSD and other areas upstream for posting to the system of record.

BOL is a FNAX developed COBOL II application using IMS DB/DC (Fastpath) Release 3.1.

It supports 157 transactions, 62 Teller, 70 Admin(which includes the CSD transactions) and 25 support functions.

There are eleven application strip files plus 29 other data bases supported by the BOL application. The eleven application strip files are:

1. INV - Investments - mutual fund, Annuity, Brokerage
2. RLS - Retail Loans
3. SBS - FNAX Bankcard Services
4. RET - Retirement
5. TSP - Trade Services
6. SDB - Safe Deposit Box
7. CDA - Certificates of Deposit
8. SAV - Savings

361

9. DDA - Demand Deposits
10. FLP - Lease
11. OTP - Outstanding Transaction Processing (overdraft and NSF history)
BOL interfaces with 25 other production applications.

There are 10 million BOL Teller transaction per month and 4 million BOL Admin transactions per month. (CSD transactions are a subset of these BOL transactions)

It is linked to an Electronic Journal archive of 62 days of on-line monetary transactions. (See below Host Electronic Journal for more detail)

Accounts in Jeopardy (AIJ) is a subset of the BOL system that acts as an account closing warning system. It is an in-house developed IMS DB/DC, COBOL on-line and batch processing application. AIJ records account closing information that is then passed onto personal bankers for follow-up. There are 2 on-line transactions that are used by the AIJ sub-system.

The Inbound Telemarketing (ITM) application is also a sub-system of BOL. It contains product information. This is an in-house developed IMS DC application. The product data is stored in DB2 tables. ITM contains updatable product information and simple scripts to support the sales effort.

Customer Information System (CIS)

This is a Hogan Release 8306 IMS DB/DC.

Customer Information file (CIF) containing 4.8 million records.

Account Information File (AIF) containing 7.9 million records.

Address Information File (ADF) containing 3.5 million records.

CSD performs approximately 20,000 IMS transactions to retrieve CIS each day.

Automated Teller Machine (ATM) / Point of Sale (POS)

The switch and drive for the ATM / POS application is resides on the Tandem mainframe.

Host authorization, logging, network settlement and ATM balancing all reside on the IBM mainframe. There applications are FNAX developed, IMS DB/DC applications.

Currently, ATM / POS transaction detail comes to the mainframe through a batch process. The detail is stored in multiple flat files daily processing.

The ATM / POS environment handles over 11 million transactions per month.

CSD does not directly access either application at this time. The limited ATM

/ POS information currently available is viewed through the BOL application.

Credit Card / Merchant System

Cardpac Release 5.0 CICS / VSAM application.

Personal account volumes of 1.1 million, other account volumes in excess of 150,000 accounts.

CSD does not directly access the Cardpac application. Credit card transactions are processed through the BOL application.

Host Electronic Journal (Host EJ, HEJ)

Host EJ contains an on-line record of all BOL monetary transactions.

It is a FNAX developed COBOL II, DB2 application.

The application load to Host EJ is performed in the IMS environment, the on-line inquiry capability is through CICS /ESA.

The application process approximately 400,000 incoming transactions per day through IMS.

Currently, CICS transactions are running at approximately 20,000 per day.

There are approximately 70 different screens used by the HEJ application.

Demand Deposit Accounts Application (DMS)

Hogan application Release 8603

IMS DB/DC

967,000 DDA accounts

Information for this application is stored on BOL strip files. Inquiries are processed by the BOL system. Updates from CSD are sent through BOL to the system of record each night in a batch process.

Each day after batch processing BOL strip files are updated account by account.

BOL keeps memo-post information for DDA accounts on the strip files.

Memo-posts that occur after the batch processing cutoff are reapplied to the strip file account records after an account is refreshed.

OD protection via credit card or line of credit, on-line exception processing and restraint processing for stops, hard holds and reserved/pledged funds are all sub-systems of the DMS (DDA) application..

Savings Application (SAV)

Hogan application Release 8603

IMS DB/DC

910,000 open linked savings accounts

364,000 open unlinked savings accounts

Information for this application is stored on BOL strip files. Inquiries are processed by the BOL system. Updates from CSD are sent to the system of record each night in a batch process.

Each day after batch processing BOL strip files are updated account by account

BOL keeps memo-post information for SAV accounts on the strip files.

Memo-posts that occur after batch processing begins are reapplied to the strip file account records after an account is refreshed.

CD processing, on-line exception processing and restraint processing are all sub-systems of the Hogan Time deposit application..

Microfilm Services Automation (MSA)

FNAX developed IMS DB/DC (Telon generated)COBOL application.

Interfaces with the Procars a PC based microfilm application.

MSA passes requests to the Procars application three times a day using a batch process.

MSA keeps a record of requests made but not the detail for the request.

Procars operates in a standalone DOS environment. It is not tied to a LAN and communicates to the host through 3270 emulation.

Procars groups the requests in batches, creates pull lists and detail pick lists, contains status information about each request, and contains three months of history of each request at the detail level.

Clarke American check vendor information

FNAX is currently developing a pass through access to the Clarke American check order on-line application.

Electronic Mail (eMail)

CA eMail release 3.2.

Application runs in the CICS environment

Support the X.400 protocol (note: the product as installed at FNAX does not support this protocol even though the vendor product does)

TCS Telecenter workload projection and scheduling system.

A windows application written in FoxPro running on a PC resident in CSD.

Schedules over 500 individual staff members.

Interfaces with the Aspect ACD and uses call volumes to project future staffing needs.

1.5.7 Application processing cutoffs

BOL, the primary application supporting CSD, is a 24 hour, 7 day a week application. Batch processing for most applications occurs on a Monday through Friday schedule. To support timely batch processing applications have standard times to cutoff that day's work. After a cutoff the transactions for the application are recorded as transactions for the next day. For example, the credit card application may cutoff work around 5:00 PM while BOL transactions may not cutoff that day's work until after 10:00 PM. That means that throughout a 24 hour period applications will be processing work for different business days. It is possible that the data source for application information may change throughout the day due to business day cutoff criteria.

In addition, account refreshes are performed on an application by application, account by account basis. It is possible to inquire on a checking account that has been refreshed or updated from batch processing yet inquire on that customer's savings account that has not been updated. For that reason, refresh dates are stored at the account level. It is important to track the work-of date for each application throughout the day when providing customers with account activity and status information.

1.5.8 Session concept

There are two aspects to the session concept with regard to CSD. First, each 3270 type terminal (the basic CSD workstation) has the capability to handle four host based sessions or logons. Second, each customer interaction is considered a session.

3270 sessions

Each station or terminal has the ability to handle four separate host logons.

As configured, the first session (session A) is the only session defined to IMS and therefore is the only session that can access the Branch On-line (BOL) application. This also means that to access any other IMS application the CSR must log off of BOL and onto the other IMS application. CICS applications can be accessed in any session but cannot be accessed through a session already logged onto IMS. All four sessions are used. The applications most commonly in use would be:

1. BOL - Branch On-line, this is the application under which CSD functions.
2. eMail - The electronic mail system, a CICS application originally the ADR eMail, now a CA product, is used for written communication throughout the bank.
3. Access - is the FNAX application that allows the CSR to locate information about staff members, departments and branches throughout the bank.
4. Host EJ - this is a record of the transactions performed by the branches and the memo-post transactions performed by CSD.

Customer event session

A customer session originates with the TS07 CSD Identification screen.

The account number entered for identification purposes is carried and pre-filled where appropriate throughout the session. (Only one account number is carried throughout the session)

Sessions can contain a number of individual application transactions. A session continues as long as the CSR is dealing with activity for the same primary account. If you want to perform transactions for a different account number you must begin a new session.

At the end of a session, the decision is made to charge or not charge the client for the activity requested. Either charge choice brings the session to an end.

The decision to charge or not charge is up to the CSR in most cases. (Some micro-film requests result in automatic charges, some items such as statements are manually charged). The CSR is provided with guidelines to follow in making the decision which is processed by pressing a function key for "charge" or "no charge".

1.5.9   Host applications and userids
Each CSR has a Host defined userid. All userids for IBM mainframe access are defined through the RACF security application. This userid is used to logon to a Network Director screen with pre-defined CSD appropriate applications. IMS and CICS applications can be entered by screen selection or using the appropriate function key, an additional userid is not required. eMail requires a different userid and password in order to be accessed. The same userid can be used to logon to each of the four sessions but each of the four sessions can have a different user logged on. CSRs may logon to multiple CICS applications but only one IMS application since only one session is defined to IMS.

2.0   Requirements
All requirements listed in this Section must be addressed. The vendor is to respond to the contents of this section in the same order, and numbered in the

same manner, as set forth herein. Where a list of requirements or questions occur, the vendor should respond in the same sequence. Functionality is listed in either the form of a description or a question. In either case, the vendor should indicate how the feature is supported and avoid 'yes' or 'no' answers.

List any ways in which the vendor product(s) exceeds the requirements or has capabilities not directly addressed by the questions. Include these additional capabilities directly after the most closely related requirements listed below.

The response must clearly indicate which functional requirements are:

1) Deliverable on a turnkey basis (no development or customization required)
2) Deliverable on an end-user definable basis (no technical resources required to adapt to FNAX requirements). Differentiate between functionality which can be implemented by the end-user without technical support and that which requires more formal procedures or control for implementing.
3) Deliverable using standard vendor provided tools and pre-built modules which can be adapted. List which tools or modules will be used and/or what modifications must be made to pre-built modules. The level of effort to adapt the tool or module to the FNAX requirement must be described.
4) Deliverable only with new development.
5) Is a deliverable of a planned release not currently in production.
6) Not deliverable.

Tools and pre-built modules which are identified in this Section should cross reference to full descriptions of those specific tools/modules in Section 4.2 (General Package Information).

Vendor must indicate specifically if a requirement can be supported but is not being used in production by any clients at this time. Vendor must be prepared to provide references, including contacts, for sites who are using any functionality which is in production today.

If the vendor considers a requirement to be a future enhancement which will not be deliverable with the current product(s), then the proposed availability date must be specified.

The vendor may propose various alternative solutions for meeting any of the requirements. The determination of the suitability of an alternative will be solely at the discretion of FNBA.

The appendices contain the following documents which will assist you in understanding our direction and system requirements.

- Appendix B contains a 'Key Concepts' document describing some of the major issues that we are facing and our high level approaches to solving those problems.

- Appendix C contains a description of the 'Top 15 Events' (most frequent) which occur in the Customer Services area and the types of changes we would like to make to the way those events are handled today.
- Appendix E contains a description of one particular event and a detailing of the requirements for one specific path down to a fine level of detail. This is intended to provide vendors with a sense of the relative complexity and functionality that are likely to be built into final requirements.

Where 'Supplemental Information' is requested, the questions do not indicate requirements for that section but indicate an interest in understanding the vendor's ability to support the functionality or feature discussed.

2.1   Functional requirements

2.1.1   Event and activity tracking
Event and Activity Tracking must be supported.

2.1.1.1 Contact History Database
1. A Contact History database must be maintained for all customer calls. The file would contain a record of all events and transactions which occur during a call. Information maintained in this file must be accessible by customer, account, or reference number (a unique number assigned to each call which is provided to the customer on monetary transactions).

Logging of all key data required to trace which individual performed a transaction needs to be supported. Logging may be selective based upon the transaction or event being performed. Logging should include information on the operator ID, a time stamp, etc. As this is required for audit and control purposes, logging must be performed with a high degree of data integrity.

2.1.1.2 Activity Tracking
1. During the course of an event with a customer, specific follow-up activities may be identified which need to be tracked. For example, microfilm copy requests, statement copy requests, closing account checks mailed, etc. may occur during a customer call. Each of these action items may require work be performed after the call is completed. The ability to track, update, and manage these tasks from initiation to completion needs to be supported. Specifically, describe how you would support each of the following capabilities:
2. Provide on-line status of any follow-up activities which are required due to an event.

Allow viewing of the items in a 'tickler' file by each CSR or by their supervisor or surrogate. Should allow the viewing to be filtered by date so that only today's events are reflected.

Allow reporting of overdue activities to Supervisors for their group.

The ability to review the history of an activity including who updated the status or completed various portions of the work associated with it.

3. Scheduling of future activities (e.g. a customer callback) which may occur.
4. Automatic entries based upon call events. For example, should a photocopy of a check be ordered (this would also apply to a referral), an item should be entered into the activity file on the expected day of arrival of the photocopy. The expected day should be calculated by the system based upon standard lead times.
5. The initial status (e.g. "Copy requested") of an activity generated by the system should be automatically entered into the Contact History database.
6. The ability for a CSR to exit an event or transaction before it is complete (e.g. a customer letter, a photo request, etc.) and have the transaction automatically logged to a pending file for future completion must be supported. CSRs must be reminded of open activities requiring completion from the pending items list (e.g. when logging on or off). Recall of the transaction in its stored state must occur when that item is selected from the pending list.
7. Allow updates of transactions from different customer events or transactions. For example, should a customer call result in the resolution of a pending activity, the ability to note that the activity is complete during the call should result in the activity being updated and dropped from the active list. Essentially, we do not want to perform separate maintenance of the status of customer events and the outstanding activity files.
8. Automatically remind CSRs of current items in the tickler file. Reminder could be by a message which pops up on their screen during the day or at logoff times.
9. Allow updating of the status of activity items by the CSR or their supervisor. Activities marked as complete should be dropped from the tickler file.
10. Restrict access to certain types of activities, i.e. fraud research items.
11. Identify 'surrogate' CSRs who can access and update the Tickler file should the primary CSR not be present.

2.1.1.3 Workflow management
1. Activities which are currently manual such as printing Cashier's Checks, ordering microfilm copies, or creating time cards may be automated in the

new environment, as note elsewhere in this RFI. Describe overall workflow management capabilities supported by your system and the process for tailoring those capabilities to our environment.

2. Open activities may result in work being performed by persons other than the CSR. This would include the Research department in CSD. Research personnel are responsible for processing the activities from all CSRs. To support this, the following capabilities must be supported:

3. The ability for a CSR to forward a 'form' to a research area for action needs to be supported. This 'form' would need to include all relevant call and action data so that the research area would know what action to take and have the information they need to process it. Content of this 'form' could be generated automatically by the system in some events while in other situations the CSR would manually enter information such as the task description and customer data. Describe how you would support this capability. For example, this could involve:

- A message being generated from the event and placed in a 'Research Queue'.
- A file being created during the event and placed in a 'Research Folder'.

4. Allow open activities to be prioritized by the personnel in Research based upon activity type or an indicator set by the CSR (such as a date or 'high priority' indicator).

2.1.1.4 Event recall and correction

1. The ability to recall data about specific transactions for a customer from the Contact History database must be supported. Recall may occur based upon customer name, representative ID, time of day, reference number, or account number. The system may return a qualifying set of transactions (e.g. all transactions for a customer) and provide an ability to select one for research or processing. Once selected, the system must support the ability to 'delete', modify, or 're-enter' the transaction; this may require host updates.

2.1.2 Computer/Telephony Integration capabilities

FNAX is currently uses an Aspect ACD and does not intend to switch ACDs as part of this automation effort. In this section, note any features or capabilities which would be available to FNAX (or be substantially easier to deliver) should FNAX switch to another ACD and specific switches these additional features apply to.

1. Screen pop with call, customer, or contact history prior to CSR receiving call must be supported. Customer identification will be based upon the entry of

an account number and privacy code by the customer onto the VRU which is validated by the host mainframe.

2. Describe the different options which you support for controlling what screen appears on the recipient workstation. Criteria must include, at minimum control over screens based upon what group the receiver is in, source of call (e.g. a Bank of Axxum, Idaho customer versus a FNAX customer), the call event.

3. Upon attempting to transfer a call or contact an individual within CSD, should they not be available, return the status of that person to the operator, e.g. on phone, on break, not at work (see Section 2.1.12, Schedule Adherence). Provide options to forward the call to a surrogate CSR or to add a message to the original CSR activities file to return the customer call.

4. Ability to control phone from PC including:
   - Answer call
   - Select phone number from context sensitive pick list
   - Transfer calls
   - Conferencing others into call
   - Other phone commands such as wrap-up, take message, etc.
   - A single sign-on to the workstation which automatically enables a sign-on to the ACD.

5. Support of coordinated voice/data for call conferences or transfers. This must include the capability to:
   - Provide the receiving person with the active screen from the senders workstation.
   - Provide the receiving person with a different screen. For example, when transferring a call to the Sales area on a referral, the initial screen may be the client relationship screen regardless of the original event.

6. When a call transfer is performed, the current status of the customer event in progress must 'transfer' with the call to the new representative. For example:
   - If a customer has already been identified to a specific level, then the system should continue to manage the event and screens for that transaction as appropriate.
   - The screens appropriate for Bank of Axxum, Idaho customer must be presented to the new CSR if the call originated from an Idaho 800 number and screens appropriate for FNAX customer must be presented to the new CSR if the call originated from a Washington 800 number.

7. Support for conferences with VRU while having visibility to what the customer does on the VRU (e.g. have customer entry of numbers displayed on the CSR screen).

8. Calls may be transferred back to VRU.

9. Multiple screens or windows may be brought up as part of screen pop. List any restrictions you may have on the number of windows.

10. Describe capability to perform call blending during low volume call periods such as the ability to mix inbound and outbound calls. Include discussion of:

   • The ability for the system to control whether call blending occurs based upon incoming call queue volumes.

   • The ability to select which CSRs are selected for alternative activities (e.g. CBT) based upon flexible parameters.

   • The ability of the system to provide Computer Based Training (see Section 2.1.11) to selected staff (as opposed to making outbound calls) during slack periods. Must include the ability to identify to the system who is eligible for CBT and which modules they should review.

11. Support of electronic FAX direct from workstations to client FAX numbers.

Supplemental information

S1. Describe your capability to determine call routing based upon previous customer contacts? Parameters should include option for alternatives should primary receiver be busy, date of last contact (if old may be irrelevant), how busy the current queue is, etc.

S2. Can you support multiple switches or switch types simultaneously (e.g. Aspect and Northern Telecom)

S3. Can the end of a call (e.g. a hangup) result in the system automatically providing the CSD the session ending screen (e.g. call wrap up, whether to charge the customer or not, etc.)?

S4. Do you support 'expert agent selection' to help select the recipient of an incoming call based upon CSR skills, customer profile, source of call, queue volume, etc?

S5. Do you have the capability to intercept customer calls when the queue is heavy and provide an option for the customer to leave a message?

S6. List ACD switches that your software currently interfaces with in production at this time. List additional ACD switches that your software supports although not in production at this time.

S7. Do you have a 'Snoop' feature allowing supervisors to view the screen that the CSR is viewing, including any data entry and updates the CSR may be making, in conjunction with listening to the client phone call, with or without the CSR being notified?

2.1.3   Check vendor interfaces

1. The system must include an interface with Clarke American, Deluxe, and

Harland for check ordering and performing inquiries as to order status and history. Describe the interfaces and functionality supported for each of these check vendors.

2. The check vendor screens may be reformatted to present within the standard Graphical User Interface (GUI) look and feel that will be built for this project. In addition, the ability to capture information about check ordering transactions into the Contact History database (see Section 2.1.1.1) needs to be supported.

2.1.4   Letter generation

The capability to generate customer correspondence must be supported.

1. Standard letter templates will be created for various call types. These letters will include the ability to:
   - Pre-fill caller/account information from current call data
   - Limit staff from editing the letter except for specific 'comments' sections for CSR input to the content. Staff editing of letter must be controllable by user class.

2. Access to letters will be imbedded in the event flows. Depending upon the event, the letter may be either required or optional for the CSR. More than one letter option may be available for the CSR for specific events.

3. Once a CSR has created a letter and is ready to send it to the recipient, the system must control the output. Based upon the letter selected, one of the following scenarios will occur. Describe the development necessary to support each scenario.

   3.1) The letter is immediately printed on a central printer.

   3.2) The letter is batched (and sorted by type of letter or zip code of the recipient) for printing on special paper stock. The ability to include supplemental materials to be printed with the letter must be supported.

   3.3) The CSR may automatically fax the letter to the customer from the workstation. The ability to restrict faxing to specific correspondence must be supported (e.g. correspondence not requiring enclosures).

   3.4) The letter is electronically routed to the Research area along with notification of a request for off-line information. The Research area prints the letter, provides the appropriate enclosure(s), and forwards to the customer.

4. The ability to incorporate graphics images such as the corporate logo into documents must be supported. Describe embedded graphics capabilities including what type of graphical objects may be included.

5. Non-technical (user area) staff must be able to create or modify standard letter templates and to place templates into production without a systems out-

age. The ability to change or create letters must be limited to appropriate staff. Describe how staff would perform maintenance to correspondence.

6. All customer correspondence must be maintained on-line for an archiving period of approximately two weeks (and reprintable during that period).

7. The batch processing capabilities of your tools must include the ability to:
   - Reprint specific letters which may have quality problems upon output (e.g. wrinkled paper)
   - Restart production at the point of failure upon print failure, etc.
   - Automatically update a central customer database that the letter has been printed and sent

8. The system must support multiple fonts including multiple fonts on a single letter.

Supplemental information:

S1. Describe how you would provide the ability for CSRs to enter text while keeping the method simple.

S2. Does your software permit the use of Microsoft Word for Windows to create letters and correspondence? Are there any other standard word processing or forms packages which can be used? How would an external letter creation package interface with your software?

S3. How would you support automatic spell check for all CSR entered text (e.g. it is automatically invoked when they update correspondence)?

2.1.5 Screen navigation and context sensitive events

2.1.5.1 Event control and transaction paths

1. Each event may take diverse paths being taken depending upon customer input, data received from the mainframe, contact history, or CSR judgment. For example, the 'security screen' described in Key Concepts (Appendix B) identifying the customer should be skipped if a CSR has already performed a security check on the customer at the appropriate level for the current transaction. Describe the manner in which event and transaction paths are created and controlled.

2. CSRs (or other users) will not be expected to input all data to complete events or transactions. If the data is available on the Contact History Database or from the mainframe and does not need modification or validation by the user, then that data should be used to complete the transaction (e.g. with the mainframe) without user involvement and with or without display to the user.

3. The system must be able to complete a series of automated transactions or

activities based upon a single CSR action. For example, a CSR completing a particular screen might result in the system automatically executing multiple host transactions, performing updates to the contact history database, and generating a customer letter.

2.1.5.2 Automated context sensitive actions

1. Pre fill of data onto screens or forms must be supported. Data must be available for pre fill from earlier in the call including VRU or CSR input, the Contact History database, or from the FNAX mainframe.
2. Should a call be received from a customer regarding an account which has certain pending action, e.g. an outstanding photocopy request, then a 'pending action' flag may be set in the initial screen provided to the CSR about that customer or account (see Section 2.1.1.1, Contact History Database).
3. A customer indicator may be set which results in a specific message being presented to any CSR who accesses that customer's data on an incoming call. The message could be generic or tailored for a particular customer.
4. Where possible, call wrap up codes should automatically be determined based upon event path. The option of having the operator override the codes or add their own input where automation is not feasible needs to be supported.
5. The ability to translate data from other sources such as the VRU or host into different presentation data (e.g. in a more friendly form) must be supported. For example, account type codes might be converted into easy to understand account descriptions.
6. The ability to identify the results of a host update transaction to a CSR needs to be supported in easy to understand terms. For example, was the transaction accepted on-line or off-line? Was it rejected? The response format such as highlighting, flashing, color, etc. may vary based upon the result.

2.1.5.3 Navigation

1. Standard tools should support:
   - The ability for a user to navigate among different screen fields via the forward and backward tab keys.
   - Full point and click Mouse capabilities
   - Function buttons and icons must be accessible via the keyboard, i.e. function keys or key stroke combinations (two keystrokes maximum).
2. Transactions, field values, pick lists, and product or sales scripts should be variable based upon:
   - The source of the incoming call (e.g. origination city or state); or
   - Call data from current or previous customer calls and mainframe data (e.g. customer account list based upon our Customer Information File data)

3. Two different paths for the same event must be supported: 1) 'Fast-track events' for senior representatives who understand transaction requirements and can utilize faster paths even if it involves more complicated screens. 2) 'Walkthrough' events for novice staff who may need assistance in completing transactions. Describe your support for the following:
   - The ability to control the set of screens and transaction paths a CSR utilizes based on user class.
   - Requiring supervisor approval for changes to user profiles (e.g. which class they are a member of).
   - Development of the event paths that are assigned to user classes, e.g. specifically outlining which screens will be viewed in which sequence for which events. Describe how this control is provided to the user area.
4. The completion of a transaction by a CSR (e.g. by pressing the enter key or clicking on a 'complete' icon) may occur while the cursor is in any position on the screen.

2.1.5.4 Miscellaneous

1. The ability for CSRs to dynamically sort and filter a list needs to be supported. For example, checking account activity may be sorted in check number order, date of posting order, and/or may not include specific transaction types. Each option may be selected through menu lists or on-screen icons.
2. The ability to select multiple items from a list to be acted upon needs to be supported. Items may not be contiguous in the list. This may require host updates be made for each item selected. For example:
   - Selection of multiple checks and a photo command could result in multiple mainframe photos requests being generated.
   - Selecting multiple overdraft fees to be reversed may results in all fees being reversed (conversely, the selected fees may be aggregated and reversed as a group).
3. The ability to recall (and fill into the 'active' field) previously used account numbers such as:
   - The session account number
   - Last used account number
   - Previous session account number
4. The ability to easily develop different sets of screens and scripts based upon which state a call originated in. Can scripts be developed with variables which are dynamically changed in real time during the call based upon source? Differences could include:
   - Different identifier for the source in all screens, e.g. 'Bank of Axxum,

Idaho' or 'FNAX'. The identifier could be text or graphics (e.g. a corporate logo).
- Different product scripts or product information.
- Different event flows based upon state requirements.

5. FNAX may choose to not re-engineer all transactions and activities performed within CSD. This would likely be true of small volume transactions where there will not likely be significant savings due to redesign. Thus, many current mainframe transactions will be 'passed through' to the CSR with limited or no modification. These transactions would have to be seamlessly handled by the CSR as part of a customer call without having to 'switch' to another host session or 3270 emulation. Describe alternatives, involving minimal development effort, for:
  - Incorporating these 'passthrough' transactions into the standard look, touch, and feel of the overall CSD system that you would provide. This could involve insuring that similar keystrokes and navigation techniques are used by the end-user as well as providing them with a similar visual feel.
  - Including statistics and information about these transactions into the Contact History database.

2.1.6 On-line user help or reference information

1. User help and reference information such as procedures manuals or product information needs to be available on-line.
2. Describe how the following user help capabilities are supported:
   - Context sensitive help dependent upon event, transaction, or the active field.
   - Placement and size options of help windows at user area discretion.
   - Instant 'pop up' of help screens (e.g. avoid the slow speed typical of MS Windows help screens).
   - Automatically invoke user help, e.g. through CSR errors, based upon specific events or transactions, etc.
   - Control by supervisors as to who automatically sees help windows based upon user group parameters and the ability to assign staff to different user groups.
3. For help and reference screens, describe the following capabilities of your system:
   - Alternatives for accessing help or reference information which minimize keystrokes.
   - Help or reference screens content and parameters that are maintainable by end-users.

377

- Quick access to help or reference screens by topic.

2.1.7 Scripting (dialogue)

1. The ability to provide CSRs Scripts which help them service or sell to customers must be supported. Include descriptions of how your software supports the following features.
   - The ability to create context sensitive scripts which are context sensitive, event specific, or dependent upon the customer data from various host or CSD databases.
   - Displayed optionally or automatically based upon parameters.
   - Indicating to a CSR that a script is available for their use if not automatically displayed.
   - Placement and size options of script windows at user area discretion.
   - Linking multiple, interactive script windows together and support of multiple screen paths available to the CSR based upon their input or selection.
   - Supervisor control over who automatically sees which scripts screens based upon User Group parameters and the ability to assign staff to different user groups.
   - Maintenance of script screens content, flow, and parameters by end-users.
   - Alternative formatting options for script text or windows such as automatic scrolling across the screen, flashing of script, etc. must be available depending upon call or customer data. Describe options.

2.1.8 Screen design capabilities

2.1.8.1 Menus

1. Menus must be customizable by user class.
2. The ability to ghost menu options based upon event or call data must be supported.

2.1.8.2 Pick lists

1. Describe your software's ability to support the following regarding pick list options:
   - Ability to vary content based upon prior call data
   - Function key activated
   - Error activated
   - Scrollable/pageable
   - Size options
   - Placed by program at user discretion.

- Minimal key strokes to select value
- Easily maintained and modified in production by end-user.
- Automatic display based upon prior call, customer, or mainframe data,
- Long pick list. Identify any limits.
- Describe methods for easing navigation of a long list.
- Embedded pick lists within a hierarchy.
- Ability to assign specific keystrokes (e.g. numbers or letters) to allow Pick List items to be selected.

2.1.8.3 Fields

1. Describe your software's ability to support the following field alternatives:
   - Highlight or change field format (e.g. highlight, bold, flash, etc.) based upon call or customer data, information received in a response from the host, or whether data had been entered in error, etc.
   - Blocks or methods of showing CSR the field size by a visual clue.
   - Ability to ghost fields, determined by event selection and customer or call data.
   - The ability to AutoTab to the next field upon completion of the current field.
   - Edits based upon co-dependent fields.
   - Edit against pick list.
   - Defaults can be determined by other field entries.
   - Default position of cursor varied by customer data, call data, or event path.
   - Ability to place cursor in first field in error if multiple errors. Tab should result in cursor placement at next error field.
   - Ability to edit each character of data entered into a field, e.g. use the mouse or standard edit keys such as arrow, delete, insert, etc. to change any character in a field.,
   - Varied fonts and font sizes within same window.
   - Field protection permitting specific fields to be blocked from change. Protection may be dependent upon customer data, call data, or event path.
   - Provide an indicator to the user, such as an auditory signal or flashing, when the maximum number of characters for a field as been entered. Retain the data entered in the field when maximum characters entered.

Supplemental Information

S1. Describe cursor options supported which enable the cursor to be easily spotted (e.g. varying sizes, blinking, the active cell, etc.)

2.1.9 Department message broadcasting

The following department messaging capabilities are required:

1. The capability to broadcast messages to CSD groups (or sub-groups such as 'special queue' personnel) or individuals which 'interrupts' call flow and forces staff to review before proceeding with calls. Note: messages should be targeted at specific CSD groups of personnel, not workstations.
2. The message system should be easy to use. Describe what makes your message system easy to use such as easy editing, easy to understand commands, etc.
3. Capability to broadcast to only those currently logged on (or scheduled in the current shift. (See Section 2.1.12, Schedule Adherence)
4. Alternatives for placement and formatting of messages when received such as flashing, scrolling, etc. Ability to create these different formats must be easy to develop by the user area based upon standard formatting selection options.
5. Store and forward of messages (e.g. should the recipient be logged off or the Workstation is off-line).
6. Standard template e-mail messages must be available for either repeat messages or easy modification before broadcasting.
7. Supervisor review of routing list and whether CSRs have reviewed messages.
8. The ability to 'pace' messages so that not all CSRs receive the message simultaneously.
9. Directory of past messages for retention, review, and possibly re-send.

2.1.10 ATM and branch location capabilities

1. The ability to present CSRs on-line maps of ATM and branch locations needs to be supported. Verbiage would be presented with the map such as:
   - Cross streets and nearby landmarks
   - Directions from major freeways or arterial routes.
   - Addresses
   - Unusual hours of the ATM
   - Branch phone number.
2. There must be an ability to search the location database by alternatives such as zip code, street name, town or city, branch name, etc. Maps will be supported for three or more states.

2.1.11 Computer Based Training

1. The ability to develop on-line training and skill certification courses is desired. Capabilities of the authoring tool and training environment must include:
2. Support of automated scoring logic for tests, certification, etc.

3. Tracking of results by user (by group, supervisor, site, total) including:
   - scoring
   - completion - partial or full
   - certification
   - module use
   - error statistics by module for all students
   - clock time spent in the training and individual modules.
4. Remedial pathing, i.e. the ability to affect pathway based upon results of test. For example, if a student answers one or more basic questions incorrectly, force them to return to a more remedial section.
5. Computer Based Training (CBT) availability 24 hours/day without impacting system performance of live operations.
6. CBT deliverable to CSR workstations.
7. Visual development tools appropriate for course designers, not technical staff. Describe the tools.
8. Ability to run two versions concurrently, e.g. training modules based upon upcoming system changes versus modules based upon the current release.
9. Ability to assign modules to specific staff by supervisor.
10. Support of multimedia including sound, video, etc. Describe multi-media tools.
11. Ability to utilize production screens and business logic from the CSD automation system (including functionality such as access to production help screens).
12. Ability to simulate the production environment including VRU input, responses from host, field edits, etc.
13. Specific mainframe user ID's have been reserved for training. These IDs, when used, allow full navigation among system screens used by CSD but result in no changes being made to our production systems. They are 'dummy' IDs. Use of the CBT system must use these IDs when communicating with the mainframe. In addition, no logging to the production Contact History Database (logging may occur as long as it is clearly separated from production data and not included in production reports and statistics) should occur.

2.1.12   Personnel schedule adherence

FNAX utilizes the TCS Management Group's TeleCenter workload projection and scheduling system in CSD. This system generates a daily schedule for all representatives including start and finish times and break schedules. These schedules are distributed to supervisors at the beginning of each shift. Based upon this data, FNAX would like a 'schedule adherence' system which helps insure that customer service representatives work according to their assigned schedules. Functionality would include:

1. The ability to generate warnings for CSRs of upcoming breaks. Warnings would include flexible parameters as to the amount of advance notice given, the number of advance notices provided, the wording to be used for each notice, where and how it appears on the representative screen.
2. The ability to automatically log a CSR off the system at the beginning of each break period. If the representative is on the phone at the scheduled beginning then the system will log the CSR off at the completion of the phone call. The representative will only be able to continue phone calls during the scheduled break if they 'override' the logoff with a special logon sequence. The originally planned break period for a CSR should be adjusted if delayed due to a customer phone call which was in progress when the break was scheduled to begin (this would allow proper reporting on exception reports).
3. Tracking of all CSR time which is listed as breaks, lunches, research time, etc. This could be captured via specific logoff codes or the selection of a logoff type from a screen. Exception reporting should be made of each adherence exception.
4. The ability to dynamically change break periods for each representative. Access to this function must be limited to supervisors and managers.
5. The ability to import the scheduling data from the TCS scheduling system into the CSD automation system to update staff scheduling data must be supported. This file must be uploaded daily at user request. The PC containing the TCS software will be connected to the CSD network. All weekend days may be uploaded on Friday before close of business. This file may be uploaded more than once, should updates occur. The system must be able to override previously entered data without interrupting functionality to representatives who may be already at work and logged on. Note any assumptions you make in your ability to support this and alternatives for importing the file (e.g. can you read the native TCS DBF files or would you need us to create a special export file of the scheduling data).
6. Provide ability for CSRs or supervisors to interrogate specific CSR schedules.

2.1.13 Time card system

Current time card system: Time cards are currently completed manually by staff. Both the staff member and their supervisor must sign the card. Cards indicate time at work, absence reasons and hours, date of absences, etc. Time cards are scanned into a corporate time card system which calculates payroll hours and overtime and interfaces with the Payroll system.

Automation of the CSD timecard system should be supported including:

1. Data generated from the Schedule Adherence system (see Section 2.1.12) regarding when staff log-on and off the system should be used as the basis for time cards.

2. An on-line substitute for the time card should be developed to allow staff members to review and approve their hours worked. Employees should be able to indicate times which are in error (e.g. perhaps they forgot to log on or off as appropriate on a particular day) and note the correct hours or time.

3. The supervisor must review and approve the employees on-line time card. Supervisors can override hours as logged by the system and approve employee corrections.

4. A electronic file equivalent of the time cards should be created weekly for input into the corporate time card system.

5. Complete audit trails of all actions taken in conjunction with creating electronic time cards including any changes or updates made. This must include the user ID of whomever made a change. Audit trails will be either electronically archived or offloaded to Fiche as it ages for long term storage.

6. Viewing of time cards must be limited to appropriate staff. The ability to affect the on-line file of time card data must be limited to appropriate staff.

2.1.14 Telemarketing/sales specific requirements

Describe the use of your system to support a telephone sales organization or branch platform sales system. Clearly differentiate between functionality which has been implemented at a current site and that which has not. Include aspects which support:

1. Maintaining detailed customer call and history information including customer financial profiles.

2. Tracking sales, sales vs. goals, and who referred the sale.

3. Interfacing with automated dialing systems.

4. Building and maintaining complex sales and product scripts which have alternative pathing possible based upon customer responses. Scripts should be maintainable by end-users.

5. Ability to download data from other sources including the mainframe such as client data, competitor rates, FNAX rates, etc.

6. Outbound sales calls such as:
   - Building call lists based upon flexible criteria (such as customer income, product interest, etc.) and using those lists to automatically dial numbers. Include the ability to track the outcome and effectiveness of use of these call lists.

383

- Interfacing with auto-dialers.
7. Interface with an Official Check Printer system to prepare Cashiers Checks on loan proceeds, closed accounts, or CD redemption's. Describe any experience you have with interfacing with Official Check Printing systems.
8. Creation of very complex forms (e.g. loan documents) which have the ability to include or exclude sections based upon customer data.
9. Fulfill sales by sending data to host systems for boarding of the new account(s).
10. Create 'What If' scenarios, including graphs, of:
    - FNAX vs. competitor products.
    - Results of alternative investment decisions.
    - Results of borrowing based upon different loans products or loan amounts.
11. Ability to print graphs (locally, remotely, FAX) which might be generated from the 'What If' scenarios.

Supplemental information

S1. Describe your ability to provide Auto, RV, or Boat loan blue book rates online. Include listing of any interfaces to third party providers of this data you support today.

2.1.15   Imaging support
Describe your products ability to handle document images including what graphical and imaging /compression standards are currently supported by your software. Specifically, can your software:

1. Interface with an external DBMS which contains images of documents (such as checks)
2. Receive one or more images as part of the data returned in the response to a query or transaction sent to that DBMS.
3. Display the image(s) received from a DBMS or image system.
4. Does your product interface with any Check Imaging systems? Which ones? Is this capability in production at any site?
5. Provide any future plans that you have for interfacing with imaging or Check Imaging systems in your product line and when that support is estimated to occur.

2.1.16   Miscellaneous
1. A basic business on-line calculator should be available on a pop-up screen via quick keystroke or an on-screen icon.
2. Reference numbers uniquely identifying an event or customer call for audit

and recall purposes must be created. These reference numbers are provided to customers at the end of each call.

3. Develop an interface to the FNAX Microfilm systems to support the ordering of check copies. This would include forwarding a file of check copy requests to the MSA system and the ability to look up the status of check orders in the Procar microfilm system.

Supplemental Information

S1. Does your software provide the ability to perform screen prints to a printer connected to the local network?

S2 Can you support flexible keyboard arrangements and varying use of keys?

2.1.17   Reporting and data availability

1. The software environment must capture information which will allow the following reports and on-line queries to be supported. Note: the data for calls received in both the Xzzum and Axxum locations must be consolidated.

2. All terminals, based upon user class, must have the ability to display in real time the following:
   - System status
   - Agent Status
   - Queue status
   - Trunk status

3. When customer calls are received by a representative, the ability to display the following information must be supported:
   - Transfer path (e.g. VRU to CSR to second CSR).
   - Total call time
   - Wait time

4. Reporting and tracking by event such as:
   - Type of event
   - Length of calls for an event such as fee reversals or refunds by operator, group, time frame, department, etc. If multiple events occur during a call then the tracking may be requested for the length of time spent during that event within the overall call time. If multiple transactions occur within an event (e.g. a letter is generated, a fee is charged, etc.) then data about the individual transaction must be tracked.
   - Tracking of the path an operator took to complete an event. Describe alternatives for this tracking ability and how you would recommend displaying this type of information.
   - Statistical correlation of events to determine what types of events tend to occur in the same phone calls and in what order.

5. Productivity reporting by operator, group, location, or department including:
   - Average length of call broken out by VRU, CSR, and individual queue time
   - Average length of call wrap-up
   - Idle times
   - # of events/call
   - Pathing of event navigation
   - Specific actions taken by operators during phone calls such as fee waivers, reversals of fees, referrals, etc.
   - # of calls/hour (day, week, month, etc.)
   - # of calls/hour vs. number of calls in queue
   - # of transfers
   - Outbound call statistics
   - All time spent by CSRs including lunch, breaks, training, etc.
6. Reporting and queries of customer contact history including:
   - date of contact
   - events during contact
   - staff contacted
   - status: (closed/pending) for each event, tickler items, etc.
   - notes regarding customer contact such as whether they tried to obtain a fee reversal
7. Tracking of customers who hang up while in queue. Tracking of customers who hang up while using the VRU in the middle of a transaction (e.g. they do not complete the activity). Keep track of ANI data for analysis of these types of calls.
8. Lists of customers who have multiple or repeat problems to aid in retention.
9. Describe how your report writing tool supports each of the following features:
   - Selection of data across multiple tables. Hides database structure from user area.
   - Building of multiple table reports.
   - Filtering of data.
   - Ability to build standard filters for dynamic application to query results.
   - Sorting of data.
   - Ability to build standard sorts for dynamic application to query results.
   - Automatic totals on breaks.
   - Labeling of data.
   - Ability to easily set performance goals for unit (group, department, division) and build exception reports of events, service levels, productivity statistics which do not meet those goals.
10. Describe reports that are delivered with your solution on a turnkey basis.

11. Must support the ability to upload all data from the Contact History Database to external databases.

2.2 Sample event flow development

Appendix E contains a detailed description of potential requirements for an event. Provide a detailed description of the development steps which would be taken, tools which would be utilized, the amount of effort it would take, etc. to develop the event per the requirements.

Also describe any suggested design criteria for screen flow, screen design, etc. based upon the detailed event description.

2.3 Overall requirements

2.3.1 System and application security

2.3.1.1 Security overview

1. Describe how you would support single sign-on by user which enables access to multiple host applications or sessions and the ACD.

2. Security architecture should be compliant with standard Distributed Computing Environment (DCE) security services, as defined by the Kerberos standard. If your security architecture is not in compliance, list your plans for becoming compliant. If you do not have plans to become compliant, describe your reasons for not doing so.
3. Describe your software's abilities to support the following and how you would implement these features:
   - Limiting physical access to your environment including workstations, servers, etc.
   - Limiting direct logons to the server (e.g. support use of a NTAS security server or equivalent to perform logons). List options currently supported.
   - Security software provided by the server application or database software and not the operating system.
   - Access RACF to authenticate users. Keep passwords in synch with RACF.
Note: existing mainframe systems such as IBM and Tandem will retain their internal security controls. Native security services will continue to control access to mainframe resources, i.e. RACF and Tandem.

2.3.1.2 Client Workstation Security

Indicate your ability to meet each of the following workstation security standards. If you would address these requirements through the use of third party software, indicate your recommendations for such software.

1. Logon should require authentication with a password of at least 7 alpha numeric characters.
2. There should be a logon failure process that prevents logon after a predetermined number of attempts (maximum of 5 to 10 logon attempts).
3. Initial passwords set by administrators should be pre-expired, requiring the owner to change default passwords at first logon.
4. Passwords expiration should be definable by the security administrators (30 to 90 days).
5. Passwords should be encrypted for storage.
6. Passwords should be masked at entry.
7. There should be the ability to authorize more than one employee to the work station without sharing an individual userid.
8. There should be a way for a guest to use a workstation for network access without accessing the owner's data files and proprietary programs.
9. Access control should provide for the protection of data files and programs by individuals and groups.
10. Security software should be enabled at start-up and should be protected from alteration without prior security administrator authorization and prevent boot from removable media.
11. There should be the ability to perform password resets from a remote location.
12. The software must support pre-configuration.
13. Software should provide a quick lock feature that allows for temporary absence from the workstation.
14. The software should provide for locking of an unattended/inactive workstation.
15. The software should support single sign-on across the network.
16. Product should allow for multiple administrator ID's.
17. Password history should be maintained at least for 20 generations to prevent password reuse.
18. Password changes can be performed by the ID owner for routine password changes.
19. Product keeps an audit trail that can be accessed remotely for reporting and investigation.
20. Encryption support for both fixed drives and removable media (floppies and tape).
21. Controls for deciphering data.
22. Supports centralized distribution and minimizes user interaction.

2.3.1.3 Virus Protection Standards

Indicate your ability to meet each of the following virus detection standards. If you would address these requirements through the use of third party software, indicate your recommendations for such software.

1. Software must run in an active mode that provides constant screening.
2. Software must detect known viruses and support cleanup of identified viruses.
3. Software should have a feature that will allow it to identify suspicious activity.
4. The software should be tamper resistant.
5. Supports centralized distribution and minimizes user interaction.

2.3.1.4 File Server Security

The server operating system is viewed as the primary means to implement data security objectives over and above complementary functionality found in network, application and database products.

Indicate your ability to meet each of the following file server security standards. If you would address these requirements through the use of third party software, indicate your recommendations for such software.

1. Logon to any department server from the server keyboard should require authentication with a password of at least 7 alpha/numeric characters.
2. There should be a logon failure process that prevents logon after a predetermined number of attempts (maximum of 5 to 10 logon attempts).
3. Initial passwords set by administrators should be pre-expired, requiring the owner to change default password a first logon.
4. Passwords expiration should be definable by the security administrators (30 to 90 days).
5. Passwords should be encrypted for storage.
6. Passwords should be masked at entry.
7. Access control should provide for the protection of data files and programs by individuals and groups.
8. Security software should be enabled at start-up and should be protected from alteration without prior security administrator authorization and prevent boot from removable media.
9. There should be the ability to perform password resets for administrators and users from a remote location.
10. The software must support pre-configuration.
11. The software should provide for locking of the server keyboard if left unattended/inactive.
12. The software should support single sign-on by accepting pre-authenticated users from an external source.

2.3.1.5 Network Security

Network Security will be provided by software that will support network encryption and user authentication. Components of the software will be housed on each workstation and server in the environment. Indicate your ability to meet each of the following network security standards. If you would address these requirements through the use of third party software, indicate your recommendations for such software.

1. The software should provide for integrity and/or the privacy of a message (message authentication/encryption).
2. The software should provide for transport of a users credentials (single sign on support).
3. Software should support extended authentication.

2.3.1.6 Application security

1. Discuss application level security that your system provides. Include, at minimum, the following abilities:

- Controlling access to transactions or procedures by individual or group.
- Controlling the ability to complete transactions based upon criteria such as dollar amount or customer value by individual or group. Customer value may be determined from a mainframe based indicator.
- A history of access levels for each individual or group and audit trails as to who affected changes to access levels.

Supplemental information

S1. What physical workstation security products is your software currently running in production?

S2. Do you have the ability support single sign-on by allowing the use of an external security software product for user authentication? Which software packages does your software currently interface with?

S3. How do you prevent CSRs or other end users from using the workstation for any applications other than the system you provide? How do you prevent CSRs from by-passing this and possibly harming the system during system failures, e.g. during restarts?

2.3.2 System reliability

2.3.2.1 System Uptime

[CSD is open for customer calls 24 hours a day/7 days a week. Any downtime has a significant impact on our customers.]

1. Describe your ability to support the following requirements.

- System administrative and maintenance activities must not take the system down.
- Ability to make changes to software and system without taking down whole system. Specifically, list which features and maintenance activities can be performed while the system is live. Include listing of what user parameters can be changed such as scripts, help screens, pick list tables, etc.

2. For software updates which are not user controlled (e.g. involve code changes), changes must minimize impact on production. Describe how you would implement changes without impacting the rest of the system.
3. Can the system support two different Client systems running simultaneously while downloads of new client software are being made (assume that the interface between the client and server may change as part of the new software)?
4. Describe any tools which support notice of possible impending failures (e.g. monitoring of network or disk access failures which exceed certain parameters).

2.3.2.2 Restart/recovery
Provide complete descriptions of how your system would support the following requirements:

1. The ability to recover from various component crashes including server, workstation, power failures, disk crashes, host, ACD. Include ability to switch to alternate devices. Describe the events that take place upon a failure of each component.
2. System crash recovery processes must allow for reconstruction of all files to the last checkpoint.
3. Should a server fail, automatic switching to a backup server must occur. How long does it take for the backup system to become operational? Does this take manual intervention?
4. The system must have the capability of handling damaged files gracefully. Can it continue to function while providing notice to a system administrator that a problem exist? For which files? Which files would cause a system failure if broken?
5. The phones and VRU must continue to function even though the network or server is down. Describe how you would accomplish this and any ACD or VRU functionality lost as a result of such a failure. What incoming phone data (e.g. source of call) is available when the server/CSD automation system is down?
6. Support for remote recovery must be provided, if appropriate. Support for all Xzzum systems occurs from Axxum.

Supplemental Information

S1. Explain operational and system recovery procedures for when the power or equipment fails during operation.

S2. Explain redundant hardware capabilities and how these capabilities are utilized.

2.3.2.3 Store and Forward and off-line Capabilities

1. What are the off-line capabilities of each component of your architecture? Separately describe the ability of the server(s), workstation, or other architectural components to function when:
   - The Host is unavailable.
   - The host is only partially available (e.g. specific applications are down but the host is generally functioning).
   - The Server is unavailable.

2. Describe your software capability to (and the development required to support):

3. Recognize that the host has not responded on a particular transaction and based upon the event being processed, backout other related transactions which may have been completed with the host and provide unique error messages back to the agent for the particular failure which has occurred. Describe the development tools used to customize this capability.

4. Describe store and forward capabilities including:
   - Practical limits to the length of time the system can store and forward.
   - Retry controls and capabilities upon multiple transaction (full or partial) failures.
   - The ability to manage system workload upon store and forward. For example, pacing transactions a few at a time to the host or at controllable intervals in order to prevent degradation in system performance upon reinstatement of mainframe availability.
   - The ability to run without manual intervention except in exception conditions.
   - The ability to inform designated areas or personnel of an off-line condition.
   - Workstations must run independently of server including the ability to perform store and forward. What capabilities would they have?
   - The ability to inform a designated area of off-line transaction exceptions (e.g. store and forward transactions which are rejected by the mainframe) so that they can research and process accordingly.

Supplemental information

S1. If a workstation crashes while it has store forward transactions queued up, how are these transactions accounted for?

S2. Provide alternatives and recommendations which you can support for insuring that the host has received all activity which is performed by CSRs.

S3. Provide alternatives and recommendations which you can support for insuring that the host does not process duplicate activity (consider issues with store and forward transactions or retries upon no response from host).

S4. Do you support the ability to access data (e.g. a daily download of customer identification data such as personal security codes which has been loaded onto the server) when the host is down in order to complete a transaction?

2.3.3   Disaster recovery requirements

1. What are your backup and disaster recovery recommendations?

2.3.4   Volume requirements

The following projected transaction daily volumes must be supported by the CSD automation system:

Activity

1995

1996

1997

1998

Total Call Volume

100,000

120,000

145,000

175,000

Calls handled by VRU

70,000

88,000

110,000

137,000

Calls handled by CSR

30,000

32,000

34,000

38,000

Host transactions generated by Calls

415,000

500,000

600,000

720,000

Host transactions generated by VRU

120,000

150,000

190,000

240,000

Transferred calls (within CSD)

8,000

9,000

10,000

11,000

Transferred calls (total)

47,000

51,000

55,000

60,000

Notes:

- 60% of the volume is at the Axxum Site vs. 40% at the Xzzum site.
- The system will need to handle no less than 300 simultaneous opera-tors in production.
- The above represents nearly 14 host transactions for each customer call han-dled by a representative today. VRU usage results in approximately 2 host transactions being generated per call.

Based upon these numbers, a peak 1/2 hour call volume is projected to be:

1995

1996

1997

1998

Peak Volume

6,200

7,500

9,000

11,000

Peak Mainframe transaction volume

30,000

35,000

42,000

50,000

1. Describe your system's ability to process the above volumes.
2. The most difficult system load may be after a system recovery when as many as 300 CSRs may attempt a logon at in an extremely short period of time. Describe how your system would handle an extremely high peak transaction load and what the client would experience should the system capacity be exceeded for a short period of time.

See Appendix A for detailed volume statistics.

2.3.5   Response time requirements

1. Current response time for host access by CSD representative is .8 seconds including host and network time. No degradation in this response time is acceptable. Describe the expected response time overhead added to host access by a properly configured system using your architecture. Include screen refresh time.
2. For a transaction which requires no host update (but will require accessing the Contact History Database, identify the amount of time utilized between the time a request is generated by a client workstation (e.g. a CSR presses 'enter') and the CSR receives a fully refreshed response screen back from the server. Assume a Contact History database storing 6,000,000 calls for 1,000,000 customers.
3. Describe how you would configure the Computer Based Training environment to avoid conflict with the production environment.

4. Describe how you would configure the report writing and query access on the CSD automation system database to avoid conflict with the production environment.

5. Describe how you would configure the CSD automation database (e.g. for customer contact history) to provide adequate response time to CSR staff.

2.3.6 System management, monitoring, and support
Describe the tools that you provide and their functionality for the following:

1. Network management
2. Configuration management
3. Performance monitoring and system tuning tools including identification of software, network, or hardware bottlenecks.
4. Full system backup while on-line.
5. Hardware monitoring

If you don't provide system management tools, which tools do you recommend?

2.3.7 Multiple site support
FNAX has Customer Service sites in Axxum and in the Xzzum area. Information about customers calls and contacts, is required in an integrated fashion for both sites. This means that even if a prior call originated in Xzzum, a Axxum CSR must have transparent and complete visibility to that call data in real time. Customer and call statistics need to be integrated across both sites, as well. Note: a call may arrive in the Xzzum location, be answered in Axxum, and transferred back to the Xzzum.

1. Provide explicit detail on how your system will manage call and customer data in such a way that both sites will have access to all Contact History data including that from the current call (e.g. in a transfer situation across sites).
2. Describe how your system will handle call transfers and call Conferences (with associated data) which are cross-site in nature.
3. Describe any distance specific issues such as server location in relation to client workstations, e.g. is their a maximum distance that your recommended configuration will support (keep in mind our dual site situation).

2.3.8 Other functionality supported
1. The vendor should describe other significant application functionality not covered above which may help differentiate their product.

2.4 Implementation

2.4.1 Implementation strategy

1. The vendor's implementation proposal is to begin with a brief narrative on the strategy recommended for installing a CSD automation solution. Alternatives can be recommended on all items. It is important that the following information be part of the vendor's response:
   - Methodology to be used for all aspects of the project from requirements through implementation and a description of each phase.
   - Migration procedures the vendor would use for converting the current CSD environment to the new one.
   - What portions of the implementation can be or should be   phased
Note: FNAX will expect to pilot functionality for 4-8 weeks.

2.4.2   Project methodology
1. Describe each phase and the deliverables to be included. Include a sample project plan and a typical list of deliverables. Include an example of well written requirements documentation.

2.4.3   Quality assurance strategies
Describe what strategies you use for insuring that your delivered product will be error free. High quality includes issues of meeting our objectives. Include a discussion on:

1) Approach to testing (during each phase).
2) Error log/tracking methodology.
3) Change control methodology.
4) Performance and volume testing.
5) Quality assurance strategies before you deliver code or product.
6) Automated testing tools.

2.4.4   Expected FNBA system changes
1. Describe the possible changes that we might have to make to our systems to support this project. Although our intent is to minimize changes to our ACD, VRU, and mainframe environments, what types of changes do you recommend?
Include changes to the ACD, VRU, network, mainframe, etc. Describe the possible extent of those changes. In the case of issues such as 3270 emulation vs. LU 6.2, describe what conditions would cause us to want to convert to LU 6.2.

2.4.5   Recommended training for FNAX staff
1. Please describe the training program, and what classes would be required to allow FNAX to be fully independent from a customization and development standpoint. Include recommended classes, whether or not your company performs the course in-house. Include video training or equivalent to reduce

training cost. Include specific identification of the pre-requisites for staff attending the courses (technical or business backgrounds).

2. Clearly identify training which can be provided to client areas (the system owners) which would allow them to make changes independent of FNAX technical staff. Insure that this information cross references clearly to Section 4.2, General Package Information.

3. Include a full course listing.

4. Describe the extent and type of training that you will recommend for our Customer Service Representatives to adapt to use of the new system. Describe the training required to 'train the trainer' should we decide to perform the training in-house.

3.0 Contractual issues and expectations

The contract governing this project will be developed specifically to encompass this RFI and your Response. Standard contracts may be used as reference, but will not be the basis for negotiations. The General Terms and Conditions included in Appendix D forms the basis upon which the Special Terms and Conditions (project-specific) will be attached. Please review and note any changes you would propose or exceptions you have to those General Terms and Conditions.

Your Response you present will be made a basis for an exhibit to the contract in the event FNAX elects to proceed and you are the vendor of choice. As contracts typically state, the final contract encompasses the complete and exclusive agreement as to the products and services being acquired and the contract will supersede all communications, oral or written, between the parties. In support of this clause, FNAX Bank may require your Response or sections thereof and any additional documentation which describes the project framework, products, services, guarantees, warranties, response times, support, etc., to be made a part of the contract.

The following sections are key components of a contract that would be addressed in any subsequent agreements that we would make with your firm. Please indicate your firm's position on each of these issues and respond directly to any questions.

1. Pricing and Payments: Prices are subject to discounting. Wherever possible, to affect more attractive pricing, you are encouraged to be creative, including price bundling. Payments shall be tied to project milestones, with final payment of not less than 40% of the contract due after final acceptance.

2. Development, Installation, Testing, and Training: These tasks may be scheduled for off-hours. No incremental charges will be assessed.

3. Scope Changes: There will be a formal section dealing with scope changes

covering the opportunity to re-price in the event the scope is reduced or expanded. Scope changes must address not only price but also project impact and deliverable dates. How does your firm formally incorporate changes into a project?

4. Warranty: 3 year on-site warranty is required at no additional charge.

5. Maintenance: You must be willing and capable to contract for maintenance and support for both hardware (if you provide) and software. (If approved in advance by FNAX Bank, a qualified subcontractor may be used.) Due to the critical nature of this installation, response time from the time the problem call is received to the time a service technician arrives on FNAX's site must be not more than 2 hours. Additionally, escalation procedures for unresolved critical problems will be addressed.

6. Escalation: There will be a formal escalation procedure to be followed during the project and subsequent to acceptance. Problem logs, status meetings, and open communication will be a part of this section.

7. Acceptance Testing: This will be developed prior to the installation of the products. You will be required to comply with the acceptance testing. In the event you will not contract for acceptance testing, an indication on your Proposal must clearly state why and what you propose as an alternative to acceptance testing. No acceptance test may disqualify you as a supplier under this project.

8. Documentation: Manufacturer standard manuals will be required. In addition, you may be required to develop a "project specific" user manual. The contract will not be considered complete without proper documentation.

9. Trade-in policy: You may propose a framework for trade-ins for any hardware included as part of your proposal, particularly if a manufacturer discontinues a product line.

10. Liquidated damages: Once the project milestones have been established, you must agree to liquidated damages if you're solely responsible for missing or delaying a milestone.

11. Performance Bond: Indicate your willingness to post an adequate performance bond to ensure that the project will be completed in the event you withdraw or are unable to complete the installation.

12. Performance Warranty: Indicate what performance warranty would remain in place after your proposed system had been installed and accepted by FNAX (i.e. uptime, response time for inquiries, reliability etc.)

13. Third Party Agreements: Provide samples of any third party agreements that FNAX would need to execute, as part of your response, such as third party software or hardware maintenance agreements.

399

14. Bank of Axxum Agreements: Are any or all of the components contained in your proposed system currently being utilized by Bank of Axxum? Please provide copies of any Bank of Axxum contracts related to your proposed system. In addition, it is expected that any terms and conditions reached with FNAX as a part of an agreement will also apply to Bank of Axxum should they decide to utilize your product at a future time.

15. Configuration Errors and Omissions: Are your prepared to offer a warranty that all the necessary components (hardware, software, network cards, interfaces, etc.) that will be required for FNAX to test and implement your proposed solution will be included in the contracted price [or if not included in the contracted price, identified as required elements in your formal response?].

16. Software Please describe whether each software component that you are proposing would be delivered to FNAX in Object and/or Source Code.

17. Escrow Agreement: Describe any existing Escrow agreement for your proposed software that would be incorporated into overall agreement with FNAX.

18. Ownership of Custom Developed Applications: If your proposal involves the development of a custom application by your organization then described who would retain the rights to re-license this application and how do you envision supporting it.

4.0   Vendor information & questions
Provide detailed responses to each of the following subject areas.

4.1   Corporate information
1. Provide a brief narrative about your company. This narrative should contain information about the company's background, its current size and the products and services it offers.
2. Provide information that demonstrates the expertise of the company and its personnel, including a discussion of the size and expertise of those technical staff whom would be candidates to support an implementation of your product at FNAX.
3. Provide information that demonstrates the company's financial stability. As part of this information, we request a copy of the company's most recent audited financial statement.

4.2   General package information
In addition to responding to the following questions, specifically identify whether any feature, development capability, or characteristic of your environ-

ment would need development in order to be supported. Identify whether or not that development is part of the quoted price estimates to follow in Section 5. If in any case, support for a function is provided by third party products, identify which products you recommend.

4.2.1   System architecture

The vendor is to provide an introduction to their system including descriptions of all significant system components.

1. Include pictures which describe all hardware and software aspects of the proposed system including those functions for which FNAX is responsible (mainframe, VRU, etc.)
2. Describe the role of each component and the specific functionality implemented as a part of that component. Components should include any significant hardware, software packages or subsystems; especially any component requiring independent development to customize or integrate. Describe the standards and manner in which each component communicates and interacts. Note any limits to expansion or issues associated with scale.
3. Describe the development tools used to customize each aspect of the system. What does the development package provide without having to incorporate the use of user exits or 3GL's? What types of functionality typically require the use of a 3GL or user exit? Do you have a visual development tool and/or 4GL and what are their capabilities?
4. Explain the mechanics of how your system manages the Computer Telephony Integration (CTI) including Screen Pop, Conferences of calls and data, etc. Describe the different components that help manage screen pop, their role, how these components would interact with our ACD, etc. If your CTI capability can function as a separate module, i.e. interfacing with other third party software, describe the functionality which would be provided by the CTI module alone.
5. Identify any third party tool sets which are included in your tool set.
6. Describe what you consider to be strengths and weaknesses of your current offering.
7. Explain exactly why use of your product set would make us more productive than developing the CSD automation system in house using tools such as:
   • Standard visual development tools, i.e. Visual Basic.
   • A standard middleware package such as MDI.
   • Sybase as a local database engine.
   The ideal solution must integrate with FNAX's standard distributed computing environment (see Section 4.2.5) and internally developed

client/server applications. These applications are currently targeted for development using Visual Basic and Sybase System 10.

8. Can your system use Visual Basic for the presentation tool and Sybase System 10 as the server database? If not, why not? What is the impact of using either of these tools? If you would recommend against use of either of these tools, please explain?

9. Describe how your system will integrate within our client/server application environment with the following capabilities:
   - Share data
   - Share objects and modules
   - Call or launch one application from another
   - Implement Computer Telephony Integration across applications (e.g. transfer the voice and data to a telemarketing representative who is using a FNAX developed application from a CSR using the CSD automation system.)
   - Implement a common look and feel
   - Run under common platform

4.2.2 Documentation

What kinds of software documentation standards are used in the following areas:

1. User manuals.
2. Module or subroutine headers.
3. Comments in the source code.
4. Program design languages.
5. Audit trails of modifications to code.
6. Include a section of a sample user manual in your response for our review. Include any other samples which would provide us with a feel for the quality of your documentation.

4.2.3 Software standards

1. Describe how your application design supports portability across platforms. For example, does the application make direct calls to the Operating System (O/S)? Are calls to the O/S isolated in separate modules?
2. What compiler is used by your development tools to generate executable code? What platforms does this compiler run on?
3. What functions must run on client, server, host, other? How much flexibility is there in where application modules operate?
4. Are your applications developed in a modular basis? Which of them function independently and can interact with other vendor applications? What are the

interfaces and are they documented for third parties?

5. What operating systems does the product work with today?

6. Does the product conform to the OSF standard? What other open software standards does the product conform to? Assuming that your products are only partially compatible with these standards, how do they vary?

7. Which network drivers does your product work with (e.g. Named Pipes, TCP/IP)?

4.2.4   Hardware standards

1. What type of hardware does your software currently run on? What hardware, in your experience, provides the best price/performance?

2. Describe in detail benchmarking results (and source) which support identification of appropriate processing power required to run your software based upon varied volumes. Identify the highest volume production shop using your software in a fashion similar to FNAX's direction, the volume of transactions currently being supported, the platform being used, and response time to the client for representative transactions.

4.2.5   Compliance with FNAX distributed computing standards

For consistency in our computing environment and to leverage our existing support resources, we prefer that the vendor solution conforms as much as possible to existing Bank computing standards. If the solution recommended by the vendor departs from the standards listed below, the vendor should describe plans for bringing the system into conformance. If the vendor has no plans for bringing their product into performance, describe the reasons for not doing so.

Existing distributed standards or preferred solutions:

1. Server Operating System: MS Windows NT Server
   (The Bank also has an installed base of NetWare 3.12 systems; if a UNIX solution is required by vendor, HP-UX is preferred solution)

2. Server Hardware: Compaq Proliant
   (Proliant 2000 is typical; 4000 is acceptable; should be rack mount)

3. Network Protocols: TCP/IP on Token Ring (16Mb/s)
   (IPX is alternative; NetBEUI is not permitted; DLC for direct FEP access is available)

4. Workstation Operating System: MS Windows 3.11
   (Win NT Workstation, Windows 95, or OS/2 3.0 will be considered)

5. Workstation Hardware: Minimum P111 333 Mhz with planned migration to PIII 900 Mhz.
   (Compaq XLs are preferred. Dell are alternative; amount of RAM or video

subsystems are optional; TR NIC is the Madge card)
6. Server Database: Sybase System 10 or most recent version (Microsoft SQL version 4.9 or most recent version will be considered)

4.2.6 Database specifics

4.2.6.1 Data Architecture

The new teleservicing environment will require the development of a number of new data repositories. For example, new data repositories will be required for contact history, pending activity, future activity and payroll information. There is the possibility that data can be stored in the mainframe environment, at the server, at multiple servers and at multiple servers in multiple physical locations.

1. Does your existing architecture specify the location of new data repositories?
2. What strategy do you use and what factors do you consider when you determine where databases should be located?
3. How would you fulfill the requirement to have all data available to all CSRs in both locations.?

4.2.6.2 Database management system

1. Does your product include or require a proprietary database management system (DBMS)?
2. What third party DBMS products do you support? Given our requirements, what DBMS would you recommend? What experience do you have with the recommended product? Specifically, how many production implementations have you completed?
3. Can you support Sybase System 10 if that were to be a project requirement? What impact would that requirement have on your architecture, application functionality, application development time and cost?

Respond directly to the following questions if your product includes a database engine. If your product utilizes third party database engines, then identify whether any of the following capabilities are precluded from use by your system.

4. Does the database support DBMS conversion, including automated assimilation of data structures and data? For which DBMS'?
5. Are there any restrictions to the size and/or number of records that can effectively be used in your database engine without affecting performance?
6. Does the Database support (yes or no answers are permissible unless otherwise indicated):
   Synonyms or aliases
   Headings, titles, or labels

Comments or remarks
Date formats
Time formats
Money formats
Variable-length fields
Keyed or indexed files
Data structure management
Ability to read, query, or report from ASCII files
Ability to import ASCII files
Ability to export ASCII files
Ability to translate EBCDIC files to ASCII and vice versa
Ability to preserve versions of data model and applications
Cross reference for dictionary objects
Ability to read, query, or report from formatted files
Ability to import formatted files (what types?)
Ability to export formatted files (what types?)
Ability to modify files while loading
File check when loading
File modification

BLOBS
Parallel query
Provides Parallel Data Loader or a high performance loader
Parallel index builder
Non-Blocking Sequence Generator.
Adheres to ANSI SQL, which version (extensions are to be described)
Control access by user ID and password and user groups.
Data dictionary based triggers.
"Like" SQL statements.
Non-Blocking queries.
Unique key generator.
Parallel sort
Cascading updates and deletes.
Referential integrity

ODBC

DRDA
7. Describe your backup and recovery capability including:
    Roll back capability.

On line (during regular processing) backup and recovery.
Unattended backups.
Automatic database recovery (automatic, manual, semi-automatic)
Mirroring regardless of hardware and UNIX operating system variations.
On-line recovery
Database re-organization
Archiving capabilities
Compression
Remote recovery, such as recovering a database in the Xzzum remotely from Axxum.

Archive medium.
8. Describe the support of performance management including:
Ability to support a 24 hours a day 7 days a week availability.
  Resource limiter
  Performance monitoring tools
Describe alternatives for minimizing network utilization which your DBMS supports.

Path analysis
9. Describe facilities and strategies for aging information off the database and the decision criteria that can be used.

4.2.6.3 Database development
1. Do you provide or recommend a data modeling tool?   Does your modeling tool interface with your preferred DBMS?
2. Do you provide or recommend a Data Dictionary? Is there connectivity between your recommended modeling tool, Data Dictionary and DBMS?
3. What features does the Data Dictionary include? Include its capabilities at defining or capturing:
      logical and physical characteristic of the data
      relationships amongst data
      business rules
      blobs
      error handling and messaging
4. Is the Data Dictionary centrally stored and accessible by more than one developer? Can it control access to all data?
5. Can the Data Dictionary activate changes to data definitions during run time?

6. Describe your database application development tools, including support for the following:

Thorough 4GL that conforms to modular programming techniques.

Quick prototyping capability

Graphical user interface which insulates developers and end users from native SQL.

Utility to display relationships, dependencies, triggers, procedures, and views.

7. What other tools do you provide or recommend that facilitate database development?

4.2.6.4 Database performance

Detail performance characteristics of your database environment. Include responses to the following questions:

1. Assume that you must deal with a database that must store 100,000 new 2K transactions per day. How many days worth of data can your system maintain in the database without affecting performance given your recommended hardware configuration (see Section 5)? What factors would affect this?

Respond to questions 2-6 for the following scenario:
   - 100,000 new 2K transactions each day.
   - Storage of those transactions for 60 days.
   - Data stored on multiple tables (assume four tables updated for each transaction).
   - 20,000 - 30,000 inquiries against a combination of the four tables each day.
   - The inquiries are on the call critical path, for example, part of the screen pop and intelligent routing process.
   - Data must be accessible by at least three criteria groups such as: CSR identifier, client account number and activity or event type.
2. What effect would this have on your recommended architecture?
3. What would be your concerns?
4. What restrictions would you want to apply in order to provide acceptable performance?
5. Using your recommended backup facility, how long would it take to create a backup of the database previously described?
6. How long would it take to recover?
7. What strategies do you recommend for insuring that simultaneous logging, queries and reporting do not affect OLTP performance?
8. Describe the largest database environment you currently have in production? How many rows of data does that database contain? What is the transaction volume, in terms of logging and inquiring, for that database environment?

4.2.4.5 Database Access
If your software can access multiple database backends, what access standards does it support such as ODBC and DRDA?

4.2.7   Development tools and integration

4.2.7.1 Development tools description
1. Describe tools used for customizing the system to FNAX' needs. For each tool, clearly identify the pre-requisite technical expertise, training, and time required for FNAX staff to become fully productive in its use. Specifically identify development tools and capabilities which are expected to be usable by the user area. Specifically describe tools required for the following (in sequence, if applicable):
   - Forms development and management
   - General development tool support
   - Computer based training
   - Reporting:
   - Work flow

Presentation tools (graphical user interface)Database access

Development of scripts and user help screens

2. Which of your tools are proprietary? How long have they been in   production?
3. Describe what functionality specific to a bank call center has been pre-built. What objects are currently supported and in production that would be applicable to our use?

4.2.7.2 Graphical user interface
1. What options to you support for adding color to the user interface? How many colors do you support and what color parameters (e.g. text vs. background) can be controlled? What are the limiting factors and are there any performance considerations?
2. Are colors customizable by the user area and if so, for any screens and windows features? How does the user area maintain screen colors? Can colors be reset to default schemes?
3. List all controls supported as objects by your development environment.
4.   Does the product support the following User Interface features (yes or no answers are permissible unless otherwise noted):

      Graphical Debugger
      Moveable forms

      Control Panel (Separate windows of buttons for common commands)
      Definable Control Panels
      Push buttons
      Check buttons
      Large graphic images
      Icons
      User defined data types:
      Standard data types (list):
      Boiler Plate definitions
      Active field indicator
      Intuitive date/time constructs
      Keystroke macros

5. How does your forms development tools support multiple table forms including multiple rows within an embedded form? Does this include the following:
      Automatic table:table coordination
      Automatic referential integrity enforcement
      Automatic cascade processing
      Automatic restrict processing
      Automatic nullify processing

4.2.7.3 Development language support:

1. Does the product support the following language features:
      SQL Implementation
      Simple queries on one file
      AND/OR logic - one file
      Relational join
      AND/OR logic - two files
      Complex Boolean logic
      Traps command or statement errors
      Compiles code
      Interprets code
      Condition logic
      While constructs
      Repeat Until constructs
      Labels
      Field Layout definitions
      Relational and arithmetic operators
      Intrinsic date/time functions
      Intrinsic numeric functions

Statistical functions

Ability to process missing data

Procedural capabilities/constructs

3GL interaction allowing write to application fields from 3GL

4.2.7.4 General development environment support

1. Does your development environment support "Self documenting" or "automated documentation"
2. How do your tools support multi-threading?
3. Is there a central repository accessible by all developers?
4. Describe the ability to develop on one platform/operating system and deploy on another and how it would be accomplished.
5. Describe the ability to develop using one DBMS and deploy on another and how it would be accomplished.
6. Can we utilize our own text editor for development? Are there any restrictions?
7. How do your tools support tracking and documenting of objects?

4.2.7.5 Integration with external software packages/tools

1. Describe your support of OLE 2.0 and DDE.
2. Describe other API's that your product provides or supports.
3. Does your product provide a VBX(s) to allow features to be used within Visual Basic?
4. Describe any features that your system has for integrating foreign packages into your system such as word processor, spreadsheets, or custom applications to include menuing, launching, and sharing of data.

4.2.7.6 Application development features

1. Does your system support the following application development environment features? (yes or no answers are permissible unless otherwise noted)

Interface to other languages

Interface to operating system

POSIX Compatibility

On-line HELP

Clear and concise documentation

Error messages and warnings

Product standard

User defined error messages and warnings

Ability to call a 3GL including C and COBOL

Ability to be called from a 3GL including C and COBOL

Ability to pass parameters to a 3GL
Ability to receive parameters from a 3GL
National Language Support (list):
Branching logic
Looping logic
Language integration
Ability to call subroutines
Ability to imbed queries in procedures
Ability to share temporary data
Ability to assign global variables
Ability to create local variables
Ability to define calculations using field values
Module re-usability via sharable libraries
Cross reference of common modules used in different applications
Ability to simply define keyboards by capturing key escape sequences
and prompting for key function
Ability to switch keyboard definitions on demand
Ability to restrict field entry or modification based upon data values
Query by Example
Query by Form
Data location transparency
Pre-defined event triggers
       Field level event triggers
       Form level event triggers
       Entity level event triggers
Ability to redefine event triggers to place objects in libraries and reuse as
a library
Ability to automatically generate stored procedures

4.2.7.7 Debugging Tools

1. Describe tools for helping debug problems after the system has been installed
2. Describe the product's debugging tools including:
    step capability
    step forward
    step backward
    step in steps
    set a break point at module on line
    cancel a break point
    set break points on call instruction

cancel break points on call instruction

allow messages to accumulate in a known area

return from a 4GL procedure without executing any further instructions

dump statements of a current procedure module

dump the procedure statements of a named central procedure library

display contents of the next register/variable

display ;contents of a named register/variable

display contents of a field

set contents of a register/variable

display break point settings

send performance benchmarks to the message area

examination of functions

extend trace capability, showing:

> trigger procedure is in
>
> module
>
> any arguments given to procedure statement
>
> Dial-in capability for deployed applications

4.2.7.8 Computer Telephone Integration scripting tools

1. Provide a detailed description of how CTI related scripts are developed which control the disposition of phone calls. It is expected that control over call routing, alternative routing for problem resolution, which queue receive incoming calls (e.g. based upon source of call), etc. are all user area controllable dynamically. Include a sample of a script and explain what it controls. Explain how control over phone calls in your system interacts with control over calls inherent in the scripting capabilities of our current ACD. Include consideration for our VRU environment.

4.2.8 Computer based training

(See functionality described in Section 2.1.11.)

1. Describe the technology (hardware and software) used for your Computer Based Training product.

4.2.9 Report and query writing tool

1. Describe the tools which support end-user creation of queries and reports.

4.2.10 User flexibility and control

1. Clearly identify those functional changes which can be made without programming, specific tools utilized for those changes, and the method for putting such changes into production. This category will exclude any changes

requiring use of a procedural language (e.g. 3GL) or changes that if implemented incorrectly could significantly affect the stability of the production environment. Differentiate between changes that the end-user (e.g. CSR) can make versus those changes that are global in nature and would be changed by a central staff person in the user area.

4.2.11 Communications

1. What remote communication line speeds are accommodated by the system?
2. What communication protocols are supported by your product and which ones are in production today?
3. Which ACD interfaces are supported (e.g. TSAPI, TAPI, CT Connect, TServer)? If you support a proprietary interface then describe the reason and the incremental functionality that it allows.
4. Describe alternatives for communicating with the Intervoice VRU. List other supported VRU platforms.

4.3   List of current package users

1. Provide a list of all current institutions using the proposed system. Provide name, address, telephone number, and contact name for at least three financial institutions who are using your product in a call center of at least 200 workstations. Please select institutions who have deployed the product in a similar fashion as FNAX is considering.

4.4   Maintenance support description

1. Describe your release methodology for updated or new products. Include:
   - How long you will commit to supporting old releases past the introduction of a new release. What is the impact to us of not upgrading?
   - Whether you anticipate separate releases of individual components (such as visual front end tools, CTI tools, etc.) of your software or whether we will be required to update all software simultaneously.
   - What is your typical release frequency? List dates of your previous three releases.
2. Do you provide a single point of contact for failure? Does this include hardware and software should you provide the hardware? If you do not provide the hardware, what expectations do you have regarding our role in determining whether the problem is a hardware or software problem, e.g. for performance problems?
3. Will you provide a dedicated account representative assigned to FNAX? Does the customer have input as to whom that representative will be?
4. Describe the trouble-recording, personnel-dispatching procedures which will

be used in servicing your proposed products(s) at our facility.

5. Describe the following company procedures and services:
   - Telephone support
   - Installation/upgrade support
   - Escalation procedures for problem resolution

6. Describe how workstation software is updated when changes are made, e.g. is it automated? Is downloading supported from a central server?

7. Do you support a CompuServe or Internet link for fixes?

8. If a problem involves a third party provider (such as language compilers or NTAS solutions), can we obtain the fix directly from the third party or from your firm? For which tools?

4.5   Architecture and/or vendor future direction

1. Describe the technology direction in which your company is headed. Include portability to different platforms, database technologies, open standards, connectivity, development tools, etc. in your response. Be as specific as possible.

2. What are the major elements of planned product release(s)? When are they due?

4.6   Other

1. Describe any additional features or information about your company or products which would make your proposal more advantageous than others.

5.0   Proposal

Descriptions for varying scenarios and vendor options follow. The vendor should provide estimates for each of the following combinations of these variations. Sections 5.1 - 5.9 should be responded to for each Option.

Option 1:	Scenario 1 with Vendor Option 1
Option 2:	Scenario 1 with Vendor Option 2
Option 3:	Scenario 2 with Vendor Option 1
Option 4:	Scenario 2 with Vendor Option 2
Option 5:	Vendor Option 3

In all estimates, include recommended development environment and testing laboratory hardware and software in your responses. Appendix E, the sample detailed transaction, is intended to provide the vendor with a sense of the level of complexity and functionality which will be expected in each transaction.

Scenario 1

Scenario 1 includes all functionality included in the following sections only:

Section 2.1.1	Event and activity tracking

Section 2.1.4	Letter Generation
Section 2.1.5	Screen navigation and Context Sensitive events
Section 2.1.6	On-line User Help or reference information
Section 2.1.7	Scripting (Dialogue)
Section 2.1.8	Screen design capabilities
Section 2.1.10	ATM and branch location capabilities (does not include on-line graphical map capability)
Section 2.1.16	Miscellaneous
Section 2.1.17	Reporting

All Overall Requirements described in section 2.3 except Store and Forward

In addition, the estimate should include the above functionality only for the following 'Top Events' as described in Appendix C:

- Verify account activity
- Overdraft research
- Changes to service charges/ product type
- Copies of checks/statements
- Stop payments
- Balances
- Error resolution
- Provide Branch/ATM information

For further clarification, see pages 1-6 of Appendix B, Key Concepts.

Scenario 2

Scenario 2 includes all functionality in Scenario 1 plus:

Section 2.1.2	Computer Telephony Integration capabilities
Section 2.1.3	Check vendor interfaces
Section 2.1.9	Department message broadcasting
Section 2.1.10	FNAX ATM and branch location capabilities (with on-line graphical map capability)
Section 2.1.11	Computer Based Training
Section 2.1.12	Personnel scheduling adherence
Section 2.1.13	Time card system
Section 2.1.14	Telemarketing and sales specific requirements
Section 2.1.15	Imaging
Section 2.3.2.3	Store and forward

This functionality should be estimated for a project which would include all 'Top Events' in scenario 1 plus:

- Support calls (SBS, RLSC, branches)

415

- Check/Deposit slip orders
- Address changes (personal)

Remaining Top Events do not need to be estimated.

For further clarification of the remaining functionality in this section, see pages 7-10 of Appendix B, Key Concepts.

Vendor Option 1

1) The vendor performs all customization, development, and testing for software they provide. FNAX would:
   - Manage the project
   - Define requirements (the vendor would be expected to provide detailed product and teleservicing expertise during this phase).
   - Develop any required changes in our own environment such as VRU or host.
   - Manage and staff acceptance testing.

Describe how your bid would be affected if we required development using Visual Basic and Sybase.

Vendor Option 2

2) FNAX performs all customization, development, testing, and implementation of the product in addition to the functions assumed in Option 1. The vendor would only provide necessary training of the technology being licensed and respond to all technical questions throughout the life cycle of the initial implementation. For this scenario, provide a high level estimate as to the development time, in hours, FNAX technical development staff could be expected to spend in learning and customizing your software.

Describe how your bid would be affected if we required development using Visual Basic and Sybase.

Vendor Option 3

FNAX will consider alternatives which involve combining tools from multiple vendors with the idea of:

- selecting 'best of class' tools
- leveraging off of current FNAX initiatives, technical standards, and expertise.

Please describe alternatives which you would support which involve participating with other vendors as part of an overall CSD automation solution. Include options providing individual components such as a Front End - GUI, Local Database, CTI, Middleware, Computer Based Training, etc.

Include the following in your response:

- What role will you consider playing in these scenarios? Prime contractor? Sub-contractor?
- What experiences do you have with alliances with other vendors (list vendors)
- Which aspects of your software and services were used?
- Estimates ($'s, time) for each of the alternatives that you would support.
- Describe where your strengths and weaknesses would be for this type of partnership.

In particular, can you provide CTI functionality only? Would you be willing to participate as the provider of that portion of the project? What would be the estimated cost of this alternative?

5.1   Equipment (hardware)

Describe recommended hardware specifications.

Include separate specifications for client workstations and servers. Include memory, processor speed, video configuration, size of monitor, bus requirements, etc.

Include recommended configuration for all hardware including backup server processors and network components.

Include recommendations for all peripheral devices (printers, CRT's, tape drives, disk units, etc.) which can be attached to the proposed system.

Describe any limitations or special restrictions in connecting multiple peripherals.

Insure that you account for ability to use the PC (workstation) as a phone device.

Identify recommended level of hardware redundancy and identify where that has an impact on our costs.

Provide pricing for any hardware components listed above for which you are a supplier or estimates for the above if you are not a supplier. Include maintenance expenses for years 1-5 for each of these components.

5.2   Vendor software

1. Describe expenses associated with the licensing and support of all software included in your response which is proprietary to your firm.
2. Should any specific functional requirements listed in Section 2 require substantial development, list the development cost for that portion as a separate line item. The intent is for us to understand if any specific requirements which we are requesting significantly affect the price of a delivered product.

Include maintenance and support expenses for years 1-5.

5.3   Non-vendor software

Identify operating systems, network software, or development environment components that are not proprietary to your firm which you support or are embedded in your product. Identify those software components you recommend or are required. Indicate whether or not you are a supplier of that software and provide an estimated cost for each component. Provide the number of installations you currently support for each configuration.

5.4   Other vendor expenses and services

Identify vendor expenses and services other than hardware or software which may be included in a proposal. Include in this section:

- Development services
- Integration services
- Installation services
- Training
  - Technical Staff
  - User Area Staff
  - Training for FNAX staff responsible for CSD training programs
  - Full initial training for all CSRs and end users (by your firm)
- Other services or expenses

Under each category, describe your services in detail and the cost of your different support options. Include expertise of staff and estimate headcount of staff you would provide for each of these services, should FNAX contract for them. Differentiate between staff that are full time employees of your company and those with whom you would contract to service FNAX. If you include any of these services as part of overall product pricing, describe what is included. The limits to your commitment (number of hours, months, etc.) should be identified along with the costs to FNAX of exceeding those parameters. Include hourly rates for any category, as appropriate.

5.5   Other non-vendor expenses

1. Identify any expenses that we might incur that are not included in your service or identified above. Include an estimate of the percentage increase in ACD utilization we might experience as a result of this project and the factors which will affect that percentage.

5.6   Upgrade and expansion expenses

Address costs relating to expansion or upgrades after the initial implementation (additional or upgraded equipment, software, documentation, etc.) At minimum, describe and list estimates for expenses which we are likely to incur related to your proposal if:

1. The number of data centers increases from two to four.
2. The volume of transactions triples
3. The number of workstations doubles

5.7   Leasing
1. FNAX would be interested in exploring the idea of leasing the system from you, including customization expenses. Provide us with information regarding your willingness to explore this option and what types of terms and conditions you would consider.

5.8   Expected implementation schedule
1. Provide an estimated implementation schedule to final product including time frames for each phase and milestone for each Scenario listed in Section 5.0.
2. Identify at what point in the project you would be willing to commit to a specific schedule and budget for completion. Describe what information and level of detail you will require before such a commitment. Provide an estimate of the cost to perform the work necessary to develop a firm budget for completion.
3. If you believe that there will be differences in project time frames depending on which portions of the project are contracted to you, describe those differences and why they exist.
4. State any restrictions on when we would be able to start this project with your firm, or limits to your ability to fully staff the project.

5.9   FNAX resource requirements
1. Describe the level of involvement which you will expect from the business client at FNAX especially focusing on up front requirements and the testing phases. Provide an estimate of the time that will be expected from FNAX staff to support requirements and acceptance testing.
2. Include estimates for the time required from FNAX technical staff to support this project. Assume minimal system changes on our part (we will estimate the time for any such changes). Rather, focus on the needed expertise to describe technical interfaces and to participate in the testing phases of the project.

5.10   Laboratory evaluation
Upon selection of finalist vendors, FNAX would like to perform laboratory work with each of the vendors remaining under consideration. The purpose of this

effort would be to obtain a hands on feel for the tools and software of each vendor and to understand how difficult it will be to customize that package to our
environment. The effort will likely be based upon the 'Account Activity' transaction detailed in the Appendix E.

1. Please indicate your willingness to support such a laboratory effort at your
   cost and your conditions for support (e.g. where, how long you would recommend, etc.).
2. FNAX would prefer this laboratory work be performed on our premises.
   Describe the needed hardware configuration at our location required to support such an effort.

Appendix II

Request for Proposal
ACD System

INSTRUCTIONS TO VENDORS

GENERAL GUIDELINES — Companies bidding on this RFP are encouraged to submit ideas for optional configurations and systems enhancements as appropriate. The requirements as stated in this document must be substantially met. Any alternative ideas should be included and discussed as secondary items to the primary ACD requirements. Alternate proposals and feature strategies will be accepted and evaluated as secondary to responses to meeting the business objective envisioned in the stated requirements. Responses to all proposals (five copies) are due by 4:00 PM ON FRIDAY, 01/XX/2001, and should be addressed to,

Ms. Lois Lane
Call Center Manager Response Center
12 West 21st
Street New York, New York 10010

If your system meets most of the requirements specified, but there are instances where physical or feature requirements are not fully met (such as proposed system maximum capacity is 350 lines vs. 356), your company is encouraged to respond with proposal since minor deviation from minimum requirements may be considered acceptable by I/O Marketing following comparative analyses.

PROPOSAL STRUCTURE — Vendors are requested to follow the request and response format in submitting your proposal. For your convenience, our request is being provided to you on disk in a flat ASCI file. Appending your answers to each request will ease the evaluation and comparative analysis of your systems. Please keep references to supporting documentation to a minimum as this greatly complicates the evaluation process.

You are requested to follow the outline below with appended material as appropriate.

1. Introduction
2. Overview of Company and Products Lines
3. ACD Operation
    A. System Functions
    B. Agent Position Features
    C. Supervisor Position Features
    D. CTI Integration
    E. Universal Customer Contact Integration
    F. Central Customer Contact Status Reporting
4. Management Information System Reporting
    A. System Reports
    B. Channel Reports
    B. Gate and Trunk Utilization Reporting
    C. Agent Activity Report
    D. Contact Outcome and Status
5. User Programmability Requirements
6. Voice Response System Interfaces
7. Operating Environment
8. Training
9. Maintenance Support
10. System Costs
11. Installation and Cutover

The capabilities summary matrix forms contained in Appendix A must be completed by your firm to summarize proposal offerings based on requirements specified.

1. INTRODUCTION

A discussion of the business objectives of this acquisition project. This discussion should cover current operations and the planned outcome. Best estimates of call center or business growth should be included.

2. CORPORATE OVERVIEW AND FINANCIAL PERFORMANCE

Please provide a short overview of your company and product lines. The most recent financial performance should be included in the form of the latest quarterly or annual report for publicly held companies. If your company is privately held, please provide financial references that we may follow up on.

3. ACD OPERATING REQUIREMENTS

The following discussion covers operating requirements for the automatic call distributor system in the categories of system, operator and supervisor functions. The requirements listed are considered MINIMUM FUNCTIONS to be delivered by the selected vendor. This request is not designed to preclude proposal of other functions or alternative methods that are perceived to be achieve the same operation and management objectives. Vendors are encouraged to comment on functions listed as well as recommend optional functions that may be available to further support the performance goals of I/O Marketing, Inc.

A. SYSTEM FUNCTIONS

GATE/GROUP ASSIGNMENTS

The ACD shall have the ability to be functionally divided with trunk groups assigned to one of a number of functional gates or group assignments. A total of sixty-four voice gates are required. The ability to intermix various types of cicuits, including WATS, foreign exchange, off-premise extensions, direct inward dial, T-1 and local circuits, etc., shall be provided for voice each gate.

423

PRIORITY QUEUING

Each trunk groups will use this feature to have calls placed in a predetermined order for answer by an agent. Priority trunk assignment will be by physical or logical trunk definition within a gate, (Physical trunk termination or in the case of DID, DNIS or ANI identification, by logical identifier). The priority rules shall have the ability to intermix functionally similar calls arriving on different circuit types. Alternatively, give functionally similar calls arriving on different types of circuits different treatment priorities. Examples: setting a higher priority for WATS circuits over local lines; Customer Service calls receiving lower priority than callers wishing to place an order.

ORDER-OF-ARRIVAL QUEUING

For each functional gate, calls waiting in queue will be assigned to the next available agent based on earliest order-of-arrival. LOG-ON IDENTIFICATION Each agent and supervisor should have an individually assigned log-on identification number that permits individual statistics to be collected by the ACD management information system. Multiple log on events by the same individual during a work period at different terminals should be tracked as one ""shift."

ABANDONED CALL SEARCH

The system shall have a feature to monitor all calls in queue to insure abandoned calls (caller hangs up prior to operator answering) are removed from the queue. Statistics should be gathered to provide caller delay tolerance as described in the management information system reporting section of this document.

INCOMING CALL ROUTING
The system shall automatically route incoming calls to alternate groups within the ACD or to specific trunk group assignments. Capability for overflow to a voice response system should be provided based on a predetermined threshold associated with delay time of the longest call waiting in queue for operator servicing within the ACD. Each gate and/or telephone line shall have the ability to overflow uniquely according to predefined parameters. DNIS codes will be the basis for determining overflow to the voice response system.

DYNAMIC CALL ANSWERING
The system shall search for an available agent in the primary group before returning supervision to the serving telephone company central office. This shall occur on a call by call basis until a time out or scheduled answer point is encountered in the call processing table, or an agent becomes available.

OVERFLOW TO SECONDARY AND SUBSEQUENT AGENT GROUPS
When this occurs according to the call processing table rules, the system must allow for simultaneous scanning or "look back" to previous groups.

ALERT WITH CUSTOMER CALL
The system shall provide an audible signal to indicate an incoming call to the agent. The call should then be connected to the agent without further console operation.

OPERATOR CALL ASSIGNMENT
The ACD shall attempt to distribute calls equally to agents across assigned gates. A first-in first-out basis shall be used by the ACD to distribute calls to agents who have been idle longest.

CALL TRANSFER
The ACD shall have the capability to transfer a call to designated gates, OPX's, administrative extensions, etc., for selective handling inside or outside of the system.

DELAY ANNOUNCEMENT
The system shall provide a queue with delay announcement capability for the

purpose of indicating service delays, posting general information and providing other instructions for callers when no agents are available.

A total of XX DELAY ANNOUNCEMENT units (minimum length - 30 seconds) shall be provided by the system with the ability to distribute these across gates as desired. I/O Marketing Inc. will not consider tape-messaging technology.

It is I/O MARKETING,Inc.'s objective to provide callers with progressive call delay messages (up to 3). There must be no barging into partially played messages by the caller. Additionally, caller should not have to wait before hearing first message. Separate messages will be used for each of the functional groups described in Overview of Requirements. Recording source and media should be specified and included in bid.

NIGHT SERVICE

The ACD shall allow for calls to be transferred to outgoing trunks or tie lines, off-premise extensions, a voice response system and to specific gates for announcement to indicate operating hours or other instructions to callers. Treatment shall be selective by telephone line (s) and/or gate.

ADMINISTRATIVE PBX FUNCTIONALITY

In addition to agent consoles, the ACD shall have standard PBX functions.

LINE/CONSOLE CAPACITY

The ACD must be capable of handling up ;to xxx agent positions and terminating xxx lines. The initial configuration must include xxx

EQUIPPED AGENT POSITIONS AND xxx INCOMING LINES.

MUSIC-ON-HOLD

The capability for using music-on-hold should be provided.

REMOTE DIAGNOSTIC CAPABILITIES

This capability must also be available to I/O MARKETING Marketing, Inc.

OUTBOUND SALES

The system shall be capable of providing outbound calling over WATS, local lines, foreign exchange and two-way 800 lines for Sales operators. Designated gate(s) for monitoring and separating Sales function is a requirement. Station Message Detail Recording (SMDR) is needed for outbound calling to include time and date of call, length of call, operator ID, number dialed and circuit accessed.

DIALED NUMBER IDENTIFICATION SERVICES (DNIS) CODES

I/O MARKETING,Inc. currently uses a four (4) digit DNIS code identifying number dialed. This code must be provided by any new switch and displayed at agent positions. The location of the display must be easily viewed by the agents. The DNIS code must also be passed with calls overflowing to the voice response system.The ability to pass this code directly to a local computer/terminal (also being used by agents taking phone orders) through a digital/communications interface is required. This feature is planned for I/O MARKETING, Inc. use within the near future.

MOVES, ADDS and CHANGES

In addition to the ability to reconfigure lines with gates and operator to gates, the system must allow for local physical/logical reconfiguration of operator consoles,

PBX

telephones, supervisor consoles and monitoring equipment.

POWER PROTECTION

For all system components indicate memory protection and/or continuous operation plus other related capabilities of benefit to the I/O MARKETING, Inc. operating facility.

B. AGENT POSITION FEATURES

AFTER CALL WORK

Agent Operator consoles shall include a button that removes agent/console from incoming calls on queue. Time spent in this work state must be included in the individual operator statistics along with group and system statistics.

SYSTEM TRANSFERRED CALL IDENTIFICATION

Agent shall receive a distinctive tone distinguishing calls transferred internally within the ACD operating environment verses the audible tone identifying new call arriving.

INCOMING CALL IDENTIFICATION

Capability should be provided for the agent to receive pre-announced identification by trunk group or DNIS code of calls appearing at the target agent console. Indication should be both announced and LED display with up to available.

ZIP TONE

During headset operation the system shall have the feature of providing the operator with an audible indicator prior to connection of a customer call arriving at the ACD.

WORK STATE INDICATORS

In addition to ;the after-call-work button, the capability for an agent to indicate one of THREE NON-ACD WORK STATES is a requirement. The ACD shall accumulate statistics by each work state indicated by agent number. The capability shall be timed-on, by agent, and timed-off with agent return to the system regardless of position occupied.

ASSISTANCE BUTTON/DIAL CODE

Agent console shall have the capability for immediate signaling to the operation supervisor when the agent needs assistance in handling customer problems. The capability must be provided for the call to roll over to an available supervisor should the first supervisor be busy.

CALL TRANSFER TO GATE/OPX

The agent shall have the capability to transfer an ACD call to another gate within the system, to an administrative line or an off-premise extension. For more frequent transfers, an operator console shall be capable of activating this function automatically (speed dial) and release the agent to proceed in handling other incoming calls.

CALL WAITING INDICATOR

Each agent console and supervisor console shall have the ability to indicate that incoming calls are waiting for service. The indicator (light) will flash when call(s) in queue exceed the predetermined threshold.

TROUBLE REPORTING/CONDITION INDICATOR

The agent console must have the ability to indicate circuit problems or tabulate counts of certain conditions under study. Unique classification by trouble type is desirable. MIS reporting must log tabulations by agent and specific conditions indicated.

EMERGENCY INDICATOR

A rapid transfer and/or alert capability must be provided to the agent to get assistance, activate recording of conversations and transfer calls to a pre-defined number (as to the assigned supervisor).

MUTE (COUGH) BUTTON

The console should be capable of muting agent noise momentarily from customer conversation.

LOG-ON AND OFF
The agent console must be capable of allowing logging on and off with employee using a 4 digit identification number.

IN-BUTTON
Agent shall have the ability to remove himself from idle states by depressing in in-button.

HOLD BUTTON
A button used to place a call on hold shall be available to the agent plus statistics regarding time spent by the agent shall be accumulated by the ACD/MIS function. A complete log of events per operator in transaction detail for is desired.

OPERATOR HEADSETS/HANDSETS

Two headset types (earplug and overhead) plus handset options are desired by I/O MARKETING, Inc. employees and management. An initial assortment and quality, plus spares, must be provided with system comparable in quality to consoles and telephones requested. Volume controls are required on all head/handsets that are effective and easily operated. Agent consoles must be equipped with dual headset jacks to permit supervisor and/or lead operator to plug directly into the console concurrently to assist in training or counseling of an agent.

C. SUPERVISOR POSITION FEATURES

AGENT ASSISTANCE
Automatic answering of agent requests for assistance must be provided at the supervisor's console. Calls for assistance from agents shall roll over to an available supervisor's position.

BARGE IN
This feature allows a supervisor to barge into an active call for the purposes of assistance or service observation. Activation of this feature will establish a three-way conference between the supervisor, the operator and the calling party.

CONFERENCING CAPABILITY
The system will have the capability for one to three supervisors to conference call directly; conference call between one to two supervisors and an agent, and conference call between two agents and one supervisor.

MONITORING CONSOLES

Each supervisor's console will have capability to monitor agents calls in a silent mode. It is imperative that calls do not lower volume, generate clicking or any other noticeable disruption to the agent and calling party. Additionally, the system shall have the capability of up to xxx CONSOLES capable of observing agent performance. These consoles but must have the capability of monitoring all operator positions, call supervisor console and otherwise be usable as a an agent instrument. Each supervisor console and other monitoring equipment must be capable of monitoring any agent assigned to any gate. Further, more that one monitoring position must be capable of observing the same agent/call concurrently.

BUSY CONDITION

An indication must be provided to each active supervisor to observe the condition of each agent position. Indications will include all agent status conditions.An active supervisor must be capable of viewing the status of agents assigned to any gate and not be limited in number of agents permitted in total for observation (that is, a limited number of agents can now be observed by a supervisor. It is this restriction that I/O MARKETING, Inc. seeks to remove).

429

DELAY ANNOUNCEMENTS ACCESS

The supervisor's console shall have the capability of controlling delay announcements. Access to delay announcement recordings shall be made available to the supervisor to record, playback, activate, and deactivate one or all of the announcements.

NIGHT SERVICE

Supervisor consoles also have the capability of activating the Night Service feature providing callers with announcements regarding hours of operation, etc. Night service can vary by gate and/or any functional area.

NUMBER OF SUPERVISOR CONSOLES

The system shall have the capability of configuring up to xxx SUPERVISOR CONSOLE POSITIONS and xxx POSITIONS for call monitoring.

4. MANAGEMENT INFORMATION SYSTEM
REPORTING REQUIREMENTS

The ACD system should include management information reporting

which will satisfy system monitoring, operator and gate/trunk reporting. The requirements listed below for these categories are considered minimum functions needed. This data is required for administration, and to support operations and generate required payroll and accounting information. Reports are expected to be available through a cathode ray tube (CRT) and to xx supervisor positions. The system shall maintain historical report data on-line, or on magnetic media capable of being immediately reprocessed by the ACD. Downline processing by PCs or other devices is discouraged. The following reporting requirements are included with an indication of those needed through CRT display, printing capability and whether the time frame is by half-hour, daily, etc.

A. SYSTEM REPORTS

SYSTEM STATUS DISPLAY

By gate assignment, this display will show in plain language, the active agents, the positions they occupy, the instant, individual agent status, the number of positions in after-call-work state. The same screen should also provide incoming call service information for the gate, such as the oldest call waiting length (in seconds), the number of calls waiting, the number of incoming calls handled during the reporting interval, the number of calls answered after XX seconds. This display must be provided in realtime. More than a 5 second delay in screen update is considered undesirable.

INCOMING CALLS HANDLED REPORT

This report is a summary of the interval of time (in seconds) required for handling incoming calls. The time intervals included on the VDT screen display are in ten second increments ranging from 0 to 9 seconds, etc. until 140 to 149 seconds. Each gate of the ACD will show by the seconds-interval number of abandoned calls within each interval along with total number of calls received per interval. This information is displayed by each time interval as well as by each gate. Totals and averages are reported for the time period requested. These reports will typically be provided in thirty-minute intervals; however, provisions must be allowed for longer intervals as for the entire shift or for a twenty-four hour period.

EXCEPTION REPORT

The exception report will be generated with line items for parameters preassigned for established thresholds once exceeded. Example: the average- speed-of-answer exceeds twenty seconds for calls on queue. The time of the occurrence, along with the gate, and the gate assignment within the ACD will be delineated as a line item for the threshold exception. Other thresholds may apply here for similar exception reporting pertaining to operator performance, trunk utilization and/or other system related problems.

THRESHOLD DEFINITION

The system should have a screen display available to the designated supervisor to activate certain settings for various thresholds being monitored. The threshold used in exception reporting for average-speed-of- answer exceeding twenty; seconds is an example. All other threshold level settings would be maintained as part of this system VDT display in the form of a fill-in-format for ease of updating and control.

B. GATE AND TRUNK UTILIZATION REPORTING

The next set of reports is intended to provide an overview of how trunks and trunk groups are allocated and utilized throughout a reporting period. Additionally, performance by gate assignments to assess employees' and/or groups performance are delineated through the three reports listed.

GATE SUMMARY

The report recapping gate activity is listed by gate number and includes the following items in minutes: seconds.

Average number of positions and work state
Number of incoming calls during the interval displayed
Average-speed-of-answer
Number of calls exceeding threshold (e.g., call waiting on queue longer than 20 seconds)
Number of abandoned calls
Number of outgoing calls
Percent busy time reflecting proportion of all operators in agate or in work state verses idle and/or other non-ACD callhandling activity

Percent active call time which is the percent of time that acall in a gate is connected to an operator including time heldat the agent's console

Weighted Call Value which is the average time per callincluding talk time and after call work

In addition to the above information, the gate report shouldshow totals by gate assignment for the various non-ACD workstates. Further, the cumulative statistics on stroke countsindicated by operators for signaling specific conditionsshould be listed by gate. The number of troubles by categoriesassigned should also be listed by gate.

The gate summary report must be available via a VDT display forthe most recent half-hour or for any period during the day asdefined by the supervisor.

TRUNK REPORT

This report will list by each trunk group the following items:

Number of trunks in group
Number of incoming calls during interval
Abandoned calls
Average hold time on trunk per call
Average ring time, time on queue, time to operator or voiceresponse system
Average-speed-of-answer
Hundred call seconds of traffic during interval reported
Total outgoing calls
The proportion time an average trunk is busy on a call duringthe time interval reported

The call answering efficiency computed as the time a call isconnected to an operator divided by the total time in thesystem during the interval reported

Percent of time all trunks are busy

DNIS CODE REPORT

The following information should be listed by DNIS Code on thisreport: Number of incoming calls during interval

Abandoned calls
Average hold time on trunk per call
Average ring time, time on queue, time to operator or voiceresponse system
Average-speed-of-answer
Hundred call seconds of traffic during interval reportedTotal outgoing calls

The call answering efficiency computed as the time a call isconnect to an operator divided by the total time in the systemduring the interval reported

OUTGOING CALL DETAIL REPORT

This report will include the following information by operatorfor outgoing calls:

Agent name
3 digit project code
Position number
Telephone number dialed
Length of call
Date and time of day
Circuit number used
The capability for generating these reports automatically willbe every half hour

or some selected interval of time within 24hours. Cumulative totals in the various categories will beprovided through the supervisor's ability to change an intervalof reporting during the course of the day.

TROUBLE REPORT SUMMARY

This report delineates specific troubles by time-of-day asdesignated through console key indication by agents orsupervisors. The report simply lists the time, the date, theagent identification number, a trunk/line number and the typeof problem experienced (noise, hum, low volume, or abandonedcall).

C. AGENT ACTIVITY REPORT

The following items are listed for inclusion in the worksummary report by agent and assigned gate. Items listed areintended to determine workload, activity accomplished by eachagent, identify the gate assignment, measure relativeefficiency of agents and determine amount of time spent onnon-ACD work activity. This includes non-customer servicestates as seeking assistance, etc. The following items arerequired for CRT display on a half hour interval or for aninterval to be defined by the supervisor viewing the display.Additionally, the ability to print this information on a timeinitiated basis by shift, by day and by individual is likewiserequired. The following items are listed on the report by agentidentification number.

Agent Identification NumberGate Assignment
Total time logged on to gate in minutes:seconds
Number of assisted calls during reported interval
Time in active various call states (minutes:seconds)
Number of incoming calls handled
Number of outgoing calls placed
Average talk time for calls handled (minutes:seconds)
Time in each of three non-ACD work states (minutes:seconds)
Operator efficiency ratio (computed as a ratio of the totalsystem talk time divided by the number of calls featured overthe total agent talk time divided by the number of callshandled by the agent)

Active time per call (calculated by combining the average talktime with after-call work time divided by calls handled duringreporting intervals)

In addition to the above information it is necessary to have asign-on log identifying each agent with an identificationnumber, the date and time sign on and sign-off took place. Thisincludes a log of time in and out of various non-ACD workstates (break, lunch, etc.).

5. USER PROGRAMMABILITY REQUIREMENTS

In the course of conducting operations at I/O MARKETINGMarketing, Inc. it is important to allow flexibility in settingsystem parameters and in developing additional reportinformation that requires simple programmable functions.

It is the intent of features in this category to provideflexibility as opposed to defining in advance all of thereporting capabilities necessary for monitoring agents, systemoperation, agents and trunk utilization.

There are three basic areas for programmable function that areexplained as example of the capabilities. Vendors areencouraged to comment on software programmable functionsavailable in high level format, ""user friendly", which willenable staff personnel at I/O MARKETING, Inc. tooperate the system efficiently without increasing major ACDoperating expense, staffing, training or sophistication.

REPORTS

Programmable functions should permit selective sorting over aperiod of time in developing a report. An example would be tosearch out operator efficiency ratios over X period of time toassist in conducting performance reviews. Additionally, trunkreport data may need to be listed by type that may lead torequesting (Class A) treatment of circuits assigned in the ACDoperating environment.

Additionally, I/O MARKETING, Inc. is interested indeveloping operator performance data for the previous day foremployees to have access once loaded into the computer systemdata base. Programmable function here coupled with ACD datalink capability will enable this development by XXX.

GATE ASSIGNMENTS

The ability to prioritize the various trunk groups by gateassignment is required as a function of programming ACDsoftware parameters. Operating personnel need the ability toselectivity assign trunk groups to gate as needed to support promotional programs which will vary from time to time.

TRUNK PRIORITIES

Since I/O MARKETING, Inc. will be configured withvarious types of circuits (WATS, foreign exchange, and locallines) it may become necessary to give higher priority to themore costly network facilities. Additionally, sinceI/O MARKETING, Inc. provides different sales andcustomer service functions, it may also be desirable to providea higher priority of service to callers who are calling for tobuy rather than enquire about account status.

While these are representative of the typical types ofprogramming capabilities, local report generation capabilitybased on a menu driven approach should be provided. The abilityto set parameters and research historical data from reportslisted above should be provided by using the ACD processorcomplex or an auxiliary unit (e.g., personal computerattachment) to facilitate such report generation. Sinceprogramming skills will be kept at a minimum atI/O MARKETING, Inc. it is recommended that astraightforward and simple software programming capability beproposed.

6. VOICE RESPONSE SYSTEM INTERFACE REQUIREMENTS

NOTE: Here these requirements should be spelled out if they areneeded. Due to the breadth of information needed to implementthese devices this section is being intentionally treatedlightly.

7. OPERATING ENVIRONMENT

Equipment operating within environmental specifications aslisted below should be suited for installation atI/O MARKETING, Inc. The environmentalspecifications being followed by the building contractorsshould satisfy hardware, software and storage mediumrequirements when equipment operates within the specificationsbelow.

ENVIRONMENTAL TEMPERATURE:

Ambient Operating: +15 degrees to 32 degrees Centigrade (+60degrees to 90 degrees Fahrenheit)

Ambient Storage: -10 degrees to 50 degrees Centigrade (+14degrees to 120 degrees Fahrenheit)

RELATIVE HUMIDITY: 40-90% NONCONDENSING

ELECTRICAL: AC INPUT 105 - 126 VOLTS, 50 - 60 HERTZ

Complete specifications for all equipment proposed by vendorsshould be included as part of proposals tendered.

8. TRAINING REQUIREMENTS

SUPERVISOR COURSES

Training for supervisors of I/O MARKETING, Inc.Response Center should be available and provided by the vendoras part of the system bid for the ACD system. Training shouldinclude an overview of fundamentals for ACD operations withpractical instructions on the various reports, gatereconfigurations, and other aspects of managing Response Centeroperators.

OPERATOR INSTRUCTIONS

Documented training with manuals foragent operations should be provided with instructions for usingtelephone consoles and headsets along with instructions forsigning in, signing out, etc. The course should be modular andpackaged so that additional training can be provided in house,at the option of I/O MARKETING, Inc.

GENERAL TRAINING

A general management overview of all systemsproposed must be provided for administrators of the ResponseCenter. The capabilities of the equipment, intent and use insimilar environments should be made available to acquaint localmanagement and I/O MARKETING, Inc. administratorson system operational functions and benefits.

9. MAINTENANCE SUPPORT REQUIREMENTS

It is the intent of this section to solicit information fromvendors regarding all of the maintenance aspects followingacquisition by I/O MARKETING, Inc. of proposedconfigurations. Where items of importance, regarding addedadvantages to the system proposed, are not spelled out in thefollowing discussions, bidders are requested to list suchcapabilities for inclusion in the final evaluation.

MAINTENANCE RESPONSE TIME

It is a requirement of I/O MARKETING, Inc.Response Center to be operational for seven days a week,twenty-four hours a day. Vendors are expected to be capable ofproviding four-hour emergency response and/or same dayresponse, based on the nature of system problems.

REMOTE DIAGNOSTICS

While it is not a definitive requirement for a provider ofsystems to have remote diagnostic capability, such capabilitywould be considered as an extra benefit in the procurement ofthe system requested.

SERVICE LOCATIONS

It is of interest to I/O MARKETING, Inc. to knowthe precise location and number of field engineers available toservice the equipment proposed. An indication should be notedregarding the availability of service personnel, in terms ofnumber. Within the area and/or potential for using XXXfacilities as the primary office to serve both the XXX ResponseCenter and other accounts.

PREVENTIVE MAINTENANCE SCHEDULE

To avoid disrupting prime time operation between 7:00 a.m. and9:00 p.m. at the

Response Center on Mondays through Fridays,vendors should propose a preventive maintenance scheduleindicating frequency, day of the week, and time of day, as partof the response.

WARRANTY

The specific warranty period and the extent of the warrantycoverage on parts, labor, travel expenses, etc. should belisted as part of the vendor's proposal.Mean-time-between-failures (MTBF) and mean-time-to repair(MTTR) data should also be provided.

SYSTEM DOCUMENTATION AND UPDATING

The documentation on the functional and technicalspecifications should be listed as part of the proposal andprovided to I/O MARKETING, Inc. if equipment isselected. The cycle for updating such information and theprocess for providing revisions to system offered should alsobe delineated.

ONGOING SOFTWARE SUPPORT

The practice and policies of your firm in providing ongoingsoftware support, and cost, as part of the maintenancecontract, should be listed. The frequency of previouslyrevisions should also be listed to provide an indication of theextent of change that might be anticipated by I/O MARKETINGMarketing, Inc.

MAINTENANCE COSTS

The specifics costs for all of the above maintenance should belisted with variations on time to respond, day-of-the-week,and/or holidays as appropriate. Additionally, such costs wouldbe listed in Section yyy and Appendix yyy to complete the totalof all costs, for system procurement and operation.

10. SYSTEM COSTS

It is the intent of this section to provide a basis for vendorsto list all of the costs to and liabilities of I/O MARKETINGMarketing, Inc. if the equipment being proposed is selected.

ITEMIZED FEATURE/OPTION PRICES

Responses should indicate where the features specified, asminimum requirements in this document, will be pricedseparately. The basic price of this system should be included.List the cost per operator position proposed.

INSTALLATION PRICES

The charges for installing the initial configuration should bespelled out in detail to include all system components, networktermination, operator position configuration monitoringpositions, and PBX locations.

ESTIMATED OPERATING EXPENSES
The following areas and costs items should be listed by vendors

Software
Terminal/CRT/Phone Instruments
Moves, Adds and Changes
List cost per position, where appropriate, and total cost

11. INSTALLATION AND CUTOVER
It is a requirement for vendors to provide detailed schedulesfor the implementation and cutover of all proposed systems.While the requirement for expansion of the current ResponseCenter at calls for completion by XXXber, 20018, it is highlydesirable to install the system in new facilities as early asXXy 2001 to facilitate training and system check-out. Systemsproposed will be evaluated based on the reasonableness andexpediency of the proposed cutover plan. The schedule should beconsistent with orderly transition, allowing for training andacceptance of system operation by I/O MARKETING,Inc.

ACCEPTANCE TEST AND DIAGNOSTICS
The method of performing acceptance testing and demonstratingthe system as ready for cutover, should be listed. Indicate thevarious diagnostic routines conducted to assure connection ofthe network, along with operator positions, supervisor consolesand monitoring locations.

CUTOVER PLAN
The support areas and representatives from your firm that willassist in the schedule for beginning the cutover andaccommodating implementation should be discussed.

CUTOVER SUPPORT
The support areas and representatives from yourfirm that will assist in the cutover should be listed. The typeof support for correcting any network or system problem shouldbe included.

POST-CUTOVER SUPPORT
Following the cutover and system test, the followinginformation should be provided regarding post-cutover support.

Traffic Engineering Capabilities
Capacity Planning

Network Design and Engineering Support
Additional Training
System Operation support

SYSTEM FACTS AND FEATURES

This must describe the system that will be delivered. Anyplanned capability that is not currently demonstrable in aanother customer site must be clearly identified as a futurecapability. I/O MARKETING Inc. is not adverse tobeing an early release customer, but is against misleadingsales practices.

Manufacturer:
Model designation:
When first installed:

SOFTWARE

Software version (feature package, release, etc.):

Is the proposed software the latest generic offered on thesystem? Yes No. The latest Generic is....

When first installed:

SWITCHING TECHNOLOGY

Control type: Distributed , or Centralized .

Transmission: Analog , Digital , VoIP

Modulation (PAM, PCM, Delta, IP Telephony, etc.): .

COMMON CONTROL AND MEMORY CONFIGURATION OF Switch or VoIP Telephony Server?

Is the proposed controller and memory capacity adequate tosupport the proposed system?

What processor technology is used to drive the system?

Describe all computers in the proposed ""working"" system as to:

Function:
Redundancy:
Location:
What portion of the short term memory is volatile?

Is volatile memory protected?

For how long is memory protected?

Are critical electronic components redundant? Yes No .

Which system components are duplicated when redundant commoncontrols and critical electronics are provided?

What is the lowest component of the switching hierarchy that isredundant (i.e., shelf controllers are redundant, line cardsare not)?

How many calls can the system process at peak load (at the""equipped"" configuration)? calls per hour.

PHYSICAL CHARACTERISTICS

How many cabinets are provided:

In the proposed wired configuration? cabinets.

At the specified minimum acceptable expansion? cabinets.

How many modules are provided:

In the proposed wired configuration? modules.

At the minimum specified expansion? modules.

Is this configuration a single switching hierarchy.

If not, please explain.

In a multistage switching or parallel LAN systems, is there a potential for traffic blocking between switch modules or local carrier?Yes No .

If Yes please explain:

At what point (number of trunks and ACD stations) will expansion of the system require an addition shelf or LAN connectivity?

At what point (number of trunks and ACD stations) willexpansion of the system require an additional cabinet?

Provide with the Proposed Document a diagram identifying cardlocation within each cabinet of the proposed ""wired"" systemas configured for cutover. For VoIP proposals, please provide traffic guarantee for minimum of busy hour attempts processed and connected. Assume average connected and completed call length to be 3.5 minutes.

OPERATING ENVIRONMENT

Temperature High Low Optimum .

Humidity High Low Optimum .

Total BTU's per hour produced by all ACD and support equipmentin the switch room:.

The proposed wired configuration: BTU's per hour.

The minimum specified expansion: BTU's per hour.

Total power consumption of all telephone (TDM or VoIP) equipment in theswitch room at peak load:The proposed wired configuration: Kva.

The minimum specified expansion: Kva.

Specify other switch room requirements:

Electrical (number of circuits, voltage, amperage and phases ofeach, and number of each type of electrical outlet)

Lighting (type and intensity)

Backboard space (total square footage and dimensions)

Raised flooring

What environmental services (cooling, lighting, electrical andbackboards) will be required in satellite closets?

441

POWER FAILURE CONSIDERATIONS

Power failure restart — what is the system's method ofmaintaining or restoring memory in the event of a commercialpower failure?

For how long is memory protected, assuming that UPS or otherbattery plant is depleted?

How long does it take to restore the entire system (at cutoverconfiguration)? minutes.

Are recent moves and changes automatically updated when poweris restored?

Yes or No.

Power failure transfer which connects specified stationsdirectly to the telephone utilities Central Office during poweroutages are provided for stations.

In the event of total system failure, can and will the powerfailure trunks be automatically connected to their designatedstations? `

Yes or No.

Describe the batteries and associated equipment supplied:

COLD REBOOT FROM COMPLETE POWER FAILURE:

What is the time to operation for the system to return to full operation? .

Impact in reload control software? Yes   No?   Time   to   reload   _____

Mins/Seconds?
Reset trunk, line or other station cards Yes No?
Time to reset _____ Mins/Seconds?

SCHEDULED SYSTEM DOWNTIME POTENTIAL
Must the system be shut down to:

Install new software generics Yes No.
Add trunk, line or other station cards Yes No.
Add an equipment shelf Yes No.
Add a cabinet Yes No.

PROPOSED SYSTEM CONFIGURATION
Switching System:
Equipped Wired Maximum
Capacity
ACD supervisor lines?
ACD agent lines?
Trunks:
Central office
(WATS, COBWT)
DIDFX
Tie Lines (to PBX)
IP trunks
Provide with the proposal a complete list of all equipmentincluded in the basic system. On that list show ACD instrumentsand supervisor CRTs, common equipment including cabinets andline/trunk cards, software features, and other items in thebasic system.

Provide with the proposal complete lists of all items included in each

Optional Equipment requirement.

How many separate trunk groups can the ACD accommodate?

At the cutover configuration?

At the wired - for configuration?

At maximum?

CAPACITY TRADEOFFS
Standard agent station line capacity is reduced by_____lines for each _____trunks.

Standard agent station line capacity is reduced by_____lines for each _____supervisor station lines.

How many of the following types of circuits are controlled by asingle shelf/module controller, and how many circuits reside oneach card?

Circuits/card Cards/controller
Trunks
ACD supervisor lines
ACD agent lines
Dedicated Service Ports
(Service Observe or Recording ports)
Are shelf slots universal, or does each type of circuit requireits own, exclusive shelf or back plane?

Describe any limitations on the quantity of any particular cardwhich can be placed on a given shelf:

INTERFACES with OTHER EQUIPMENT
Is the proposed system compatible with other systems asspecified in section?

PBX Yes No.

Outboard Force administrative System Yes No.

Which Manufacturers.

Outboard outcall list system Yes No.

Which manufacturers.

.If the answer to any of the questions above is no, explain:

STATION EQUIPMENT — DISTANCE LIMITATION
How far can a supervisor or agent instrument be located from the control switch? Agent set feet.

SoftPhone feet.

CRT feet.

STATION EQUIPMENT — GENERAL
Proposed equipment:

Manufacturer Model Cable size

ACD agent setSoftphone

ACD supervisor CRTACD supervisory and agent instrument information:

Can ACD supervisory positions be provided with eight CRT's asrequired? Yes No.

Number of ACD supervisory CRT's operate fully andsimultaneously? Number.

What is the maximum number of ACD supervisory CRT's that canoperate fully and simultaneously at the wired — for configuration size?.

Is transmission from the agent sets to the switching systemdigital, analog or VoIP?

Do agent sets include a digital display for calls waiting or asingle lamp? Yes No.

On agent sets, list how many:

Line buttons?.

Feature buttons?.

Describe briefly the specific functionality of agent sets'feature buttons:

Describe briefly the display functionality of the agent set:

THE VOICE CCS PER LINE OR VoIP CONNECTION RATING:
Per ACD line:

At wired line size CCS

At minimum specified expansion line size CCS

SIMULTANEOUS CONVERSATIONS AVAILABLE AT:
Proposed wired configuration: conversations

Minimum specified expansion: conversations

STATION CABLING
Describe recommended station instrument cabling:

Describe recommended CRT cabling:

ACD MANAGEMENT INFORMATION SYSTEM (MIS)
Does the system collect peg count data or transaction detailrecords? Explain:

If transaction detail, please list transaction types

Maximum number of transaction records the system can store:

At the proposed configuration

At maximum capacity

Describe how the ACD supervisor CRTs are updated, payingparticular attention to the frequency of update:

Will the system allow eight supervisory CRT's to havesimultaneous access to real-time ACD information? Yes No.

Can the system automatically print cumulative reports for thefollowing periods:

Hourly
Daily
Weekly
Can the system's capacity be expanded easily and in the fieldto accommodate a minimum of 5,000 call records per day, andprovide hourly, daily, and weekly summary reports? Yes No.

What additional hardware/software would that expansion require?

List necessary items and approximate cost:

Will the system provide all ACD agent and trunk reportinformation as required? Yes No.

If No, explain:

Are sample ACD MIS reports enclosed? Yes No.

Is complete MIS system training included in the basic systemprice as required? Yes No.

ACD OR TELEPHONY SERVER CAPACITIES

How many agent groups can the ACD function accommodate?

At the cutover configuration?.

At the wired - for configuration?.

At maximum?.

Can the ACD route calls to a selected group of PBX single linetelephones and define those telephones as an agent group? Yes No.

How many trunk gates (or splits) can the ACD accommodate:

At the cutover configuration?

At the wired - for configuration?

At maximum?

How many separate ACD trunk groups can the ACD accommodate:

At the cutover configuration?

At the wired — for configuration?

At maximum?

How many agent groups can be programmed into an ACD callrouting progression for overflows?

REQUIRED and OPTIONAL FEATURES

General Instructions:

This is referenced to the specifications section. The numbersat the left of the response column below indicate the sub -paragraph number of the specifications. Indicate, by insertingthe proper letters whether the following are:

- (S) Standard in the proposed system at no additional charge;
- (OA) Optional in the proposed system and available at anextra charge which is not included in the base price. This codewould apply only to features or equipment which are bothoptional from the proposer and optional in the specifications;
- (OI) Optional in the proposed system at an additional chargewhich was included in the base price because it was specifiedas a requirement in thisRFP;
- (NA) Not available;
- (NAS) Feature is available but does not operate exactly asspecified. For example, if a feature is standard, but does notoperate exactly as specified enter the letters S-NAS.
- (FUTURE) Indicates that a feature has been announced for thesystem but that it is not yet generally in use (beta test sitesare not considered ""generally in use""). The date when thefeature will be available must be included. For example, if arequired feature, which will be available in XXXch 2001, hasbeen included in the base price of the system even though it isnormally quoted as an option by the proposer, the letters OI -FUTURE, XXXch, 2001 should be entered in the response.

A. SYSTEM FUNCTIONS

GATE/GROUP ASSIGNMENTS
PRIORITY QUEUING
ORDER-OF-ARRIVAL
QUEUING LOG-ON
IDENTIFICATION
ABANDONED CALL SEARCH
INCOMING CALL ROUTING
DYNAMIC CALL ANSWERING
OVERFLOW TO SECONDARY AND SUBSEQUENT AGENT GROUPS

ALERT WITH CUSTOMER CALL OPERATOR
CALL ASSIGNMENT
CALL TRANSFER
DELAY ANNOUNCEMENT
NIGHT SERVICE
ADMINISTRATIVE PBX FUNCTIONALITY
LINE/CONSOLE CAPACITY
MUSIC-ON-HOLD
REMOTE DIAGNOSTIC CAPABILITIES
OUTBOUND SALES
DIALED NUMBER IDENTIFICATION SERVICES (DNIS) CODES
MOVES, ADDS and CHANGES
POWER PROTECTION

B. AGENT POSITION FEATURES

AFTER CALL WORK
SYSTEM TRANSFERRED CALL IDENTIFICATION
INCOMING CALL IDENTIFICATION
ZIP TONE
WORK STATE INDICATORS
ASSISTANCE BUTTON/DIAL CODE
CALL TRANSFER TO GATE/OPX
CALL WAITING INDICATOR
TROUBLE REPORTING/CONDITION INDICATOR
EMERGENCY INDICATOR
MUTE (COUGH) BUTTON
LOG-ON AND OFF
IN-BUTTON
HOLD BUTTON
OPERATOR HEADSETS/HANDSETS

C. SUPERVISOR POSITION FEATURES

AGENT ASSISTANCE
BARGE IN
CONFERENCING CAPABILITY
MONITORING CONSOLES
BUSY CONDITION
DELAY ANNOUNCEMENTS ACCESS
NIGHT SERVICE
NUMBER OF SUPERVISOR CONSOLES

4. MANAGEMENT INFORMATION SYSTEM REPORTING REQUIREMENTS

A. SYSTEM REPORTS
SYSTEM STATUS DISPLAY
INCOMING CALLS HANDLED REPORT
EXCEPTION REPORT
THRESHOLD DEFINITION

B. GATE AND TRUNK UTILIZATION REPORTING
GATE SUMMARY
TRUNK REPORT
DNIS CODE REPORT
OUTGOING CALL DETAIL REPORT
TROUBLE REPORT SUMMARY

C. AGENT ACTIVITY REPORT

5. USER PROGRAMMABILITY REQUIREMENTS
REPORTS
GATE ASSIGNMENTS
TRUNK PRIORITIES

6. VOICE RESPONSE SYSTEM INTERFACE REQUIREMENTS

7. OPERATING ENVIRONMENT
ENVIRONMENTAL TEMPERATURE:
RELATIVE HUMIDITY: 40-90% NONCONDENSING
ELECTRICAL:

8. TRAINING REQUIREMENTS
SUPERVISOR COURSES
GENERAL TRAINING

9. MAINTENANCE SUPPORT REQUIREMENTS
MAINTENANCE RESPONSE TIME

Appendix III

Request for Proposal

Customer Experience Improvement Initiative Multimedia Recording and Analysis Solution

449

First National Bank of Axxum
National Customer Contact Centers
Confidential Reply by
November xth, 20XX to:
> Philip Smith
> Vice President
> First National Bank of Axxum
> 100 RFP Plaza
> Axxum AL 6XXXX
> Email: Philip_Smith@Axxum.com

INTRODUCTION

Objectives of RFI

The purpose of this request is to invite prospective vendors to submit a proposal to supply an automated monitoring and recording system(s) for the Customer Experience Improvement Initiative at First National Bank of Axxum. This Request for Proposal (RFP) provides potential vendors with the relevant operational, performance, application, and architectural requirements of the system. This information enables vendors to respond in a format that makes for a fair comparison and ensures that the proposed solution meets the requirements.

Overview of National Customer Service Contact Centers:

National Customer Service Contact Centers are located in both Biloxi and National Alabama. We are a retail consumer and small business service center supporting Axxum customers nationally. We have a combined workforce of approximately 2100 employees, 1700 of which are phone representatives either in service or sales related activities.

Our servicing goal is to provide a world class call center environment for our clients. As part of our requirements to be world class, we need to provide a consistent and uniform message to all callers. We have a need to increase efficiencies and to evaluate our phone associates regularly and in a consistent manner.

Background Information:

Currently, we perform monitoring functions without the aid of a formal recording and monitoring system. We have twenty-five dedicated Monitoring Representatives performing the task. In addition, all levels of management spend 1 hour each month monitoring phone associates. Frequently, it is difficult for the management group to locate the representative on an active call due to the variety of break schedules and inaccessibility of the on-line ACD or scheduling software at their desktops. The dedicated monitoring group does however, have access to these tools at their desktop.

Call Center Environment:

The hours of operation vary by department, but our intent is to seek a solution capable of handling a 24x7 operation. All inbound calls are routed to our 688 BRITE IVR, which handles slightly greater than 82%-84% of the call volume without the need to forward the call to a "live" associate. The IVR receives 300,000 to 500,000 calls per month. Therefore, the Call Center will be handling 500,000 to 750,000 per month by the end of 2001. Average talk time in the Call Center for "live" agent calls is 3.15 minutes. After call work time is 10 seconds bringing total handle time, per call, to 3.25 minutes.

Outbound call volume is primarily handled through the Avaya Mosaix 4000 or 5000 Predictive Dialer systems located in AL and in TX. The Alabama dialer group makes between 400,000 and 650,000 outbound attempts per month with a connect rate of between 24% and 28%. The Texas dialer group makes between 1,000,000 and 1,500,000 outbound attempts per month with a connect rate of between 19% and 27%. This volume is expected to triple as well.

Existing Technology Environment:
Call Center
PBX Hardware & ACD software, CTI Servers:
[provide detail]
Voicemail:
[provide detail]
Trunking:
[provide detail]
LAN type, agent & reviewer desktop computers:
[provide detail]
Non-Call Center Environment :
[provide detail]
PBX Hardware & ACD software, CTI Servers:
[provide detail]
Trunking:
[provide detail]
LAN type, agent & reviewer desktop computers:
[provide detail]

451

INSTRUCTIONS TO VENDORS

Project Schedule
| November 17, 20XX | RFI sent to all potential bidders |
| December 4, 20XX | Written & e-mail responses due |

Closing Time and Date
The response to the RFI should be submitted to Philip Smith no later than December 4th, 20XX.

Responses must be sent to:
Philip Smith
V.P. Project Management
100 RFP Plaza.
Axxum, Alabama 6XXXX
(123) 456-7890   (voice)
(123) 456-7880   (fax)
Philip_Smith @Axxum.com

Number of Copies
3 printed copy and 1 electronic copy (Microsoft(r) Word 97 document) of your response are required.

Right to Reject

Axxum reserves the right to reject any or all responses to this RFI even if all the stated requirements are met. In addition, Axxum may enter into negotiations with more than one vendor simultaneously and award the transaction to any vendor in negotiations without prior notification to any other vendor.

VENDOR PROFILE
1. State the number of years your firm has been in business.
2. Where are your head office, sales, and customer service offices located?
3. Number of employees in your company?
4. Number of employees in customer service/help desk functions?
5. Number of employees in product development?
6. Are your hardware and software dependent on third party vendors for manufacturing?
7. Provide a list of 5 references including contact names so that we may solicit feedback from them. For each reference, provide names and phone numbers of the primary contacts.
8. Provide your income statement and balance sheet for each of the two most recently completed fiscal years certified by a certified public accountant (CPA). Failure to submit a financial statement with this response may eliminate your proposal from consideration.

Future Offerings
1. What new technology does your company plan to utilize in the near future that would be an advantage to Axxum?

What are the most significant factors affecting the future success of your company and what is being planned to address it?

Market Differentiation
1. Provide a brief summary of your company's history in the marketplace. Limit response to 2 pages.
2. Please describe any features, services, or practices you provide in relation to the products requested which set you apart from your competition.

Client References
1. Provide a general list of call center clients where your systems are installed.
2. Provide any articles in call center publications that highlight your company's products/services and associated implementation.
3. Please list any pertinent industry awards your application has received.

Strategic Alliances/Partnerships

Identify strategic alliances and/or partnerships your company has with other call center technology vendors, and describe briefly the nature and objectives of the particular partnership and/or alliance. How does the alliance/partnership strengthen or differentiate your company's product/service offerings?

BUSINESS RESPONSES

Selective Recording

1. Does the proposed system support selective recording? How is this achieved?
2. Describe the systems integration and functionality with Genesis CTI?
3. Does your application support after-call or wrap-up screen/data capture via ACD?
4. Does your system have the ability to capture voice and data for both inbound/outbound calls, based on user defined business rules?
5. Does your solution provide the ability to support a free-seating agent environment in which agents can be monitored regardless of where they are physically sitting?

Total Recording

1. Does the proposed system support total recording, including e-mail recording? How is this achieved?
2. Describe the systems integration and functionality with Genesis CTI?
3. Does your system support both total recording and the quality application?
4. Dies your application support compliance recording?

Record on demand

1. Does the proposed system support agent and supervisor record-on-demand?
2. Describe the systems integration and functionality with Genesis CTI?

Multiple Recording Applications

1. Does your recording application support multiple recording solutions, i.e. total, ROD, and selective on the same platform?
2. Does your system support multiple recording solutions on voice and e-mail and is your quality application on the same platform?
3. What is the maximum amount of online storage available?

Recording e-Mail and Collaborative Chat Interactions

1. What are your organization's integration capabilities with e-mail response management systems?

2. What are your organization's integration capabilities with leading collaborative chat applications?

3. Can we define rules to trigger the recording of specific e-mail and web chat interactions? If yes, how does your application allow us to do this?

4. Does your application allow e-mail and web chat interactions to be recorded from beginning to end (i.e., until the e-mail is sent or the chat interaction is completed)? If yes, does the application also capture agent wrap up time?

5. Does your application provide the ability to compare web interactions and e-mail data with other performance metrics generated from your application (such as ACD and adherence statistics)?

6. What benefits will our organization gain by collecting and analyzing e-mail and web chat data?

General Capabilities

Does your automated digital recording system have the following features/capabilities?

- Client/Server, LAN/WAN based system
- Integrated, native ODBC/SQL Compliant database with on-line query capability for all voice recording, screen capture and data tags?
- PC-based playback applications, for voice and data, running under Windows 95/98/20XX and Windows NT
- Playback application installed on any non-dedicated workstations
- On-line help screens available to users
- An Ethernet interface for connection to the LAN/WAN
- Communication with external applications using a sockets interface over a TCP/IP protocol

Archiving Capabilities

1. What backup program is available with the proposed platform?

2. What is the maximum amount of storage space in terms of hours, contact/data recording?

3. What is the recommended length for data storage?

4. Describe playback from near-online and archived media?

5. Longest time to replay of a record?

Platform Features

1. How scalable is your platform in regards to the number of seats/workstations it can support at a single location?

2. How many playback workstations can the proposed system monitor at one time?

3. Are there any restrictions on the number of users that can simultaneously access and replay recorded voice and data files?
4. Please describe in detail the interface between the recording system and the PBX (i.e., does your system operate via analog ports, digital ports; how many ports and what is the exact type of port)?
5. Does your system use the Service Observe feature for monitoring?
6. Please describe your phone switch requirements in detail.

Quality

(a) Recording Contacts

1. Is the system able to record incoming ACD calls, as well as KANA email for a specific agent scheduled by pre-selected times, number of calls, or at random intervals throughout the day? If yes, can that schedule be repeated and/or changed?
2. What information does the software need prior to scheduling recordings of an agent? List the data fields.
3. Can the system be limited to only recording ACD calls and not the agent's private line?
4. If yes, describe how it would be configured.
5. Is the agent schedule needed to know when the agent is available to be monitored?
6. Does the system allow the creation of recording schedules so only an agent's outbound or inbound conversations are recorded but not both? If yes, describe how it would be configured within the software. Additionally, can the same be created on the KANA side? That is, can the recording schedules be broken down by outgoing e-mail versus incoming e-mail?
7. Is the system able to record an agent regardless of the location of the workstation? I.e.: How does the software know where agents are sitting in a free-seating environment?
8. If yes, how is this accomplished?
9. Does the system have the potential to capture CTI information and attach it to the call so that the calls can be retrieved using CTI information?
10. Is the system able to record a CSR on a manually initiated basis? i.e. Record on Demand. Describe how this would be accomplished. Can the user stop a recording that is occurring and begin the recording of a different agent through a manual process? Describe how this would be done with the software.
11. Can an agent be recorded in three or more time sessions? I.e.: a call at

10:00 am, 1:00 pm and 4:00 pm. Are their any limitations on the number of time sessions that can be recorded in a shift? Describe how the schedule would be created to record three or more time sessions.

12. What are the options, with respect to voice conversations if your hard drive becomes full? i.e.: archive, delete, overwrite etc.

13. What voice compression technique is used?

14. What would be the size of a one-minute, two-minute and three-minute message in kb/s?

15. What are the archiving options?

(b) Play Back Conversations & email

1. Does the system require dedicated workstations or can the reviewer use the PC at their desk to schedule and play back conversations?

2. Is the system able to allow selected voice conversations to be saved for an unlimited amount of time? Describe how the system would specify clips to be saved.

3. Is the system able to delete specific conversations on request before the retention period expires? What actions would the user take to perform this function?

4. Does the system allow the user to visually view the length and position of the silence within a voice conversation during playback? If yes, describe whether the user has the option of skipping the silence and how this would be accomplished.

5. Is the system able to convert the voice conversations into .wav files for training or playback purposes? Describe how this is accomplished.

6. When the reviewer is playing back a call does the application visually show where the playback is in relation to the full length of the voice file?

7. Does the system allow the user to play back voice files from .5 up to 2X the original recorded speed of the conversation?

8. Does the system use voice compression to minimize the amount of disk space the voice recordings consume?

9. How does your recording system protect the integrity of the recordings? What type of security is utilized to prevent unwanted access of the recordings?

10. Will the digital recording system be flexible enough to allow either dedicated quality monitors and/or team leaders and their agents to access playback functionalities, ideally from their own workstations?

11. Does the system allow the users to transfer a .wav file to their agents, ideally by e-mail, so leaders can access the file from their workstations?

12. Does the system have remote playback capabilities?

(c) Evaluation of Agents

1. Is the system able to weight each of the questions on the quality assessment form with a different value and does the software calculate an overall score for all of the questions?
2. Does the system provide the user the ability to customize on-line evaluation forms? This includes adding questions, changing weightings and changing values? Please describe your solution to this requirement and how a user would change or create an evaluation form.
3. Does your system allow you to predefine lists of data that can be linked to the evaluation form for scoring and/or training purposes?
4. Within the scoring template, does the system provide the ability to aggregate questions based upon a group of similar questions, i.e. customer Service or Opening Questions?

If yes, can this information be used to run an agent or group performance report based upon the scores of the group of questions of the header? Describe how this is accomplished with your system.

5. Does the system allow the user to create, without vendor customization, multiple grading templates using questions provided by the customer?
6. Does the system have the capability to design evaluation forms by utilizing custom wizards?
7. Does the evaluation system allow for sub totaling of the evaluation questions? Can evaluation forms have multiple pages of scoring, and if so, how many?
8. Does the system have the capability of inserting notes on a per-question basis and a summary notes into the on-line grading form? Describe how the notes are later retrieved for reporting purposes.
9. Does the system provide the user with the ability to search the database of voice conversations and then select a specific voice conversation for evaluation by using only the software application?
10. Does the system permit the user to filter out those calls that have already been reviewed? Describe how the user would filter the calls to select only non-reviewed calls for playback.
11. Does the system provide the capability to calibrate evaluations, i.e. offer a second opinion?
12. Does the system allow three or more reviewers to review the same call at completely different times for the purposes of evaluating reviewer consistency? Describe the reports that can be created to compare the reviewers.

(d) Reporting
1. Does your system provide wizards to automatically generate custom reports?
2. Can report criteria be saved for later use to maximize database storage? I.e. does the reporting function store the entire report or just the report criteria?
3. Does your reporting function offer both text and graphical reports?
4. Can the format of the graphical reports be modified?
5. Does the system provide the capability of exporting the data from all reports to a text file to allow for the manipulation of the data as desired by the users?
6. Does the system provide an agent performance report with a visual graph that shows change over time on a daily, weekly or monthly basis or for any desired timeframe? How does a supervisor use it? Provide a printed copy of the report showing agent performance on a weekly basis for a one month time period.

7. Does the solution provide a report with a visual graph that shows how well an agent or group of agents has performed over time on a per question or group of questions basis (i.e.: report the scores for questions 1-5)? How does a supervisor use it? Provide a copy of the report, as it would look printed out.
8. Does the system provide a report with a visual graph based upon a custom field, such as all agents assigned the custom field of shift or customer service call? Is this assigned to the grading template at the time of scoring the call? How does a supervisor use it? Provide a copy of the report, as it would look printed out.
9. Does the system provide a report that includes a visual graph showing agent performance based upon a subset of questions represented by a heading? How does a supervisor use it? Provide a copy of the report, as it would look printed out.
10. Does the solution provide a report that includes a visual graph that compares how multiple reviewers scored a single agent over a period of time? How does a supervisor use it? Provide a copy of the report, as it would look printed out.
11. Does the system provide the ability to run a report that shows all of the notes that have been made by the reviewers while grading agents? How does a supervisor use it? Provide a copy of the report, as it would look printed out.
12. Does the system provide a report with a visual graph that shows the reviewer consistency by day, week or month for an agent or a group?

How does a supervisor use it? Provide a copy of the report, as it would look printed out.

13. Does the system provide the capability to run a report that shows how many calls have been recorded for an agent and how many calls the agent has reviewed? How does a supervisor use it? Provide a copy of the report, as it would look printed out.

(e) Management of Voice Conversations

1. Does the proposed solution allow the user to visually select a clip, a number of clips or a group of voice clips for playback by only using the software interface? Describe how this would be accomplished via the user interface and what kind of searches can used to retrieve the calls.

2. Describe how a user would accomplish a search to retrieve a call that meets the following criteria. All calls for agent = Tina, recorded using a recording schedule called "New Employees", reviewed by agents Tim and Susan, between May 1st and 31st, 20XX and that assigned a custom field of "good call" during the review of the call.

3. Does the system provide the ability to create folders that meet a set of user defined criteria (i.e. all calls recorded by group customer service that have not been reviewed) to visually group employees voice recordings to allow for simple retrieval? Describe how this is accomplished.

4. Does the system allow the user to search and playback calls by using only the software application to retrieve a call? Describe how this would be accomplished.

5. Does the system allow calls to be searched and retrieved based upon the following criteria:
   a. recorded group
   b. recording date
   c. recording schedule
   d. recording duration
   e. review status

6. Explain the different methods that can be used to select voice conversations for playback.

7. Does the system allow the user to specify individual clips that must be archived as voice conversations for training purposes for a 1 year time period? Described how this would be accomplished.

8. Does the system provide the option of logging out the user after a predetermined time period?

9. Does your system allow the creation of custom fields that can be associated with an evaluation form?

(f) Scheduling

1. Please describe your system scheduling capabilities? Does it support scheduling for the following CTI criteria?
   - Dialed Number
   - by DNIS or VDN
   - by ANI
   - Client ID

2. Does the proposed technology have the ability to program recording by time of week, time of day, etc.? Can schedules be defined on a daily, weekly, and monthly basis?

3. Does your system have the capability to schedule the amount of "wrap-up" time that is recorded?

4. Does your system have the capability to schedule minimum and maximum call lengths?

5. Does your system have the capability to record contacts from beginning to end?

6. Does your system have the capability to schedule random, percentage of, or sequential calls?

7. Does your system have the capability to schedule voice only, data only, or a synchronous combination?

(g) Monitoring Application

1. Does the system have the capability to view agent status? Does this feature show the status of the agents (the group which the supervisor has authorization to monitor), does it show, which agents are logged on, and which agents are being recorded at the moment (Voice, screen, or Voice + Screen)?

2. Does the system have the capability to start monitoring the selected agent for either (Voice, Screen, or Voice + Screen)?

3. Does the monitoring application have the capability to perform Record on Demand functions? Does your system have the capacity to perform the following functions?
   - Records one call - record the current call. Disabled if the agent is not on a call.
   - Record next calls - record the agent's next call.
   - Record continuously - record all agent's calls, starts on the current call, if exists, or on the next call, if the agent is not on a call.

4. Does your monitoring application have a system resource view for both audio and data?

5. Does your monitoring application have the capability to determine output channels?

(h) System Administration

    1. Does your system provide multiple levels of security access?

    2. Does your system allow for the definition of Groups, if so can users belong to multiple groups?

    3. Can privileges be customized on a per user basis?

    4. Can your system administration function upload user data from current databases, i.e. WFM (Blue Pumpkin) or employee database applications?

    5. Does your system support the custom definition of user fields in the System Administration function?

    6. Will this system work with a "Power" (FAT) Client

    7. Does the proposed solution offer a certain level of security as well as levels of accessibility, to allow team leaders to proactively exchange information with their agents without allowing access to the whole monitoring platform?

Implementation Process

1. Please describe your implementation process.

2. What is the shortest length of time you have implemented your solution?

3. What is the average length of time in which you have implemented your solution?

4. Please outline the responsibilities of the vendor's team members in the implementation process.

5. Please outline the responsibilities of the prospects team members/employees in the implementation process.

Training

1. What is the background of the people responsible for training the call center on the application? Detail their experience working in a call center environment.

2. What type of user training is available and can it be customized to meet the specific needs of our call center group?

3. Do you provide hands-on on-site training?

4. How many trainers do you have dedicated just to training?

5. How much training time (hours/days) is customary?

6. Describe a typical training program?

7. Is future training or customized training available? What type is available?

8. Do you offer refresher tools we can use on site such as books and e-learning software?

9. Do you provide a training program by which those who support the technical aspects of the software can be certified?

10. Do you offer different methods of delivering training, such as video conferencing, distance e-learning or other multi-media methods?

11. What additional training or consulting is available to help us smoothly implement this process into our organization?

12. Are agent orientation classes available?

e-Learning Capabilities

1. Do you have an e-learning application?

2. Does your e-learning system support remote access?

3. Please describe the basic technological functionality of your e-learning system.

4. How does it integrate with your existing quality monitoring application?

5. How is training delivered to CSR's via your e-learning application?

6. Does the system allow a supervisor to measure an individual CSR's skill levels?

7. Does the system allow for development of individualized learning plans?

8. Does your system utilize "off-the-shelf" learning management systems?

9. Does your e-learning system support multiple course authoring tools? If so, which ones?

10. Is the learning management system designed for open/public standards (i.e., AICC, IMS or SCORM)?

11. Does your system allow for measurement of productivity improvement in the:
    - Short-term?
    - Mid-term?
    - Long-term?

12. How does a supervisor/manager extract reports from your e-Learning system?

Automated After Call Voice Documentation Processes

1. Describe any ability your system has to be applied to sales verification or collection commitments?

2. Can subsequent processes be launched from key field completion in a relationship management system. Axxum uses custom, Hogan and DST systems as the primary call center support platforms.

3. Describe any other applications your company may recommend to serve Axxum applications?

Installation & Customer Support

1. How long does a typical installation require for your solution?
2. What kinds of resources are required of our company?
3. How many people are dedicated to your customer support organization? Describe the customer support capabilities.
4. What are the escalation procedures?
5. What is your software maintenance and support policy? Are free upgrades to the application included in maintenance agreement? Please provide a copy of the agreement.
6. Is remote diagnostic support available? Describe how it occurs.
7. Do you have a user group for your automated monitoring product? How often do they meet and where was the last meeting?
8. Do customers and users groups influence release of the product? Describe how.
9. Is telephone support provided? If yes, is this a toll-free service?
10. What is your guaranteed response time for user support?
11. What support is available for implementation?
12. Where is your help desk located and do you have field offices that are available to assist in the event of a problem? Where are the field offices?

Technical Requirements

1. Describe the ability to interface to the Avaya PBX to record the CSR's conversations?
2. Describe how you are able to record only complete conversations and not blocks of time?
3. Describe the archive capabilities and back-up/recovery capabilities.
4. Describe the various levels of security for this solution.
5. Is your system scaleable? Describe how for the various components.
6. Describe the ability to determine status and location of specific customer service agents in our free seating environment.
7. Provide operating systems supported and system configuration capacities.
8. What are the PC workstation requirements?
9. What are the server requirements?
10. Does the system continue to record when a silence occurs in the conversation?
11. Does your system scale to permit full time logging and monitoring? If yes, describe how this is accomplished.
12. Does a third party vendor manufacture your hardware and software?

13. Is your company Avaya certified?
14. Describe proposed system's screen capture technology.
15. Describe the proposed system's method of screen capture. Does the system capture "screen" scrapes or use "GDI"?
16. Describe how these screen capture processes are communicated to the recording server(s.)
17. Does the proposed system permit recordings to be scheduled and recorded sequentially?
18. Please describe the integration capabilities of your proposed recording solution with predictive dialing technology. Provide examples of your integrated solution with different predictive dialing solutions.
19. Provide documentation on your proposed solution's architecture. How open is your architecture?
20. The vendor should document all aspects relating to integration with the telephone switch, the back-up functions, the different levels of security and the scalability of the proposed solution.
21. How many physical server devices does your company recommend for this application?

Pricing for the Proposed System

1. For each software product, consulting service, training class, or other item, which is part of the vendor's basic proposal, please provide the following information:
    a) The vendor must provide the per Axxum or license price of the product or service and any other one-time costs.
    b) The vendor should include any recurring charges such as maintenance or continuing training.
    c) In addition please answer the following questions:
    d) Does this pricing include any of the following:
        • Incremental costs for upgrades and expansions?
        • Maintenance costs?
        • Training materials?
        • Documentation (number of copies)?
        • On-line help desk?
2. Identify any other products that must be licensed / per Axxum for use with the package or any of its components.
3. Describe any Desktop/LAN equipment required, but not included, in the solution pricing (i.e. local disk space, memory, operating system, network

connectivity standards/protocols supported, dial-up modem for support, dedicated server, storage, etc.).

Warranty Programs

1. Please attach sample copies of the standard license and service agreements offered with your proposed solution.

Appendices

2. Any additional documents you would like to include in this response may be added in the appendices.

Glossary

A

ACD — Automatic Call Distributor — This is the device that distributes calls as work-flow tasks to specific staff members trained to serve the caller and the specific subject of the call.

ACD gateway — An interflow control or the connection between ACD systems that allows one ACD system to forward a request to another ACD. This can be inband (on the actual trunk to be used by the call) or out-of-band (a data signaling channel only) The ACD gateway is configured as a step in a routing script.

ACD Interflow — The ability to offer a waiting call at center A to any other center in the enterprise. Optimally (ACD dependent) the queued customer should be recognized by all of the centers in the network for the first appropriately skilled agent to serve.

Abandoned call — This occurs when a caller hangs up before the call is answered. Calls in which the caller hangs up almost immediately do not have to be counted as abandoned but should be tracked as potential trunk signaling problems.

About box (Windows) — A dialog box that displays general information about an application. The About box usually contains copyright and version information. In most applications, you can invoke it from the Help menu.

Activity Codes — Originally this referred to the "after call work" codes that are entered by the agent into the ACD instrument so some call type or disposition state can be tabulated. These are also called after call work, wrap up or disposition codes. More recently these codes are collected in the departmental CRM or service system.

Adherence — This is the matching of the attendance record to the employee work schedule. This refers to how well agents to follow or "adhere" to their schedules. This can include such statistics as: Hours at work, hours signed on to the ACD and how much time they were available to take calls during their shifts, including the time spent handling calls and the time spent waiting for calls to arrive (also called Availability) and when they were available to take calls (also called Compliance or Adherence).

Administrative Workstation (AW) — A PC used for system supervision to manage the call handling of the ACD, or the group or split in the ACD system. This workstation is typically configured as the administrative system for configuration of resource tables (like agent skill sets,) call processing and routing tables.

After-Call Work (ACW) — This is state following the "release" of a call. Also called Wrap-up and Post Call Processing (PCP). Work that may be necessary following an inbound transaction. This often includes entering of data, filling out forms and making outbound calls necessary to complete the transaction. During the ACW state, the agent is unavailable to receive another inbound call.

Agent — A generic term for a staff member who is tasked with answering incoming phone calls. Each agent is associated with an ACD phone terminal and can be a member of one or more groups, defined by application, call type or skill set. Airlines called them reservation sales agents thus the compression became "agents."

Agent Group — The agent group is a group of agents typically tasked with serving specific types of calls. On the trunk side of the switch trunks carrying a similar call type are called splits, gates or a queue. A collection of agents share a common set of skills, such as being able to handle customer complaints may also called a "skills" group.

Agent Out Call — An outbound call placed by an agent.

Agent Status — The mode an agent is in (available, talk, after-call work, unavailable).

All Trunks Busy (ATB) — The designated state of a trunk group when all trunks are in use. The trunk group cannot accept any new inbound or outbound calls in this state. The ACD tracks the amount of time during which all trunks in a trunk group are busy. From this snap shot, frequency and length of ATB, the required trunks that should be added can be estimated. This information is available historically or in real-time.

Analog — A term applied to the first generation of transmission of the original energy wave format of telephone connection. Signals are analogous to the original signal.

The second generation was defined as digital, or the digitizing of these waves.

Announcement — A recorded announcement played to an inbound caller as a delay message.

Answer Supervision — The answer acknowledgment signal (voltage reverse in an analog world) sent by the ACD or other device to the local or long distance carrier to accept a call. This is when billing for either the caller or the call center will begin, if long distance charges apply.

Answer delay — The elapsed time the point of inbound supervision (request for connection) or the call being offered to the switch by the CO at the Customer Premise Equipment (Key, PBX or ACD device) till return supervision or answer.

Answered call — A call is counted as answered when it reaches an agent or IVR. This does not necessarily mean served. A call is not counted as handled until it is finished. Therefore, the number of answered calls and handled calls during an interval is not necessarily the same, as they could still be abandoned before reaching a point of completion.

API or Applications Program Interface — Essentially a set common of common rules that allow software to work together as if it has a "plug-in" interface. Two API compatible programs expect common control signals and data.

Application Based Routing & Reporting — The ACD capacity to route and track transactions by type of call or application e.g. sales, service etc. versus the traditional method of routing and tracking by trunk group and agent group.

Application Bridge (AB) — The Application Bridge interface (proprietary) from an Aspect ACD allows other applications to connect monitor and or control the ACD. More widely adopted proprietary links are CT Connect (Dialogic) and T-Server (Genysis).

Application Gateway Software — Application that allows an ICR routing script to invoke an external application.

Application Generator — A tool that generates application software based on the problem or desired solution.

Applications Service Provider — an outsourcing company that provides access to complex software applications via the Internet or a VPN.

Architecture — The basic design of a system. Determines how the components work together, system capability, ability to upgrade and the ability to integrate with other systems.

Area code — A three-digit (North America) prefix used to indicate the destination area for long distance calls. Can be used to classify calls into call origination types by geography.

Audiotext — A voice processing capability that enables callers to automatically access pre-recorded announcements.

Auto Available — An ACD feature whereby the ACD is programmed to automatically put agents back into the available state after they have released a call or completed predetermined after call work.

Automated Attendant — A set of applications in a voice response unit that can triage predictable calls automatically. The definition says it all. Ensure pushing "o" is the known escape to a live agent.

Automated Greeting — An Agent specific pre-recorded greeting that plays automatically when a call arrives. "Thank you for calling, this is Nname. How may I help you?"

Automatic Wrap-Up — An ACD feature whereby the ACD is programmed to automatically put agents into after-call work following the release of a call (type?).

Automated Attendant — A voice processing capability that automates the attendant function. The system prompts callers to respond to choices e.g. press one for Service, two for Sales. The Automated attendant then maps the request to the ACD group and dials that group or extension number. This function can now reside on a board in the ACD, an external on-site system or in the long distance network.

Automatic Call Distributor (ACD) — A switch at a call center that routes incoming calls to targets within that call center. After the call routing rules determine the target for a call, the call is sent to the ACD group associated with that target.

Automatic Call Sequencer (ACS) — A simple system that is less sophisticated than an ACD, but provides some ACD-like reporting. The significant differentiator is the inability to force an agent into available and to answer. On an ACS, this decision remains with the agent.

Automatic Number Identification (ANI) — A feature that provides the calling phone number which a call originated the call. On some ACD systems with intelligence that can map the caller number to a database, this can effect routing to a table as specific as this customer.

Auxiliary Work State — An agent work state that is typically not associated with handling telephone calls. When agents are in auxiliary mode, they will not receive inbound calls. This is particularly important with overflow states and integrated email correspondence states.

Available State — The state where the agent is ready to accept calls, but is not currently involved in call work typically because no qualifying calls were available.

Available Time — The total time in seconds that an agent was in the Available State.

Average Delay — The average delay of calls that are delayed. It is the total delay for all calls divided by the number of calls that had to wait in queue.

Average Handle Time (AHT) — The average time it took for a call to be handled. Handle time is talk time plus after-call work time.

Average Trunk Holding Time (ATHT) — The average time inbound transactions occupy the trunks. It is (talk time + delay time)/calls received.

Average Number of Agents — The average number of agents logged into a group for a specified time period.

Average Speed of Answer (ASA) — The average answer-wait time for calls to a service or route. This should include: ring time (often not) announcement time and total queue time.

Average Time to Abandonment — The average time that callers wait in queue before abandoning. The calculation considers only the calls that abandon.

B

Base Staff — Also called basic staff complement or "butts in seats." The minimum number of agents required to deliver specific service level and response time objectives for a given period. Seated agent calculations assume that agents will be 'in their seats' for the entire period of time. Therefore schedules need to add in extra people to accommodate breaks, absenteeism and other factors that will keep agents from the phones.

Basic Rate Interface (BRI) — One of two levels of ISDN service. The BRI provides two bearer channels for voice and data and one channel for signaling (commonly expressed as 2B+D).

Beep Tone — An audible notification that a call has arrived (also called Zip Tone). Beep tone can also refer to the audible notification that a call is being monitored.

Benchmark — Historically, a term referred to as a standardized task to test capabilities of devices against each other. In quality terms, benchmarking is comparing products, services and processes with those of other organizations to identify new ideas and improvement opportunities.

Glossary

Best-in-Class — A benchmarking term to identify organizations that outperform all others in a specified category.

Blockage — A state where a call is available and cannot reach the desired destination because the resources are not available to make the connections. These may be physical or logical resources.

"Blocked Business" — This telephony concept of the blocked call is most illustrative of a simple concept that also applies to any task; that is the resources are not available to bring the transaction to a successful conclusion. Access to the business is "blocked" by a lack of resources, tools, skills or specific offer that meets with success. Understanding these issues are vital to succeeding in future opportunities.

Blocked Call — A call that cannot be connected immediately because:

a. Unintended: No resources (agent, talk path or connectivity) is available at the time the call arrives. b. Intended: The ACD is programmed to block calls from entering the queue when the queue backs up beyond a defined threshold.

Measuring both conditions is essential to understand how much business is being "blocked."

Boolean expression — An expression that evaluates to TRUE or FALSE and used in certain decision trees. If the queue is longer than five calls (True or False?) deliver a busy signal.

Busy Hour — A telephone traffic engineering term, referring to the hour of the day in which a trunk group carries the most traffic. Typically mid morning or mid afternoon in a business day. The average busy hour represents approximately 17% of the daily traffic. Busy hour is vital to call center staffing calculations, as this is the basic criterion for staffing. Lost calls in a call center means lost business. In a administrative telephone network, engineering to this is overkill but some level of network resource queuing can make this potential blockage transparent to your administrative staff.

Busy Signal — A telephone signal that indicates no connection can be made.

Busy State — The state in which agents in a skill group are busy in other skill groups i.e. in skill groups other than the one presently being examined. For example, an agent might be talking on an inbound call in one skill group while simultaneously logged on to and ready to accept calls from other skill groups. The agent can be active (talking on or handling calls) in only one skill group at a time. Therefore, while active in only one skill group, the agent is considered by the other skill groups to be in the Busy Other state.

C

CDR — Call Detail Recorder — The device that records each detail of a call, typically date, time begun, ended, digits dialed (or received,) origination and destination data, (from trunk to agent in an ACD or station and trunk in a PBX call accounting application.) Call detail is contrasted with peg count or call event summaries.

Call — Typically a term referring to telephone calls. This has been expanded to include video calls web calls and other types of contact. The UNIX OS world has bled over into telephony causing some confusion as a system request to the operating system for service is also termed a "call."

Call Blending — The blending of inbound and outbound call workflow and delivery to a single agent or group. A system that is capable of blending automatically puts agents who are making outbound calls into the inbound mode and vice versa, as required by the incoming call demand.

Call-by-call Routing — Each incoming call is individually processed to a call destination based on call criteria specific to this call. This may be network based real-time information about the state of each call center as well as CTI or VRU appended data (network, caller or database derived.) By contrast, simple call routing statistically distributes calls among call centers based on general historical patterns.

Circuit-Switched — A switching methodology where two ports are connected via a dedicated path or "circuit" for the duration of the communication, through signal and silence brought about by gaps in the conversation (people or machines.) This is not as bandwidth efficient as packet-switched systems. Traditional central offices, ACD and PBX systems are circuit switched. "Wired and tired!"

Customer Call Center — A single site at which incoming phone calls are received and answered. Typically, each call center can provide several services and is staffed by agents from one or more skill groups. This term implies call processing only and has undergone "scope creep" to the become the customer contact center include other text-based real-time or delayed customer requests such as email, web browsing, web call backs and traditional correspondence (mail or fax).

Call Control Variables — A set of variables used by a peripheral to hold information related to a call. When client is established you may pre-select how the client's call will be handled based on intelligent rout variables such as dialed number, caller-entered digits and calling line ID.

Call Detail — Data saved by the about every call. It routes and calls that terminate at each peripheral. (Route) call detail describes how intelligent call routing processed the call. Termination call detail describes how a call was handled at a peripheral.

Call Detail Recording — Data on each captured and stored by the ACD. Can include trunk used, time in queue, call duration, agent who handled the called, any numbers dialed, (inbound, outbound or wrap-up data) and other information.

Call Forcing — An ACD feature that automatically delivers calls to agents who are available and ready to take calls. They hear a notification that the call has arrived e.g. a beep tone, but do not have to actively accept the call to answer the call.

Call Load — Also referred to as Total Workload. Call load is the product of (average talk time + average after call work) x call volume, for a given period.

Call Routing — The main part of the ICR system. The Call Router receives call routing requests and determines the best destination for each call. It also collects information about the entire system.

Call type — A category of incoming calls. Calls are categorized based on dialed number (DN), caller-entered digits (CED) and calling line ID (CLI). Each call type has a schedule that determines which routing script or scripts are active for that call type at any time.

Caller ID — See ANI.

Caller-Entered Digits (CED) — Network or caller derived identification data. Digits entered by a caller on a touch-tone phone in response to prompts. Either a peripheral (ACD, PBX, or IVR) or the carrier network can prompt for CEDs.

Calling Line ID (CLI) — Information about the billing telephone number from which a call originated.

Calls abandoned — The number of calls abandoned during an interval.

Calls answered — The number of answered calls during an interval.

Calls handled — The number of handled calls during an interval.

Calls in progress — The total number of calls to a route or service that are on-line, in queue or being handled at the peripheral now.

Calls in queue — The number of calls to the service or route that are in queue at a peripheral.

Calls in queue now time — The total time spent in queue for all calls currently in queue for a service or route.

Calls incoming — The number of incoming calls during an interval.

Calls offered — The number of offered calls during an interval.

Calls routed — The total number of calls routed to a service or route during an interval. This data is also recorded for scripts as the number of calls routed by the script during a five-minute interval.

Carrier — A company that provides telecommunications circuits. Carriers include the local telephone company and companies like AT&T, MCI and Sprint.

Cause & Effect Diagram — A tool to assist in root cause identification developed by Dr Kaoru Ishikawa.

CCSS7 — Common Channel Signaling System 7. The protocol used by the Carrier networks.

CD-ROM — Compact Disc Read Only Memory. These discs hold as much as 660 megabytes of memory.

Central Controller — The computer or computers running the Call Routing and the ACD Database Server (Logger). In addition to routing calls, the Central Controller maintains a database of data collected by the Peripheral Gateways (PGs) and data that the Central Controller has accumulated about the calls it has routed.

Central Database — The relational database stored on the Central Control, which stores historical five-minute and half-hour data, call detail records, ACD configuration data and call routing scripts.

Central Office (CO) — Can refer to either a telephone company switching center or the type of telephone switch used in a telephone company switching center. The local central office receives calls from within the local area and either routes them locally or passes them to an inter-exchange carrier (IXC). On the receiving end, the local central office receives calls that originated in other areas, from the IXC.

CCS or Hundred Call Seconds — A hundred call seconds, a unit of telephone traffic measurement. The first C is the Roman numeral for 100. One hour = one Erlang = 3600 called seconds, sixty minutes = 36CCS.

CCMS — Call Management System — A reporting package used on ACDs and PBXs.

Collateral Duties — Non-phone tasks e.g. data entry, which are flexible and can be scheduled for periods when call load is slow.

Computer Telephony Integration (CTI) — The method of connecting computers and ACDs to capture call info for screen popping or to allow the computer to control the switch routing.

Conditional Routing — The capability of ACD to route calls based on current conditions. It is based on 'if-then' programming statements e.g. 'if the number of calls in agent group one exceeds ten and there are at least two available agents in group two, then route the calls to group two'.

Configuration data — The description of your enterprise telephone system. This includes, for example, the peripherals, agents, skill groups, services, call types and geographical regions that you use. The ACD must have complete and up-to-date configuration data in order to route calls properly.

Configuration lock — A construct that allows only one person to make changes to configuration data at one time. Only the holder of the lock can make changes to the configuration data or to the call organization data.

Configuration registry — A repository of configuration data maintained by the Windows NT operating system. All applications on a computer can access the registry to store and retrieve configuration information. You can use the Windows NT Registry Editor (regedt32.exe) to view or modify data in the registry.

Controlled Busies — The capability of the ACD to generate busy signals when the queue backs up beyond a programmable threshold.

Cost Center — An accounting term that refers to a department or function in the organization that does not generate profit.

Cost of Delay — The money you pay to queue callers, assuming you have toll-free service.

Customer Contact — Customer transaction.

CTI Gateway — The process that acts as a server for the CTI clients to communicate with the ACD. The CTI Gateway process may run on the same computer as the Peripheral Gateway process or on a separate computer.

Current Call — The call that the ACD (or a specific script) is currently processing. The ACD receives a routing request for a call and locates a target to which the call can be routed.

Customer — The organization whose calls the ACD routes. A customer may have several semi-autonomous units or business entities. Optionally, the ACD can be partitioned so that business entities can operate independently. In the standard configuration, an ACD routes calls for a single customer. In a service bureau configuration, a multiplex ACD system may service multiple customers.

D

Database Call Handling — A CTI application, whereby the ACD works in sync with the database computer to process calls, based on information in the database. E.g. a caller inputs digits into a voice processing system, the database retrieves information on that customer and then issues instructions to the ACD on how to handle the call, e.g. when to route the call, what priority the call should be given in the queue, the announcements to play, etc.

Database Server — The part of the ACD system that stores information about the entire system in the central database. The Database Server (maintains the data that is used in reporting and making routing decisions. The Database Server is also called the Statistics PC.

Day of Week Routing — A network service that routes to alternate locations, based on the days of the week. There are also options for day of year and time of day routing.

Default call type — The call type the ACD uses if a call's qualifier is not available.

Delay — Also called Queue time. The time a caller spends in a queue, waiting for an agent to become available. Average delay is the same thing as average speed of answer.

Delay Announcements — Recorded announcements that encourage callers to wait for an agent to become available, remind them to have their account number ready and provide information on access alternatives. In some systems, delay announcements are provided through recorded announcements routes (RANs).

Delay in Queue — The sum of time that calls spent in the queue for a route or service. Delay in queue can also take into consideration abandoned calls.

Delay time — The time spent processing a call after it arrives at a peripheral, but before it is either queued or presented to an agent.

Dialed Number (DN) — The number that a caller dialed to initiate a call e.g. 01403 214400. You can define a set of internal dialed numbers understood by the ACD and map them to actual dialed number strings passed by routing clients.

Dialed Number Identification Service (DNIS) — A string (usually three or four characters long) is indicating the number dialed by a caller and how the ACD, PBX or IVR should handle the call. The ACD uses the DNIS and trunk group to indicate the destination for a call.

Direct Inward Dialing/Dialed Number Identification Service (DID/DNIS) — When a call arrives at an ACD or PBX, the carrier sends a digital code on the trunk line. The switch can read this code to determine how it should dispatch the call. Typically, this value is the specific number dialed by the user. By mapping each possible code with an internal extension, the switch can provide direct inward dialing (DID). The ACD uses the DID/DNIS value to specify the service, skill group, or specific agent to whom the switch should route the call. The switch reads the value from the trunk line when the call arrives and dispatches the call appropriately.

DPNSS — Digital Private Network Signaling System — Allows switches to connect to other locations.

Dual-Tone Multifrequency (DTMF) — the tonal selection generated by a 12 key tone dial pad or touch tones. A signaling system that sends pairs of audio frequencies to represent digits on a telephone keypad. It is often used interchangeably with the term Touchtone (an AT&T trademark).

Dynamic Answer — An ACD feature that automatically reconfigures the number of rings before the system answers calls, based on real-time queue information. Since costs don't begin until the ACD answer calls, this feature can save callers or the call center money when long distance charges apply.

E
_____

E1 — The European equivalent of a T1. This differs from a T1 because it has 32 versus 24 conversation paths.

Enterprise — An entire company or agency, possibly spanning many call centers. Enterprise refers to the set of call centers served by the ACD.

Enterprise-wide call distribution — A strategy for allocating calls among several call centers or other answering locations based on real-time information about activity at each location. The ACD implements enterprise-wide call distribution and allows calls to be sent to any network-addressable location within, or outside of, an enterprise.

Envelope Strategy — A strategy whereby enough agents are scheduled for the day or week to handle both the inbound call load and other types of work. Priorities are based on the inbound call load. When call load is heavy, all agents handle calls, but when it is light, some agents are reassigned to work that is not as time sensitive.

Erlang — One hour of telephone traffic in an hour of time e.g. if circuits carry 120 minutes of traffic an hour, that is two Erlangs. Named after the statistician K.K. Erlang.

Erlang B — A formula developed by A K Erlang, widely used to determine the number of trunks required to handle a known calling load during a one hour period. The formula assumes that if callers get busy signals, they go away forever, never to retry ('lost calls cleared'). Since some callers retry, Erlang B can underestimate trunks required. However, Erlang B is generally accurate in situations with few busy signals.

Erlang C — Calculates predicted waiting times (delay) based on three things:

1. The number of servers (reps). 2. The number of people waiting to be served (callers). 3. The average amount of time it takes to serve each person. It can also predict the resources required to keep the waiting times within targeted limits. Erlang C assumes no lost calls or busy signals, so it has a tendency to overestimate staff required.

Error Rate — Either the number of defective transactions or the number of defective steps in a transaction.

Escalation Plan — A plan that specifies actions to be taken when the queue begins to build beyond acceptable levels.

Event Code — Also referred to as a wrap up code. This is a field that can be entered by an agent after the call to tag the call as a specific type, disposition or other outcome that requires later analysis. Originally this was a valuable ACD feature that should migrate to the department database system.

Ethernet — Local area network transmission protocol. This protocol uses an "open line" transmission concept where the "loudest" device gets priority so significant retries exists to complete a client/server exchange.

Expected delay — The ACD's predicted delay for any new call added to a service or route queue. The expected delay value is valid only if no agents are available for the route or service.

Exchange Line — local trunk service.

F

FAQ — A frequently asked question. This amounts to a predictable question that can be typically answered with a "canned " response, that can then be loaded onto a self service web site.

Fault-tolerant architecture — A design that allows a system to continue running after a component of the system has failed. Most ACDs includes several levels of fault tolerance that minimize time when the system is non-responsive to call routing requests. The ACD fault-tolerant architecture eliminates single points of failure and provides disaster protection by allowing system components to be geographically separated.

Fast Clear Down — A caller who hangs up immediately when they hear a delay announcement.

Fax on Demand — A system that enables callers to request documents using their telephone keypads. The selected documents are delivered to the fax numbers they specify.

Fiber optic Cable — Bundled glass fibers where each fiber allows the light pulses to represent a discrete transmission. In a simplified manner, each discrete packet of light pulses (on/off = 1/0) on each fiber is a device-to-device digital transmission. Optical data needs to be converted to electronic (digital) data then an analog signal to be a useful voice call. Lots of magic is used here.

Five-minute interval — Certain statistics within the ICR database are updated at five-minute intervals. The first such interval for each day begins at 12:00 midnight and ends at 12:05 A.M. The date and time at the start of the five-minute interval is saved with the data. This allows you to look back at data from previous five-minute intervals.

Forecast — An expected or desired set of standards for a service over a specified period of time. This includes the number of calls offered, the number of calls handled, average talk time, and so forth.

FreePhone Service — The British equivalent of North American 800 service.

Full-time equivalent (FTE) — The number of full-time agents that would be required during a period to perform the work done during that period. To calculate the FTE, divide the number of seconds of work performed by the number of seconds in the period. For example, if agents spent a total of 7200 seconds handling calls during a half-hour (1800 second) interval, the FTE for call handling during the interval is 7200 person-seconds / 1800 seconds = 4 persons.

This means that if all agents spent full-time handling calls during the interval, the work could have been done by four agents.

G

GUI — Graphical User Interface, pronounced "Gooey." Windows or Mac OS are GUIs that remove the user from the complexity of a C···> prompt world of textual computer commands.

Gateway — A server dedicated to providing access to a network.

Grade of Service — The probability that a call will not be connected to a system because all trunks are busy. Grade of service is often expressed as "p.01" meaning

1% of call will be "blocked". Sometimes, grade of service is used interchangeably with service level, but the two terms have different meanings.

H

H.323 — The H.323 spec defines packet standards, which provide a foundation for audio, video and data communications across IP-based networks, including the Internet. Gateway systems that convert data voice to PCM voice are often termed "H.323 gateways".

HTML — Hypertext Markup Language — An early web page generation language/tool. Predecessor is XML or Extensible Markup Language.

Half-hour interval — Half-hour statistics within the ICR database are updated at 30-minute intervals. The first such interval for each day begins at 12:00 midnight and ends at 12:30 A.M. The date and time at the start of the 30-minute interval is saved with the data. This allows you to look back at data from previous 30-minute intervals.

Handle time — The time an agent spends talking on an inbound call and performing after-call work.

Handled calls — A call is counted as handled when the call is finished. The calls might have been answered before the interval began. By contrast, a call is counted as answered as soon as it reaches an agent or IVR. Therefore, the number of handled calls and answered calls during an interval is not necessarily the same, but eventually each answered call is counted in both categories.

Help Desk — A term that generally refers to a call center set up to handle queries about product installation, usage or problems. The term is most often used in the context of computer software and hardware support centers.

Historical data — Data collected at five-minute and half-hour intervals and stored in the ACD central database.

Historical Reports — Reports that track call center and agent performance over a period of time. Historical reports are generated by ACDs, third party ACD software packages, and peripherals such as VRUs and call detail recording systems. The amount of history that a system can store varies by system.

Holding Time — Most correctly, the duration of a hold state during a call, ("May I put you on hold?") This is often confused with the time a caller is "held" in a queue.

I

Idle — Another name for the Not Ready state.

Incoming call — A call offered to a route or service from an external carrier.

Incoming Call Center Management — The art of having the right number of skilled people and supporting resources in place at the right time to handle an accurately forecasted workload at service level and with quality.

Incremental Revenue (Value) Analysis — A methodology that estimated the value (cost and revenue) of adding or subtracting an agent.

Index Factor — In forecasting, a proportion used as a multiplier to adjust another number.

Inter Exchange Carrier (IXC) — A long-distance telephone company.

Interflow — Intelligent overflow between remote ACD customer contact centers or systems.

Internal Help Desk — A group that supports other internal agent groups (e.g. for complex or escalation of calls).

Internal Response Time — The time it takes an agent group that supports other internal groups (e.g., for complex or escalated tasks) to respond to transactions that do not have to be handled when they arrive (e.g. correspondence or e-mail).

Integrated Services Digital Network (ISDN) — An international standard for telephone transmission. ISDN provides an end-to-end digital network and provides a standard for out-of-band signaling. It also provides greater bandwidth than older telephone services. The two standard levels of ISDN are the Basic Rate Interface (BRI) and the Primary Rate Interface (PRI).

Intelligent Call Processing (ICP) — The facility that allows third-party products such as Intelligent Call routing to pre-route calls.

Intelligent Call Routing (ICR) — The System that implements enterprise-wide call distribution across call centers. The ICR provides Pre-Routing, Post-Routing, and performance monitoring capabilities.

Interactive Voice Response Unit (IVR) — A telecommunications computer, also called a Voice Response Unit (VRU) that responds to caller entered touch-tone digits. The IVR responds to caller entered digits in much the same way that a conventional computer responds to keystrokes or a click of the mouse. The IVR uses a digitized voice to read menu selections to the caller. The caller then enters the touch-tone digits that correspond to the desired menu selection. The caller entered digits can invoke options as varied as looking up account balances, moving the call within or to another ACD, or playing a pre-recorded announcement for the caller.

Internet Protocol (IP) — The connectionless-mode network service protocol of TCP/IP. IP enables the entities in a network to communicate by providing IP addresses and by numbering and sending TCP data packets over the network. NICs, PGs, and Admin Workstations in the ICR system use IP to communicate over a wide area network.

Internet "Call Me" Transaction — A transaction that allows a user to request a callback from the call center, while exploring a Web Page.

Internet "Call Through" Transaction — The ability for callers to click a button on a Web site and be directly connected to an agent while viewing the site.

Internet Phone — Technology that enables users of the Internet's World Wide Web to place voice telephone calls through the Internet, thus by-passing the long distance network.

Invisible Queue — When callers do not know how long the queue is or how fast it is moving.

J

Jitter — This is the term used to describe what a user sees on a data screen. Packets arriving in a "disordered" sequence cause this. Crudely, "-lo arrives before "Hel-." The precise addressing and clocking of packets, in a transmission takes time, as does reassembly as a continuous voice or video call. This accounts for some delay and lower fidelity.

Job Scheduler — A tool that allows you to set up specific commands to be executed automatically at given dates and times. You can schedule a command to execute once, on several specific days, or regularly on a weekly or monthly schedule.

Judgmental Forecasting — Goes beyond purely statistical techniques and encompasses what people believe is going to happen. It is in the realm of intuition, inter-departmental, committees, market research and executive opinion.

K

Key — An entry in a database index. Each key in the index corresponds to a table row and is composed of specific column values from that row.

Glossary

L

Latency — This refers to the delay that is imposed by the distance or route (or a combination thereof) that a transmission encounters. This is critical in real-time transactions.

Local Area Network (LAN) — The connection of several computers within a building, usually using dedicated lines.

Local database — A database on the Admin Workstation that contains information copied from the central database.

Local Exchange — The switching center of the local telephone company. The local exchange receives calls from within the local area and either routes them locally or passes them to an inter exchange carrier (IXC). On the receiving end, the local exchange receives calls that originated in other areas from the IXC.

A Local Exchange trunk connects a call center directly with their Carrier.

Log file — A file used to store messages from processes within a system.

Logged On — A state in which agents have made their presence known to the system, but may or may not be ready to receive calls.

Longest Delay (Oldest Call) — The longest time a caller has waited in queue before abandoning or reaching an agent.

Longest Idle Agent — The agent that has been continuously in the Available state for the longest time. The ICR can examine services or skill groups from different peripherals and route a call to the service or group with the longest Waiting agent.

Look Ahead Queuing — The ability for a system or network to examine a secondary queue and evaluate the conditions before overflowing calls from the primary queue.

Look Back Queuing — The ability for a system or network to look back to the primary queue after the call has been overflowed to a secondary queue, and evaluate the conditions. If the congestion clears, the call can be sent back to the initial queue.

M

MIS or Management Information Systems — a system that provides real-time and historical data about some systems' performance. This may be about the employment and performance of resources attached to that system such as an automatic call distributor: Trunks, queue slots, IVRs and people.

Metcalf's Law — Named after the inventor of the LAN. The law is the value of a network is proportional to the number of participants.

Middleware — Software that mediated between different types of hardware and software on a network, so that they can function together.

Mirroring — An arrangement by which changes to one storage device are automatically written to a similar device. For example, you can set up a disk as a mirror of another disk so that all writes to one disk are also automatically written to the other. This allows for recovery from media failure. Mirrored disks are an alternative to a RAID configuration.

Monitor ACD — A tool that allows you to monitor real-time and historical activity within your enterprise. Monitor ACD allows you to build reports from templates or view reports you have saved previously. You can also view events generated by ACD processes.

Monitoring — Also call Position Monitoring or Service Observing. The process of listening to agents' telephone calls for the purpose of maintaining quality. Monitoring can be; A) silent, where agents don't know when they are being monitored, B) side by side, where the person monitoring sits next to the agent and observes calls or C) record and review, where calls are recorded and then later played back and accessed.

N

Network Interface Card (NIC) — The card inside a PC allowing it tot be connected to a local area network LAN.

Network Control Center — Also called Traffic Control Center. In a networked call center environment, where people and equipment monitor real-time conditions across sites, change routing thresholds as necessary, and coordinate events that will impact base staffing levels.

Network Interflow — A technology used in multi-site call center environments to create a more efficient distribution of calls between sites, Through integration of sites using network circuits (such as T1 circuits) and ACD software, calls routed to one site may be queued simultaneously for agent groups in remote sites.

Network trunk group — A group of trunks to which a routing client can direct calls. A peripheral may divide its trunks into trunk groups differently than the routing client does. Simple trunk groups describe the peripheral's view of the trunks; network trunk groups describe the routing client's view of the trunks.

Next Available Agent (NAA) — A strategy for selecting an agent to handle a call. The strategy seeks to maintain an equal load across skill groups or services.

Noise Canceling Headset — Headsets equipped with technology that reduced background noise.

Non ACD in Calls — Inbound calls that are directed to an agent's extension rather than to a general group. These may be personal calls or calls from customers who dial the agents' extension numbers.

Not Ready state — A state in which agents are logged on but are neither involved in call handling activity nor available to handle a call.

0

Occupancy — Also referred to as agent utilization. The percentage of time agents handle calls versus wait for calls to arrive. For a half-hour, the calculation is: (call volume x average handling time in seconds) / (number of agents x 1800 seconds).

Offered call — An incoming call or internal call sent to a specific route or service. In real-time data, a call is counted as offered as soon as it is sent to the route or service. However, if the caller hangs up before the abandoned call wait time has elapsed, that call is not counted as offered in the historical (five-minute and half-hour) data. This ensures that the number of calls offered is the same as the number answered plus the number abandoned.

Off-Peak — Periods of time other than the call center's busiest periods. Also a term to describe periods of time when long distance carriers provide lower rates.

Open Database Connectivity (ODBC) — A standard application programming interface (API) that allows a single client application to access any of several databases. Because most ACD databases support ODBC, you can query them from any third-party client tool that also supports ODBC.

Open System — A term used to describe a manufacturer independent system. Buy the appropriate hardware type from any source and run the standards based application on your preferred brand or choice.

Open Ticket — A customer contact (transaction) that has not been completed or resolved (closed).

Outbound Agent — An agent that only makes outbound calls.

Outsourcing — Contracting some or all call center services to an outside company.

Overflow — When a trunk group or agent group is busy the ACD will pass an offered call to the next group in line.

P

PCM — Pulse Code Modulation. The most common method of encoding an analog voice signal into a digital bit stream.

Packet-Switched — This is the opposite of circuit switching. Now each packet is "smart" enough to find its way across the network (potentially thousands of alternate paths) from one IP address to any other. The packet has address, clock and origination/destination data attached to it, to allow routing and arrival at the intended device. This is the magic of the Internet.

Peak Call Arrival — A surge of traffic beyond random variation. It is a spike within a short period of time.

Percent Allocation — A call routing strategy sometimes used in multi-site call center environments. Calls received in the network are allocated across sites based on user-defined percentages.

Percent utilization — The percent utilization is computed by dividing the total time agents spent handling calls by the total time agents were ready. (The ready time is calculated by subtracting the Not Ready time from the total time that agents were logged on.) The value is expressed as a percentage.

Peripheral — A switch, such as an ACD, PBX, or IVR, that receives calls that have been routed by the ACD.

Poisson — A formula sometime used for calculating trunks. Assumes that if callers get busy signals, they keep trying until they successfully get through. Since some callers won't keep retrying, Poisson can overestimate trunks required.

Pooling Principle — The pooling principle states: any movement in the direction of consolidation of resources will result in improved traffic-carrying efficiency. Conversely, any movement away from consolidation or resources will result in reduced traffic carrying efficiency.

Post Call Processing — After call work.

Power Dialer — automatically dials a sequence of numbers and presents the live answers to agents. The system can sort out and discard any other outcome such as a busy, no-answer, answering machine etc, for later retry. When a live person answers and an agent is not available, a "nuisance call" is said to occur.

Predictive Dialer — Is a more sophisticated system than a power dialer automatically places outbound calls and delivers answered calls to agents based on the predicted agent availability. The "nuisance rate" is calculated to as near as zero as possible.

Preview Dialing — screen based dialing with the ability of the agent to preview the screen before launching the call.

Progressive Dialing — the system dials progressively, launching the next dialing event when a certain closing stage and screen field is reached in the current call.

Private Automatic Branch Exchange (PABX) or Private Branch Exchange (PBX) — A telephone system located at a customer's site that handles incoming and outgoing calls. ACD software can provide PBXs with ACD functionality. Also called Private Automatic Branch Exchange.

Profit Center — An accounting term that refers to a department or function in the organization that generated profit.

Peripheral Gateway (PG) — The computer and process within the ICR system that communicates directly with the ACD, PBX, or IVR at the call center. The Peripheral Gateway reads status information from the peripheral and sends it to the Central Controller. In a private network configuration, the Peripheral Gateway sends routing requests to the Central Controller and receives routing information in return.

Post-Routing — The routing concept that allows the ICR to make secondary routing decisions after a call has been initially processed at a call center. Post-Routing enables the ICR to process calls when an ACD, VRU, or PBX generates a route request. This directs the ACD to send the call to an agent, skill group, or service in the same call center or at a different call center. In making a Post-Routing decision, the ICR can use all the same information and scripts used in Pre-Routing.

Pre-Routing — The routing concept that enables the ICR to execute routing decisions before a call terminates at a call center. With Pre-Routing, the ACD receives the route request from the IXC and passes the call information along to the ICR. The ICR is then able to process the IXC route request through a call routing script, which defines how calls should be routed. In making its Pre-Routing decisions, the ICR uses real-time data on the status of call centers. This data is gathered by Peripheral Gateways at different call center sites and passed back to the ICR.

Prefix — The leading digits of a telephone number. When defining a call type, you can specify a prefix of any length to match calling line ID values. You can also define a region that is a collection of prefixes. If a calling line ID value matches multiple pre-fixes, the longest matching prefix prevails.

Primary Rate Interface (PRI) — One of two levels of ISDN service. In the UK, the PRI typically provides 30 bearer channels for voice and data and one channel for sig-naling information (commonly expressed as 30B+D). Sometimes known as DASSII, I421 or Q931.

Private Branch Exchange (PBX) — A device located at a customer's site that switches incoming calls to extensions within that site. A PBX can be used to implement direct inward dialing.

Private network — A network made up of circuits for the exclusive use of one cus-tomer. Private networks can be nationwide in scope. They typically serve large cor-porations or government agencies.

Private network routing — A configuration in which the ACD sends routing requests to the ICR through the Peripheral Gateway. This is a type of Post-Routing.

Public Switched Network (PSTN) — The public telephone network. The PSTN provides the capability of interconnecting any home or office in the country with any other.

Protocol — These are the rules for transmission in a network. Commonly acceptable rules allow an open system like the Internet to succeed.

489

Q

Quantitative Forecasting — Using statistical techniques to forecast future events. The major categories of quantitative forecasting include time series and explanatory approaches. Time series techniques use past trends to forecast future events. Explanatory techniques attempt to reveal linkages between two or more variables.

Query — The act of requesting information from a database, or the statement used to request that information.

Queue — A place to wait or line up awaiting something. Callers are frequently placed in queues awaiting specific service resources, such as an agent.

Queue time — The time a call spends queued at a peripheral waiting for an agent to become available. Queue time occurs after delay time and before ring time.

Queued call — A call that has arrived at a peripheral, but that is being held until an agent or other resource becomes available to handle the call.

Glossary

R

Random Call Arrival — The normal, random variation in how incoming calls arrive.

Ready state — A state in which an agent is logged on to the system and is currently available to handle a call, is talking on a call, or is involved in after-call work and presumed to be available to handle another call when done.

Real-time data — Real-time information about certain entities within the ACD system is updated continuously. Real-time data includes data accumulated since the end of the last five-minute interval. Real-time records are usually stored in the local database on the Admin Workstation.

Redundant Array of Inexpensive Disks (RAID) — A storage device that provides fault-tolerance through redundant physical disks. A RAID system is an alternative to mirrored disks.

Relational database — The database model used in the ICR central and local databases. The relational database model portrays data as being stored in tables (or relations). The associations between pieces of data are implicit in the data themselves rather than being stored externally.

Reader boards — Also call Display boards or Wall Displays. A visual display, usually mounted on the wall or ceiling, that provides real-time and historical information on queue conditions, agent status and call center performance.

Real-Time Adherence Software — Software that tracks how closely agents conform to their schedules. Ass Adherence to Schedule.

Real-Time Data — Information on current conditions. Some "real-time" information is real-time in the strictest sense (e.g., calls in queue and current longest wait). Some real-time reports require some history (e.g., the last call x calls or x minutes) in order to make a calculation (e.g., service level and average speed of answer).

Received Calls — A call detected and seized by a trunk. Received calls will either abandon or be answered by an agent.

Response Time — The time it takes the call center to respond to transactions that do not have to be handled when they arrive (e.g., correspondence or e-mail).

Retry — A caller who "retries" when they get a busy signal.

Retry Tables — Sometimes used to calculate trunks and other systems resources required. They assume that some callers will make additional attempts to reach the call center if they get busy signals.

Report — The final presentation of data, titles, dates and times, and graphic elements either printed or displayed in a Monitor window. A single report can include components generated by one or more templates. For example, one report can contain a real-time pie chart and a historical grid, each generated with a different template.

Reserved state — A state in which an agent is awaiting an inter flowed call and is unavailable to receive any incoming calls. This state applies to agents on Northern Meridian ACDs only.

Ring time — The time elapsed from when a call is presented to an agent until the agent answers it. This occurs after any delay time and queue time. This value is stored in the Termination Call Detail table in the ICR central database. Some peripherals do not track ring time. For those peripherals, ring time is included within the queue time or delay time.

Rostered Staff Factor (RSF) — Alternatively called an Overlay, Shrink Factor or Shrinkage. RSF is a numerical factory that leads to the minimum staff needed on schedule over and above base staff required to achieve your service level and response time objectives. It is calculated after base staffing is determined and before schedules are organized, and accounts for things like breaks, absenteeism and ongoing training.

Route — A value returned by a routing script that maps to a service and a specific target at a peripheral; that is, a service, skill group, agent, or translation route. The ICR converts the route to a label that is returned to the routing client. The routing client then delivers the call to a specific trunk group with a specific DNIS. The peripheral is responsible for recognizing the trunk group and DNIS and delivering the call to the appropriate target.

Route call details — Data about routing requests received by the ICR and how calls were routed (that is, the route chosen for each call). This data is stored in the Route Call Detail table in the ICR central database.

Router — The device that reads packet addresses and sends them to the next routing device up or down the network hierarchy.

S

Screen Monitoring — A system capability that enables a supervisor or manager to remotely monitor the activity on agents' computer terminals.

Screen Pop — A CTI capability. Callers' records are automatically retrieved (based on ANI or digits entered into the VRU) and delivered to agents, along with the calls.

Screen Refresh — The rate at which real-time information is updated on a display (e.g., every 5 to 15 seconds). Note, screen refresh does not correlate with the time frame used for real-time calculations.

Script — A defined procedure that the ICR can execute. The ICR supports two types of scripts: routing scripts to determine where to route a call and administrative scripts that perform background processing. A script consists of executable nodes, connections, routing targets, and comments.

Seated Agent — A literally staffed position.

Server — The computer that processes requests from a client in a client server network.

Service Bureau — A company that handles inbound or outbound calls for another organization.

Service level — The percentage of incoming calls that are answered within a specified threshold. Several slightly different calculations can be used for the service level; specifically, abandoned calls can be accounted in different ways. The ICR keeps track of two different service levels: the peripheral service level is the service level as calculated by the peripheral; the ICR service level is the service level as calculated by the ICR.

Service level calculations — The ICR typically use one of three formulas to calculate the service level for a service. The formulas differ in the way they treat calls that were abandoned before the service level threshold expired. In the database, the type of service level calculation is represented by an integer:

1. Ignore Abandoned Calls. Remove the abandoned calls from the calculation.

2. Abandoned Calls have Negative Impact. Treat the abandoned calls as though they exceeded the service level threshold.

3. Abandoned Calls have Positive Impact. Treat the abandoned calls as through they were answered within the service level threshold.

Note that regardless of which calculation you choose, the ICR always tracks separately the number of calls abandoned before the threshold expired.

Service Observation — A euphemism for agent monitoring.

Shrink Factor — The difference between planned and actual staff based on no-shows, tardiness or other human factors.

Silent Monitoring — Monitoring without any notice to the agent.

Skill group — A collection of agents that share a common set of skills, such as being able to handle customer complaints. A skill group is associated with a peripheral. An agent can be a member of zero, one, or more skill groups (depending on the peripheral).

Skill-based routing — A concept whereby calls are routed to agents based on the skills those agents have. You can construct skill groups that contain agents who share a common set of skills. You can also assign priorities to the skills in each agent profile. For example, an agent might have a priority 1 assignment for handling calls from Spanish-speaking callers; however, that same agent might have a priority 3 assignment for handling Sales calls. Calls can then be routed to the skill group that has the appropriate level of expertise to handle the call. The ICR implements skills-based routing at the network level rather than just within a peripheral. This allows the ICR to examine skill groups on all peripherals before deciding where to route the call.

Speech Recognition — The capability of a voice processing system to decipher spoken words and phrases.

493

Split — A group of trunks with some common property such as carry calls related to service requests. The Service split or trunk group. The agent side is typically called a group or team.

SQL Server — The Microsoft relational database product used for local and central databases.

Statistics PC — The process within the ACD that manages the central database. And is also called the Database Server.

Structured Query Language (SQL) — A standard database query language in which you can formulate statements that will manipulate data in a database. The statements include SELECT, for data retrieval; UPDATE, for data modification; DELETE, for data deletion; and INSERT, for data insertion. You can access the ICR databases using SQL and any client tool that supports ODBC.

System time — The time as used consistently throughout an ICR system. Although parts of the ICR system can be in different time zones, they all use the same clock time. The system time is typically the local time for Side A of the ICR Central Controller.

Supervisor Monitor — Computer monitors that enable supervisors to monitor the call handling statistics of their supervisory groups or teams.

Supervisor — A person who is in charge of or responsible for a group of agents.

T

T1 Line — A digital circuit with the bandwidth for 1,544 million bits per second that allows for 24 talk paths and a signaling channel.

Talk state — A state in which an agent is talking on an inbound call.

Talking Other state — A state in which an agent is talking on an internal (neither inbound nor outbound) call.

Talking Out state — A state in which an agent is talking on an outbound call.

Talking state — A state in which an agent is talking on a call. This includes the Talking In, Talking Out, and Talking Other states.

Talk time — The total seconds that agents in a skill group are in the Talking In, Talking Out, and Talking Other states.

T1 Circuit — A high-speed digital circuit used for voice, data or video, with a bandwidth of 1.544 megabits per second. T1 circuits offer the equivalent of twenty-four (24) analog voice trunks.

Telecommuting — Using telecommunications to work from home or other locations instead of at the organization's premises.

Telephone Service Factor (TSF) — Service level offered to callers.

Telephony Applications Programming Interface (TAPI) — A first party call control CTI protocol developed by Microsoft and Intel.

Telephony Services Application Programming Interface (TSAPI) — A first party call control CTI protocol developed by Novell and AT&T.

Termination call detail — Data that contains information about how each call was handled at a peripheral. This data includes items such as the identifiers for the agent and the peripheral that handled the call, ring time, after-call work time, and the identifier for the route where the call was sent. Termination call details are stored in the central database.

Text Chat — This is a communication where two parties can use text to conversationally communicate "over" the current applications without disturbing them. This has evolved to text chatting on the web and "chat rooms." The click-to-chat feature on web sites allows this as an alternate contact medium for web site visitors to "chat" with a live customer contact center agent.

Third party CTI — The communication between the switch and the application operates through middleware (the third party) versus being a direct request from the client desktop (1st party CTI).

Threshold — The point at which an action, change or process takes place.

Tie-line — A private trunk line that connects two ACDs or PBXs across a wide area. Mostly CAS, DPNSS or Q-SIG.

Toll-Free Service — Enables callers to reach a call center out of the local calling area without incurring charges. 800 and 888 service is toll-free. In some countries, there are other variations of toll-free service. For example with 0345 or 0645 service in the UK callers are charged local rates and the call center pays for the long distance charge.

Touchtone — A trademark for AT&T.

True Calls Per Hour — Actual calls an individual or group handled divided by occupancy for that period of time.

Transmission Control Protocol (TCP) — A connection-based Internet protocol that is responsible for packaging data into packets for transmission over the network by the IP protocol. TCP provides a reliable flow control mechanism for data in a network.

Transmission Control Protocol/Internet Protocol (TCP/IP) — The Internet suite of protocols used to connect a world-wide inter network of universities, organizations, and corporations. TCP/IP is the protocol used to communicate between the Central Controller and devices in the Intelligent Call Routing system. TCP/IP is based primarily on a connection-oriented transport service, the Transmission Control Protocol (TCP); and a connectionless-mode network service, the Internet Protocol (IP). TCP/IP provides standards for how computers and networks with different technologies communicate with each other.

Trunk — A telephone line connected to a call center and used for incoming or outgoing calls.

Trunk Group — A collection of trunks associated with a single peripheral and usually used for a common purpose. A peripheral may group its trunks one way while a routing client groups those same trunks differently. In the ICR configuration, Trunk Groups describe how a peripheral groups its trunks while Network Trunk Groups define how the routing client groups the trunks.

A network trunk group maps to one or more peripheral trunk groups.

Trunk Load — The load that trunks carry. Includes both delay and talk time.

Trunks idle — The number of trunks in a trunk group that are non-busy.

Trunks in service — The number of trunks in the trunk group that are functional.

Glossary

U

Unavailable Work State — An agent work state used to identify a mode not associated with handling telephone calls.

Uniform Call Distributor (UCD) — A simple system that distributes calls to a group of agents and provides some reports. A UCD is not as sophisticated as an ACD.

V

Variable — A named object that can hold a value.

Virtual call center — An approach to enterprise-wide call center management that treats several geographically dispersed call centers as if they were a single call center. The virtual call center expands skill-based routing from the ACD to the network level.

Visible Queue — When callers know how long the queue that they just entered is, and how fast it is moving (e.g., they hear a system announcement that relays the expected wait time).

Voice Processing — A blanket term that refers to any combination of voice processing technologies, including voice mail, automated attendant, audiotex, voice response unit (VRU) and fax on demand.

Voice Response Unit (VRU) — A telecommunications computer, also called an Interactive Voice Response unit (IVR) that responds to caller entered touch-tone digits. The VRU responds to caller entered digits in much the same way that a conventional computer responds to keystrokes or a click of the mouse. The VRU uses a digitized voice to read menu selections to the caller. The caller then enters the touch-tone digits that correspond to the desired menu selection. The caller entered digits can invoke options as varied as looking up account balances, moving the call within or to another ACD, or playing a pre-recorded announcement for the caller.

Voice over Internet Protocol (VoIP) — Using the internet to make phone calls. Packet switched Internet telephony as an alternative to circuit switched telephone service.

W

Wide Area Network (WAN) — The connection of several computers across a wide area, normally using telephone lines.

Wide Area Telecommunications Service (WATS) — A special service provided by an interexchange carrier that allows a customer to use a specific trunk to make calls to specific geographic zones or to receive calls at a specified number at a discounted price.

Workforce Management Software — Software systems that, depending on available modules, forecast call load, calculate staff requirements, organize schedules and track real-time performance of individuals and groups.

Workload — Sometimes used interchangeably with Call Load. However, workload can also refer to non-call activities.

Work Not Ready state — A state in which an agent is involved in after-call work and is presumed not to be ready to accept incoming calls when done.

Work Ready State — A state in which an agent is involved in after-call work and is presumed to be ready to accept incoming calls when done.

Wrap-up — Call-related work performed by an agent after the call is over. An agent performing wrap-up is in either the Work Ready or Work Not Ready state.

Wrap-Up-Codes — Codes agents enter into the ACD to identify the types of calls they are handling. The ACD can then generate reports on call types, by handling time, time of day etc.

497

X

X-axis — A horizontal positioning system. Any point on a two-dimensional surface (such as a video display) can be specified as a position on the x-axis and y-axis.

Y

Y-axis — A vertical positioning system. Any point on a two-dimensional surface (such as a video display) can be specified as a position on the x-axis and y-axis.

Z

Z-order — The front-to-back ordering of items in a three-dimensional space. For example, overlapping windows on a video display have a specific z-order.

Zoom — To shrink or enlarge the appearance of objects on the screen.